Atlas of Human Prenatal Craniofacial Development

Ralf Johannes Radlanski

Atlas of Human Prenatal Craniofacial Development

Springer

Ralf Johannes Radlanski
Craniofacial Developmental Biology
Charité-Universitätsmedizin Berlin
Berlin, Germany

ISBN 978-3-031-88811-3 ISBN 978-3-031-88812-0 (eBook)
https://doi.org/10.1007/978-3-031-88812-0

This Springer imprint is published by the registered company Springer Nature Switzerland AG
The registered company address is: Gewerbestrasse 11, 6330 Cham, Switzerland

Dedicated to the memory of the great anatomists and embryologists

Wilhelm His (1831–1904)

Erich Blechschmidt (1904–1992)

Gerd Steding (1936–2011)

Jochen Fanghänel (1939–2024)

Preface

Seht Ihr den Mond dort stehen?
Er ist nur halb zu sehen
und ist doch rund und schön.
So sind wohl manche Sachen,
die wir getrost belachen,
weil unsre Augen sie nicht sehn.

Behold the moon – and wonder
why half of her stands yonder,
yet she is round and fair.
We are the ones who're fooling
'cause we are ridiculing
as our minds are unaware.
Matthias Claudius (1773)

With the face, we look out of our clothes. With the face, we identify ourselves and others, and we communicate.

Of course, it is virtually interesting for anatomists, medical doctors, and actually for all the people to know why we look the way we do. Evolutionary biology cannot explain this because the development of every creature begins before birth. Therefore, developmental processes of the individual must be studied there. In the training of medical students, embryology is not always given the importance that it actually has: Every malformation that fatefully accompanies a human being throughout life has its origin in embryonic and fetal development. From animal studies, our knowledge of this is immense, but in relation to humans, there are still large gaps. There is no doubt that there are ethical limitations here, which rightly restrict work in the laboratory. However, there are also few, but all the more valuable collections of human embryos and fetuses available worldwide that are accessible for scientific research.

This atlas, you hold in your hands, is the result of about four decades of intensive study of the prenatal development of the human face. After my education at the Universities of Göttingen (Germany) and Minneapolis MN (USA), I was fortunate to receive basic scientific training in the subjects of anatomy and embryology at the Institute of Anatomy of the University of Göttingen. My teachers there were Hans-Jürg Kuhn and Gerd Steding. After a residency in orthodontics, which I completed in parallel, I was appointed Professor and Director of the Department of Craniofacial Developmental Biology at Freie Universität Berlin (Germany) in 1992 (Today: Charité—Universitätsmedizin Berlin). In this department, I was then privileged to continue my scientific investigations on the development of the face with many bright collaborators, skillful technicians, and talented graduate students up to the year 2024.

The specialty is that we were able to study human material. The embryos and fetuses, which form the basis for our investigations, come from spontaneous abortions and autopsies, and were made available for anatomical basic research within the framework of the applicable laws.

Our research has been financially supported by the Deutsche Forschungsgemeinschaft (DFG, German Research Foundation) for over three decades. This enabled us to establish a collection of section series of human embryos and fetuses covering developmental stages from week 6 to week 28. In total, about 50,000 individual sections are available, which have been digitized as the Radlanski—Collection in order to be available to a broad community of researchers https://human-embryology.org/wiki/Radlanski_Collection

Basic embryological research is technically laborious because the embryos are quite small, of course. The embryologists Wilhelm His (the elder) [1] and Gustav Born [2] described a method as early as 1883 (born 1883), which we still use today in a modernized form: The specimens were prepared as continuous histological section series, which then made it possible to identify all relevant structures under the light microscope. Their contours were then recorded magnified, in our case digitally entered into computers. Whereas a hundred years ago, cut-out wax plates or plastic plates were stacked on top of each other to form 3D models, we were now able to generate virtual models of our preparations using the appropriate software (see Chap. 3). Enlargements were possible, as was virtual sectioning, i.e., the omission of selected textures in order to be able to follow the layer-by-layer structure in depth. And because we were able to do this with 34 series of sections from different stages of development, an impressive picture of the spatiotemporal development in the form of this atlas was created.

This was only possible with the enthusiasm and help of my staff and doctoral students, whom I would like to sincerely thank once again at this point. That way, we could at least contribute a small part to the long road to understanding human prenatal craniofacial development.

References

1. His W. Unsere Körperform und das physiologische Problem ihrer Entstehung. Briefe an einen befreundeten Naturforscher. Leipzig: Verlag von F. C. W. Vogel; 1874.
2. Born G. Die Plattenmodelirmethode. Arch Mikr Anat. 1833;22:584–99.

Berlin, Germany Ralf Johannes Radlanski

Acknowledgments

The author would like to acknowledge the contributions of:

Katrin Beckert
Bettina Benedikter
Nancy Bontschev
Julia Brandt
Barbara Danielowski
Alexandra Dietz (neé Konietzny)
Esther M. Fernández Rubio
Anna R. Germelmann (neé Than)
Birgit Guizetti
Uta Kalinke
Marie C. Klarkowski
Susann Lajvardi
Skadi Lange
Sepadi Lieck
Beate Lion
Maria Memering
Ulrike Müller
Herbert Renz
Christina Reulen
Nyamdorj Tsengelsaikhan
Angela von Criegern
Joanna van Straelen (née Skulimovska)
Camilla Wille (née Zimmermann)
Matthias Konietzny
Irene Schwarz
Felix P. Schuster
Aram Tabatabai
Joachim Weinhold

Contents

About the Author

Ralf Johannes Radlanski is Professor Emeritus of Craniofacial Developmental Biology at Charité—Universitaetsmedizin Berlin. He has held guest professorships at the University of California at San Francisco, at the University of Turku in Finland, at the University of Basle, and at the University of Zurich (Switzerland). In addition to numerous scientific publications, he is the author of the basic textbook *Oral Structure and Biolo*gy and author (together with Karl Wesker) of the world bestseller *The Face. Pictorial Atlas of Clinical Anatomy*, which has been translated into 16 languages worldwide.

www.radlanski.org/ralf

Part I

Introduction

1 Introduction

Why Is this Atlas Necessary?

Embryological development is a process, which takes place not only in time, but also in space. Cell division, increase in cell number, aggregation of cells, differentiation of tissues, and formation of organs are processes, which occupy space. Therefore, if we want to scrutinize and describe prenatal craniofacial development, we must do it in 3D.

There are several atlases devoted to prenatal development of the face available: The atlas by Marjorie A. England covers the embryonic and fetal development of the whole body, and it also shows the development of the head in color photographs taken from cadaver specimens, with special focus on surface anatomy [1]. Klaus V. Hinrichsen [2] and also Gerd Steding depict the outer appearance of the human face in impressive scanning-electron images taken from stages of the development of the pharyngeal arches up to the eighth week of development [3]. An abundance of histological sections ranging from human embryos of 9 mm CRL to the newborn is presented in the work of Bertam S. Kraus, Hironori Kitamura, and Ralph A. Latham [4]. Based on 3D-reconstructions from serial sections, Erich Blechschmidt established a collection of physical models showing the development of the human embryo during the first trimester of development [5] and published an atlas, devoted to the development of the whole body [6].

Also, more modern atlases are devoted to the whole body and its development during early stages, like the MRI-Atlas of the first trimester (4.6–28.9 mm CRL) [7], and the one by Bernadette S. de Bakker and her coworkers, covering the first 8 weeks of development [8].

The development of the human face was topic of research in particular devoted to how the basic patterns are being established [9, 10], but development of specific structures in the face happens in later stages. Several topics were covered in single publications, like the description of the prenatal development of the mandible [11–14], of the maxilla [15], the development of the muscles [16, 17] and those of the floor of the mouth [12], the formation of foramina, like the mental foramen [18–20] and the incisive canal [20] and some aspects of the formation of the temporomandibular joint [14, 21–26], and the development of the articular disc in particular [27].

A sample of coherent 3D-reconstructions from histological serial sections have only once appeared in a publication, with selected developmental stages and layers of virtual preparation [28].

So, there are already some isolated publications on the prenatal development of the human head region, but a coherent monograph over several stages, for human material and then consistently in 3D—that was not to be found until now.

Therefore, I have compiled this present atlas.

How to Use this Atlas?

In the foreground stand the images—exemplary histological sectional images and, a large number of three-dimensional reconstructions created from the serial sections. We also show selected reconstructions as stereo image pairs. Practiced anatomists can see a three-dimensional image in their minds when the left eye is directed to the left image and the right eye is directed to the right image. To do this, one must first look into the distance so that the axes of the eyes are nearly parallel, but at the same time accommodate to the image on the paper. With some practice, this is accomplished, and one is rewarded with an impressive spatial impression.

All illustrations show the tissue structures with a consistent color code (see Chap. 3.2). The majority of the images is provided with a texture, and we have followed the nomina anatomica. We are aware that some of the embryological structures are not included there, so we have labeled by analogy (see Chap. 2).

Actually, in an atlas, the pictures speak for themselves. In this respect, the accompanying text is kept relatively short. It is only intended to direct the viewer's observation somewhat. In addition, the rather concise description of the findings is occasionally referred to more general contexts, which are provided

R. J. Radlanski, *Atlas of Human Prenatal Craniofacial Development*, https://doi.org/10.1007/978-3-031-88812-0_1

with the corresponding literature reference. This may be unusual for a results-section in a scientific study; however, in an atlas of this kind it may be sensible and therefore legitimate.

References

1. England MA. A colour atlas of life before birth. Normal fetal development. London: Wolfe Medical Publications Ltd; 1983.
2. Hinrichsen KV. The early development of morphology and patterns of the face in the human embryo. Berlin/Heidelberg/New York/Tokyo: Springer Verlag; 1985.
3. Steding G. The anatomy of the human embryo. A scanning electron-microscopic Atlas. Basle: Karger; 2009.
4. Kraus BS, Kitamura H, Latham RA. Atlas of developmental anatomy of the face. New York: Harper and Row; 1966.
5. Blechschmidt E. Der menschliche Embryo. Stuttgart: Schattauer; 1963.
6. Blechschmidt E. The stages of human development before birth. An introduction to human embryology. - Die vorgeburtlichen Entwicklungsstadien des Menschen. Eine Einführung in die Humanembryologie. London, New York, Basel, Freiburg i. B.: Karger; 1960.
7. Yamada S, Samtani RR, Lee ES, Lockett E, Uwabe C, Shiota K, Anderson SA, Lo CW. Developmental atlas of the early first trimester human embryo. Dev Dyn. 2010;239:1585–95.
8. de Bakker BS, de Jong KH, Hagoort J, de Bree K, Besselink CT, de Kanter FE, Veldhuis T, Bais B, Schildmeijer R, Ruijter JM, Oostra RJ, Christoffels VM, Moorman AF. An interactive three-dimensional digital atlas and quantitative database of human development. Science. 2016;354:6315.
9. Blechschmidt E. Elementarprozesse der Gesichtsbildung. IMAGE, Medizinische Bilddokumentation Roche. 1970;36:11–6.
10. Avery JA. Developmental prenatal craniofacial skeleton. Oral development and histology. Baltimore: Williams & Wilkins; 1988.
11. Lee SK, Kim YS, Oh HS, Yang KH, Kim EC, Chi JG. Prenatal development of the human mandible. Anat Rec. 2001;236:314–25.
12. Radlanski RJ, Klarkowski MC. Bone remodeling of the human mandible during prenatal development. J Orofac Orthop. 2001;62:191–201.
13. Radlanski RJ, Renz H, Klarkowski MC. Prenatal development of the human mandible. 3D reconstructions, morphometry and bone remodelling pattern, sizes 12-117 mm CRL. Anat Embryol (Berl). 2003;207:221–32.
14. Radlanski RJ. Mein Gesicht. Haare, Haut, Augen, Nase, Ohren, Knochen, Mund und Zähne. Warum sehen wir so aus? Anatomische Erklärungen. Berlin: Quintessenz; 2016.
15. Kjaer I. Prenatal skeletal maturation of the human maxilla. J Craniofac Genet Dev Biol. 1989;9:257–64.
16. Gasser RF. A description of development of facial muscles in man. Anat Rec. 1965;151:352–61.
17. Gasser RF. The development of the facial muscles in man. Am J Anat. 1967;120:357–76.
18. Kjaer I. Histochemical and radiologic studies of the human fetal mandibular condyle. Scand J Dent Res. 1978;86:279–99.
19. Radlanski RJ, Renz H, Müller U, Schneider RS, Marcucio RS, Helms JA. Prenatal morphogenesis of the human mental foramen. Eur J Oral Sci. 2002;110:452–9.
20. Radlanski RJ, Emmerich S, Renz H. Prenatal morphogenesis of the human incisive canal. Anat Embryol. 2004;208:265–71.
21. Radlanski RJ, Lieck S, Bontschev NE. Development of the human temporomandibular joint. Computer-aided 3D-reconstructions. Eur J Oral Sci. 1999;107:25–34.
22. Bach-Petersen S, Kjaer I, Fischer-Hansen B. Prenatal development of the human osseous temporomandibular joint. J Craniofac Genet Dev Biol. 1994;14:135–43.
23. Fernández Rubio EM. Concerning the lack of a muscular antagonist to the lateral pterygoid. Morphogenetic and morphological investigations of the murine temporomandibular joint and masticatory muscles from stages E13.25 to P4 [Med. dissertation]. Charité-Universitätsmedizin; 2021.
24. Fernández Rubio EM, Radlanski RJ. Hypothesis for the lack of a muscular antagonist to the lateral pterygoid. Ann Anat. 2022;239:1–8.
25. Fernández Rubio EM, Radlanski RJ. Development of the primary and secondary temporomandibular joint in the mouse. Ann Anat. 2023;249:1–15.
26. Fernández Rubio EM, Radlanski RJ. Development of the human primary and secondary jaw joint. Ann Anat. 2024;251:1–8.
27. Giambartolomei LA, Brunotto M, Gomez de Ferrais ME. Human temporomandibular joint disc: anatomy and measurements in prenatal development. Acta odontol Latino-americana. 2011;24:98–103.
28. Radlanski RJ, Renz H. Genes, forces, and forms: mechanical aspects of prenatal craniofacial development. Dev Dyn. 2006;235(5):1219–29.

Terminology

2

This atlas shows embryos and fetuses of a crown-rump length from 15 to 270 mm. In a number of instances, the anatomical conditions are very different from those found in adults—at least this is predominantly the case in the developmental period before the fetal period.

And there is another special developmental peculiarity in the head area: the anatomical conditions in relation to the teeth and the alveolar process bone are of course also different before birth because tooth eruption and therefore also the formation of the alveolar process bone only takes place after birth. In the process of tooth development, instead, we find many soft tissue structures with many anatomical characteristics that determine the shape of the dental hard tissue, but then disappear after the teeth are completed. There is a separate prenatal nomenclature for this, the terms of which have been established for over a 100 years. Research into the peridental structures found during embryonic and fetal development is quite new, and there are hardly any really suitable anatomical terms, at least not from the period of postnatal development. This was then described in more detail in the accompanying text. At least there is the Terminologica Embryologica from the Federative International Program on Anatomical Terminologies, published in 2013, which was of course used whenever possible.

All terms should of course be comprehensive and precise.

And then there is the problem that in Europe a Latin nomina anatomica is used, whereas in the USA an anglicized Latin-Greek terminology is common. In the end, I have followed the American practice in the text, which makes the English text easier to read. The illustrations, however, are largely labeled according to the Nomina Anatomica or Terminologica Anatomica commonly used in Europe, which is intended to give them a more universally applicable character. My American colleagues may forgive me for this.

Further Reading

Federative International Programme on Anatomical Terminologies: Terminologica: Terminological Anatomica/International Anatomical Terminology. Stuttgart: Thieme; 1998.

Federative International Programme on Anatomical Terminologies: Terminologica Embyologica/International Embryological Terminology. Stuttgart: Thieme; 2013.

R. J. Radlanski, *Atlas of Human Prenatal Craniofacial Development*, https://doi.org/10.1007/978-3-031-88812-0_2

Material and Methods

3

Material

Thirty-four histological serial sections from the Radlanski Collection, Charité Berlin, Germany, ranging between 15 and 270 mm CRL (weeks 6–28) were used for this atlas (Figs. 3.1 and 3.2; Table 3.1). The use of this material was approved by the ethical committee of Charité-Universitätsmedizin Berlin, EA4/017/06. The specimens were obtained from legal or spontaneous abortions and autopsies, and no part of the material gave any indication for spontaneous malformation. Most of the funding for the laboratory processes described below was provided by the Deutsche Forschungsgemeinschaft (DFG, German Research Foundation) within the framework of research projects Ra 428/1-1, 1-2, 1-3 1987–1993, Ra 428/1-9 2008–2011, and Ra 428/11-1 2012–2016.

The specimens were fixed in 4% buffered Formalin, resp. in Bouin's solution and, according to standard histological procedures, the specimens were dehydrated in alcohol with increasing concentration up to 100%. Depending on the size and gross preparation, the specimens were decalcified using RDO (Eurobio, Paris, France) or EDTA for 2–30 days. Paraffin embedding was carried out according to standard procedures, and the specimens were cut as 10 µm-thick serial sections (Leica, Reichert-Jung RM 2065, Nußloch, Germany) in frontal, horizontal, and sagittal planes. Routine staining was performed with hematoxylin and eosin. In addition, with special focus on cartilaginous tissue, staining methods according to Masson-Goldner were applied, which stain the

Fig. 3.1 All histological section series are stored protected from light on slide boards

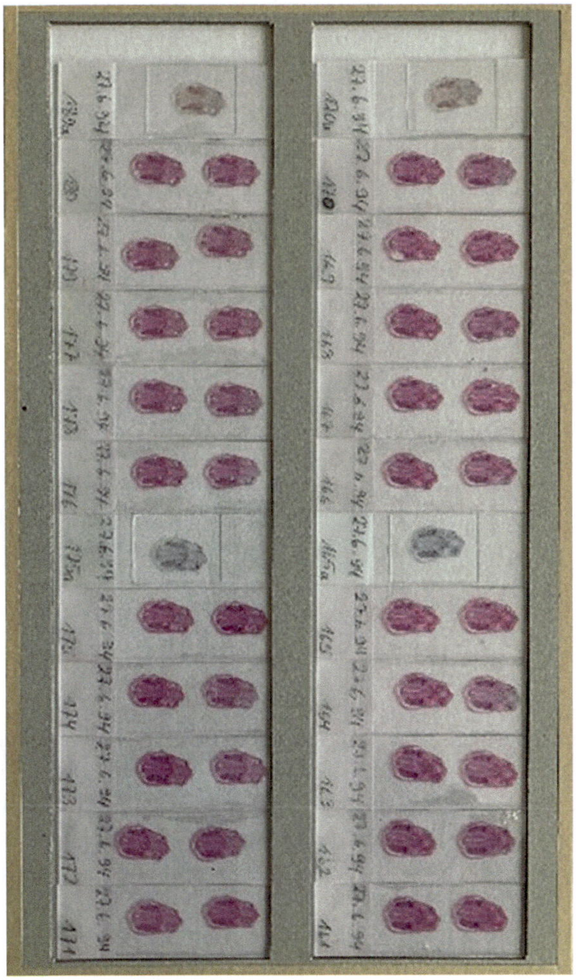

Fig. 3.2 Slides on a cardboard tray

R. J. Radlanski, *Atlas of Human Prenatal Craniofacial Development*, https://doi.org/10.1007/978-3-031-88812-0_3

Table 3.1 List of specimens

Catalog number	Size (CRL)	Age	Section plane	Staining
STE290687	15 mm	ca. 6th week	HOR	H.E.
GUS110785	18 mm	ca. 6th week	SAG	H.E.
CHR220687	19 mm	ca. 7th week	SAG	H.E.
ADE100787	20 mm	ca. 7th to 8th week	SAG	H.E.
NEL110989	21 mm	ca. 7th week	HOR	H.E.
EMM15078	22 mm	ca. 7th week	SAG	H.E.
ALI070785	22 mm	ca. 7th week	HOR	H.E.
DID110688	22 mm	ca. 7th week	HOR	H.E.
BAL110787	24 mm	ca. 8th week	HOR	H.E.
JOS080289	25 mm	ca. 8th week	HOR	H.E.
OLL0808785	29 mm	ca. 8th week	SAG	H.E.
T61_060765	30 mm	ca. 8th week	FRO	H.E., Masson-Goldner, van Giesson, mod. Domagk
PIP161189	31 mm	ca. 8th week	SAG	H.E.
DOR100785	37 mm	ca. 9th week	SAG	H.E.
KUR030389	41 mm	ca. 9th week	HOR	H.E.
HAN040389	53 mm	ca. 9th week	SAG	H.E.
THE230494	54 mm	ca. 9th week	HOR	H.E., Trichrome, Elastica-Trichrome
SIM280489	54 mm	ca. 11th week	HOR	H.E.
URS2508889	60 mm	ca. 15th week	SAG	H.E.
GIS210789	66 mm	ca. 10th week	HOR	H.E.
ART270694	68 mm	ca. 10th week	HOR	H.E., Trichrome, Elastica-Trichrome, TRAP
FUL140489	76 mm	ca. 11th week	HOR	H.E.
DES200597	87 mm	ca. 12th week	HOR	H.E., Trichrome, Elastica-Trichrome, TRAP
FLO010689	95 mm	ca. 12th to 13th week	SAG	H.E.
BEL170294	100 mm	ca. 13th week	HOR	H.E.
HUL110589	117 mm	ca. 14th week	SAG	H.E., Trichrome, Elastica-Trichrome, TRAP
T55	118 mm	ca. 14th week	FRO	H.E., Masson-Goldner, van Giesson, mod. Domagk
T26_050265	150 mm	ca. 18th week	SAG	H.E., Masson-Goldner, van Giesson, mod. Domagk
PIT2_180294	160 mm	ca. 18th week	HOR	H.E., TRAP
MIN101293	190 mm	ca. 20th week	HOR	H.E., Trichrome, Elastica-Trichrome
AMA-2_160294	225 mm	ca. 21th week	HOR	H.E., Trichrome, Elastica-Trichrome
ULF270592	230 mm	ca. 24th week	HOR	H.E.
LUI010293	250 mm	ca. 26th week	SAG	H.E., Trichrome, Elastica-Trichrome
DAG-7_020392	270 mm	ca. 28th week	HOR	H.E., Trichrome, Masson-Goldner, TRAP

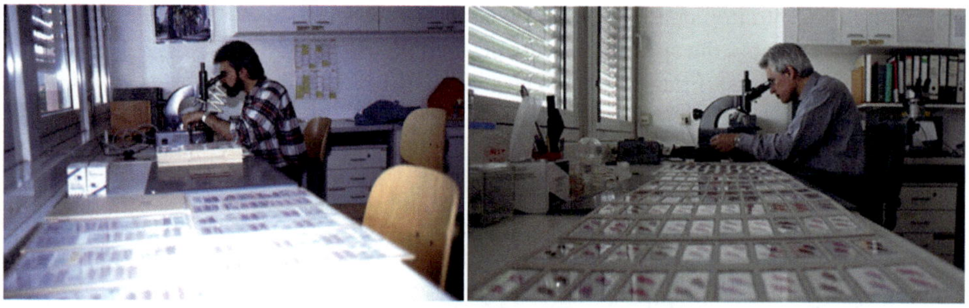

Fig. 3.3 The author has decades of experience in microscopy of histological section series

nuclei dark brown, cytoplasm and muscle fibers red, erythrocytes orange-red, and hyaline and collagen green-blue, depending on its intensity. Additionally, TRAP-Staining (Tartrate-Resistant Acid Phosphatase) was applied with osteoclasts in focus. Some sections were stained by means of a combined alcian blue and van Giesson modification according to Domagk [1]. There, nuclei stained dark brown, cytoplasm, muscle fibers, epithelium, and glia fibrils yellow, collagen fibers red, and acid mucins and other macromolecules turquoise to green.

3D Reconstruction

The sections were photographed and digitized using a VS 100 microscope (Olympus, Berlin, Germany), equipped with a motorized stage (x, y, and z axes motorized, Merzhäuser, Wetzlar, Germany) at magnifications between 1.25× and 40× (UPlanSApo objectives, Olympus, Berlin, Germany). After histological diagnosis, the different tissues were manually segmented in each section (Fig. 3.4) and the 3-D module of the software "AnalySIS" (SIS-Olympus, Münster, Germany) was used to generate 3-D images from selected stages of the serial sections. To that end, the single sections were brought into alignment (Fig. 3.5) according to the general rules of 3-D reconstructions [2]. In the orofacial region, structures like the facial contour, the form, and the symmetry of the eyes and other paired organs, and Meckel's cartilage and the course of nerves were used to check the alignment for accuracy, as it was impossible to make use of any techniques that used external markers as fiducial aids [3, 4]. The contours of all the tissues and organs of relevance for the current question were diagnosed on the screen at higher magnifications and traced using the 3-D function of the software (SIS-Olympus, Münster, Germany). All structures were reconstructed as triangular meshes by means of the 3-D function of the software (Figs. 3.6). Finally, a color code that was systematically maintained throughout all of the reconstructions was applied to the rendered 3-D models, as given in Table 3.2. In addition to plain outlines, bone remodeling activity was also recorded for selected specimens (Figs. 3.7, 3.8, and 3.9).

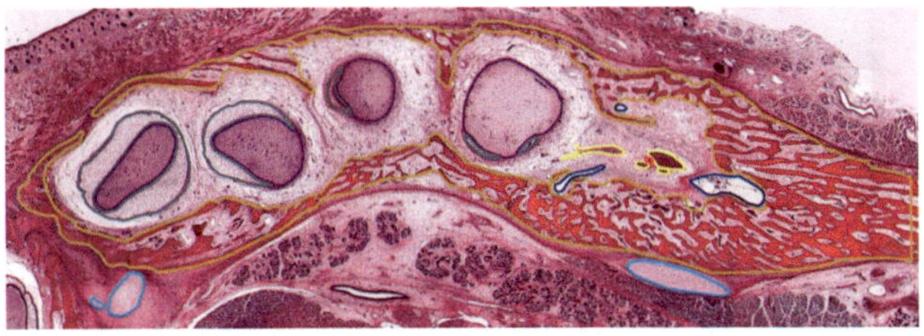

Fig. 3.4 Histological section with acquisition of the relevant contours (here dental primordia in maxillary bone, in a horizontal section). This step, known as segmentation, requires a good knowledge of histology. Until now, exact automated detection of tissue boundaries has failed due to uneven staining, artifacts in the histological section, or simply the fact that an experienced person must make a decision as to where a tissue must be delineated in order to produce a meaningful three-dimensional reconstruction

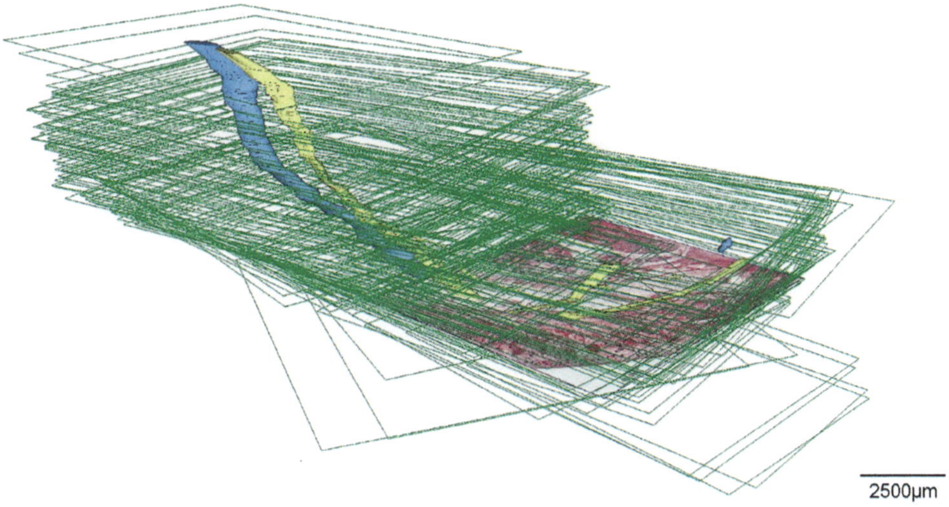

Fig. 3.5 Each individual histological section in a series must be accurately matched to its adjacent sections. This step cannot be automated either because all sections have experienced different distortions due to the forces involved in the cutting process. In this respect, an alignment of the individual sections must be carried out taking into account the individual areal expansion variations based on the known shapes of reference structures such as Meckel's cartilage, the eye or the like. Nevertheless, mainly because of the variable dimension of each cut, the surfaces of the 3D structures are so rugged in some areas

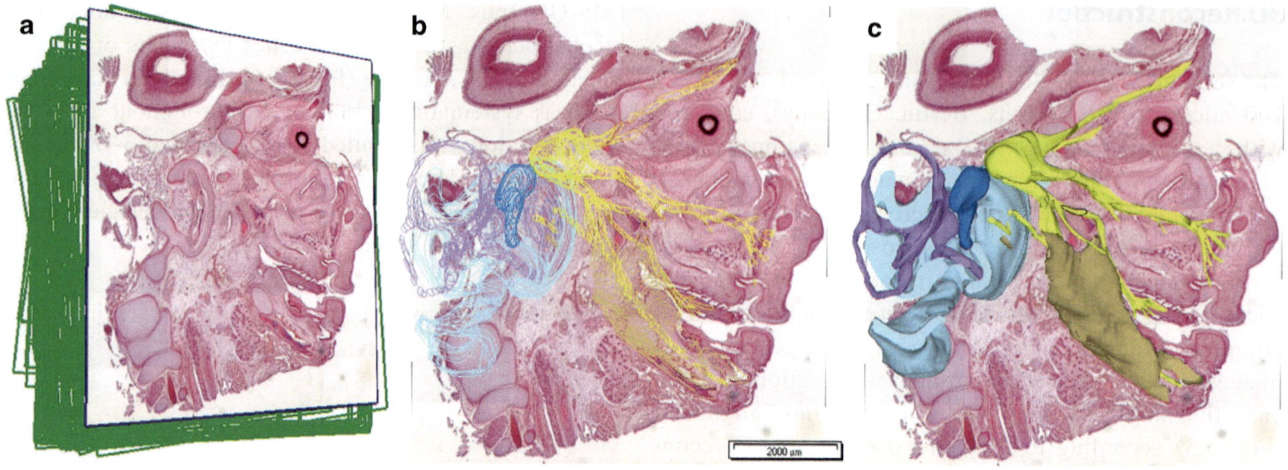

Fig. 3.6 (**a**) Example of a stack of a series of vertical sections, region of the head, correctly aligned. Staining: Hematoxylin-eosin. (**b**) The same stack, now with projected acquisition of the contours: Yellow: trigeminal nerve. Ochre: Mandible. Dark blue: Incus. Purple: semicircular canals of the inner ear. Light blue: otic capsule. (**c**) The same stack, now with now with the three-dimensionally rendered structures from **b** superimposed: Yellow: trigeminal nerve. Ochre: Mandible. Dark blue: Incus. Purple: semicircular canals of the inner ear. Light blue: otic capsule

Fig. 3.7 In addition to plain outlines, bone remodeling activity was also recorded for selected specimens. Regions where osteoblasts predominated were recorded as apposition (**a**), and where osteoclasts predominated were characterized as resorption zones (**b**)

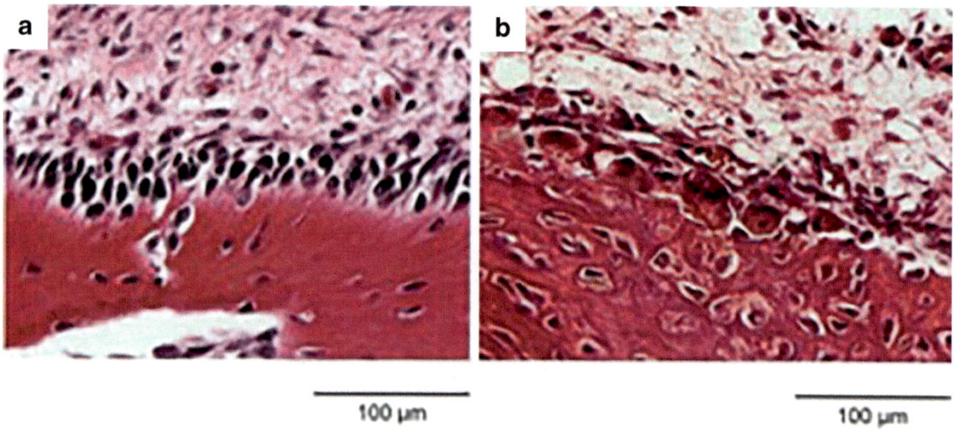

Fig. 3.8 Bone surfaces with predominantly resorption were marked in red, bone surfaces with predominantly apposition were marked in green. Regions with indifferent cells were identified as resting zones and marked in blue. Here is an example of a histological horizontal section through the dental primordia c and m_1 of the mandible with reconstructed interdental bony septa superimposed

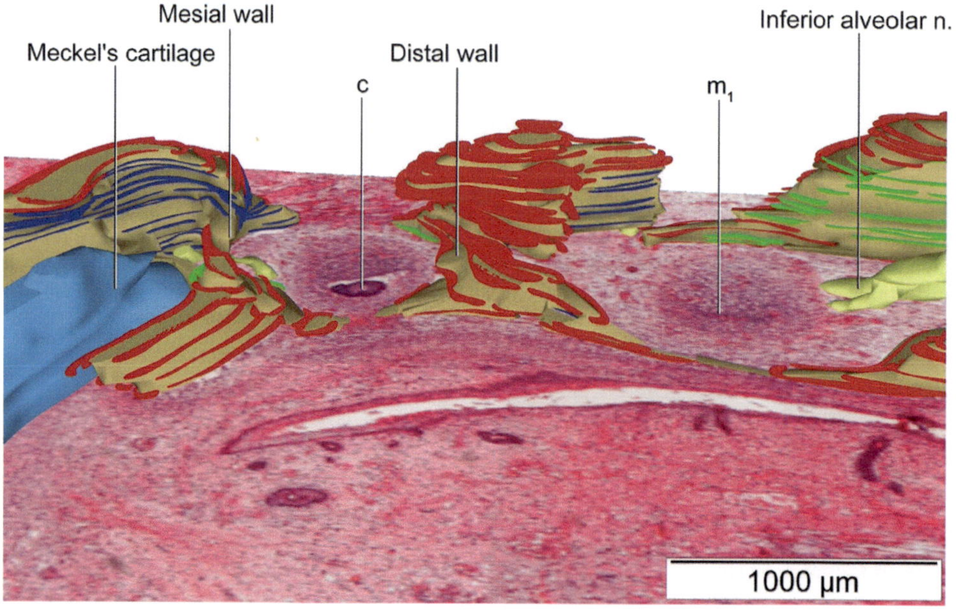

Color	Anatomy / Histology	analySIS-HSL			RGB		
		Color	Sat.	Brigh t	Red	Gree n	Blue
	Epithelium (skin and mucosa)	26	144	212	243	231	207
	Mesenchyme, condensed	200	240	225	255	223	255
	Bone general	28	146	161	222	191	120
Bone remodeling							
	Apposition				255	0	0
	Lining cells				0	0	255
	Resorption				0	255	0
Bone (desmal ossification)							
	Maxilla and Mandible	28	146	161	222	191	120
	Cartilago ossis zygomatici	133	240	160	85	200	255
	Cartilago septum nasi (Vomer, Os ethmoidale)	133	240	210	191	235	255
Bone (mixed ossification)							
	Cartilago ossis sphenoidalis	133	240	170	106	207	255
	Cartilago ossis temporalis, styloidei	137	240	160	85	183	255
	Cartilago basis cranii	133	240	184	136	217	255
Bone (cartilagineous primordium)							
	Cartilago ossis hyoidei	140	240	120	0	128	255
Cartilage							
	Nasal capsule and Meckel's cartilage	138	185	158	101	174	235
	Larynx	157	155	160	115	123	225
Ear							
	Membrana tympanica	160	0	120	127	127	127
	Tuba pharyngotympanica (Tuba auditiva Eustachii)	160	0	181	192	192	192
	Cartilago ossis cochleae (Cochlea)	133	240	142	47	187	255
	Cartilago ossis inci	138	185	125	38	142	227
	Cartilago ossis stapedis	138	184	105	26	120	198
	Cartilago manubrium mallei	138	185	158	101	174	235
Eye							
	Choroidea	160	0	120	128	128	128
	Cornea	160	0	181	192	192	192
	Lens	160	0	0	0	0	0

Table 3.2 Color code used in 3D reconstructions

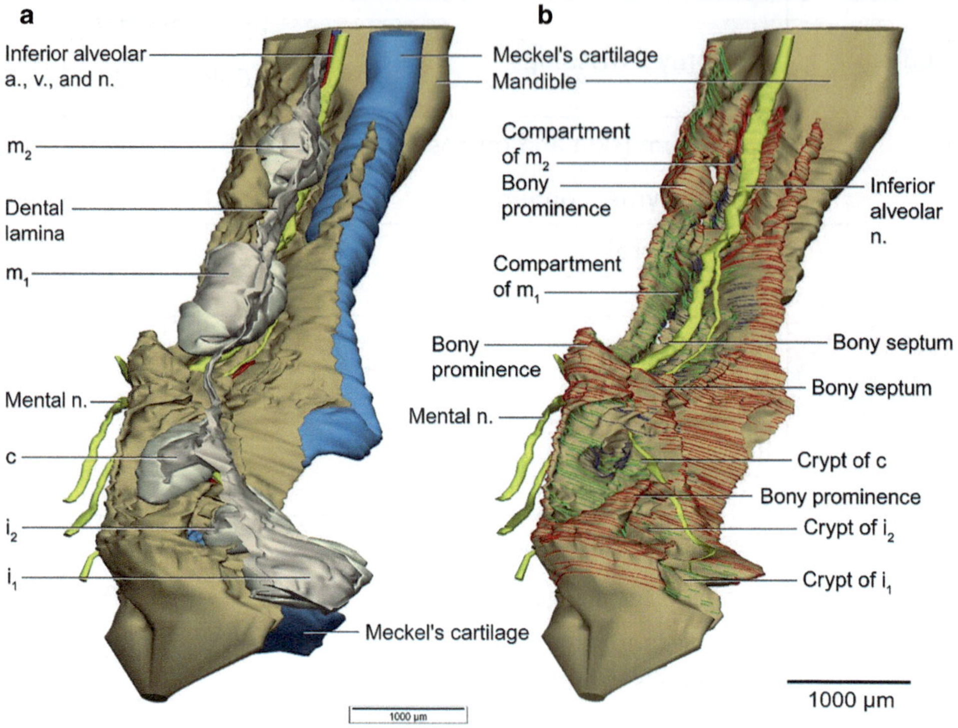

Fig. 3.9 Another example of a more complex reconstruction of a part of the mandible. (**a**) Right half of the mandible with dental primordia and Meckel's cartilage, cranial view. (**b**) The same reconstruction, dental primordia removed, bone remodeling pattern (red: apposition, green: resorption) superimposed

References

1. Romeis B. Mikroskopische Technik. 17. Aufl. München: Urban & Schwarzenberg; 1989.
2. Gaunt PN, Gaunt WA. Three dimensional reconstruction in biology. London: Pittman Medical Publications; 1978.
3. Ongaro I, Sperber GH, Machin GA, Murdoch CA. Fiducial points for three-dimensional computer-assisted reconstruction of serial light microscopic sections of umbilical cord. Anat Rec. 1991;229:285–9.
4. Knabe W, Washausen S, Brunnett G, Kuhn HJ. Use of "reference series" to realign histological serial sections for three-dimensional reconstructions of the positions of cellular events in the developing brain. J Neurosci Methods. 2002;121:169–80.

Part II

Plates

4 Embryo 15 mm CRL (ca. 6th wk): STE290687

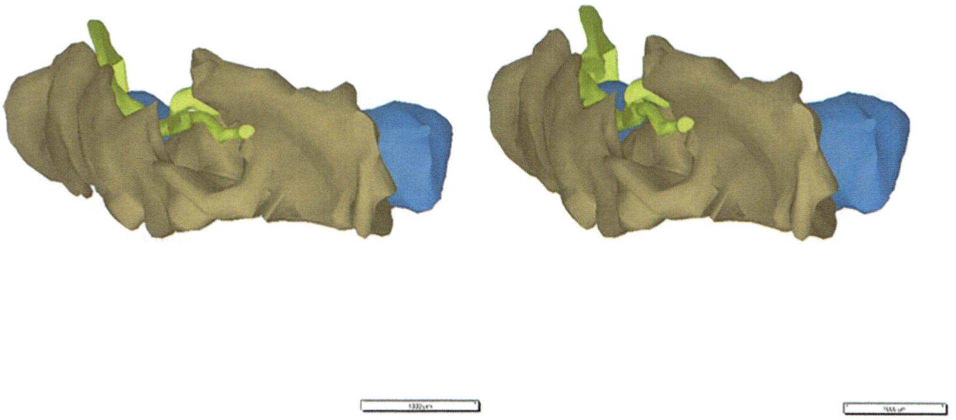

Fig. 4.1 Stereoscopical pair of images. Human embryo, 16 mm CRL. Right half of the mandible, anterior view. Ochre: bone, blue: Meckel's cartilage, yellow: inferior alveolar n., Bar: 500 µm

Mandible

The mandible is the first bone to ossify [1]. Prior to that, Meckel's cartilage had formed as a v-shaped framework within the first visceral arch in the fifth week of development [2, 3]. As Meckel's cartilage grows quickly, it exercises shearing forces for initiating and spreading bone formation in the space between the cartilage and the adjacent skin epithelium [4–6]. Bone formation starts, where the inferior alveolar nerve crosses Meckel's cartilage (Fig. 4.1) [7].

Cartilaginous Skeleton

Meckel's cartilage is running medially of the mandibular bony formation. At this stage of development, the cartilage is not embraced by bone (Figs. 4.2, 4.3, and 4.4).

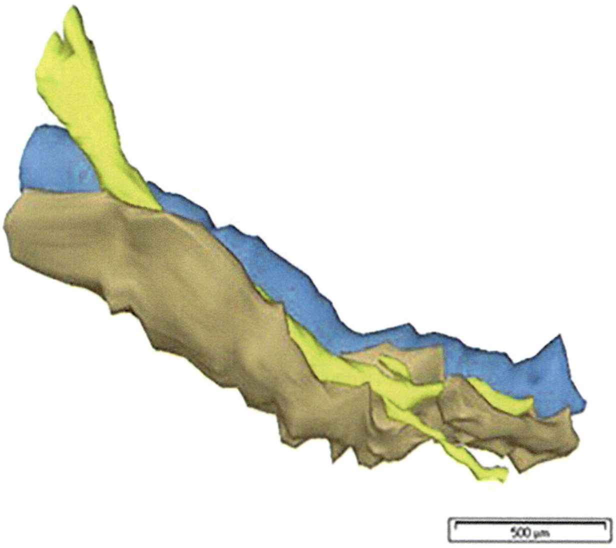

Fig. 4.2 Human embryo, 16 mm CRL. Right half of the mandible, lateral view. Ochre: bone; blue: Meckel's cartilage; yellow: inferior alveolar n.

R. J. Radlanski, *Atlas of Human Prenatal Craniofacial Development*, https://doi.org/10.1007/978-3-031-88812-0_4

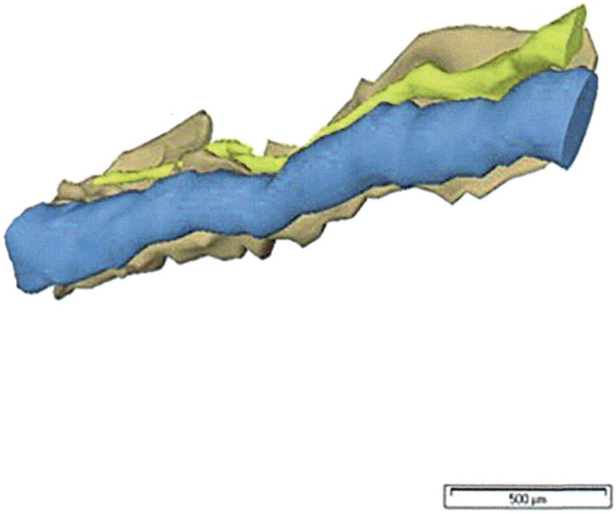

Fig. 4.3 Human embryo, 16 mm CRL. Right half of the mandible, medial view. Ochre: bone; blue: Meckel's cartilage; yellow: inferior alveolar n.

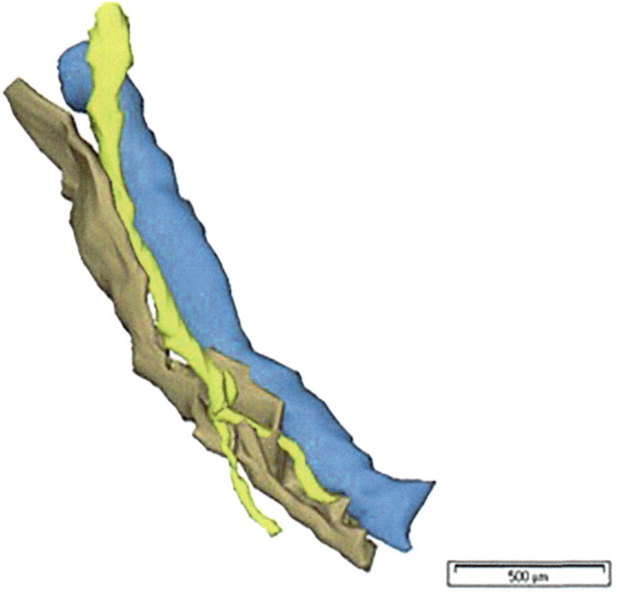

Fig. 4.4 Human embryo, 16 mm CRL. Right half of the mandible, cranial view. Ochre: bone; blue: Meckel's cartilage; yellow: inferior alveolar n.

Nerves

The alveolar inferior nerve runs in the mesenchymal space between Meckel's cartilage and mandible. At the mental foramen, the mental nerve branches off laterally. At this point, the mandible forms a small trough whose medial wall is quite close to the cartilage. Past the branch point, the inferior alveolar nerve continues in anterior direction.

Mental Foramen

Note that the orifice of the mental foramen is much wider open than the diameter of the mental nerve would require. Even if the concomitant vessels, which pass the foramen with the nerve are taken into account (not visible in this reconstruction), the foramen is much wider than actually required [8, 9]. Furthermore, at this stage of development, the bony encirclement is not complete—it is open cranially and slightly dorsally.

References

1. Patten BM. Human embryology. Philadelphia: Blakiston; 1946.
2. Radlanski RJ, Kjaer I, Vastardis H, Renz H. Morphometrische Untersuchungen zur Fetalentwicklung der menschlichen Mandibula. Fortschr Kieferorthop. 1994;55:77–83.
3. Kjaer I, Keeling JW, Fischer-Hansen B. The prenatal human cranium - normal and pathologic development. Copenhagen: Munksgaard; 1999.
4. Blechschmidt E. Mechanische Genwirkungen Funktionsentwicklung I. Göttingen: Musterschmidt; 1948.
5. Blechschmidt E. The stages of human development before birth. An introduction to human embryology. - Die vorgeburtlichen Entwicklungsstadien des Menschen. Eine Einführung in die Humanembryologie. Karger, London, New York, Basel, Freiburg i. B.: Karger; 1960.
6. Blechschmidt E. The ontogenetic basis of human anatomy: the biodynamic approach to development from conception to adulthood [edited by Brian Freeman]. Berkeley, CA: North Atlantic Books; 2004.
7. Radlanski RJ, Renz H, Klarkowski MC. Prenatal development of the human mandible. 3D reconstructions, morphometry and bone remodelling pattern, sizes 12-117 mm CRL. Anat Embryol (Berl). 2003;207:221–32.
8. Radlanski RJ, Renz H, Müller U, Schneider RS, Marcucio RS, Helms JA. Prenatal morphogenesis of the human mental foramen. Eur J Oral Sci. 2002;110:452–9.
9. Radlanski RJ, Renz H, Lajvardi S, Schneider R. A bone remodeling during prenatal morphogenesis of the human mental foramen. Eur J Oral Sci. 2004;112:301–10.

Embryo 18 mm CRL (ca. 6th wk): GUS110785

5

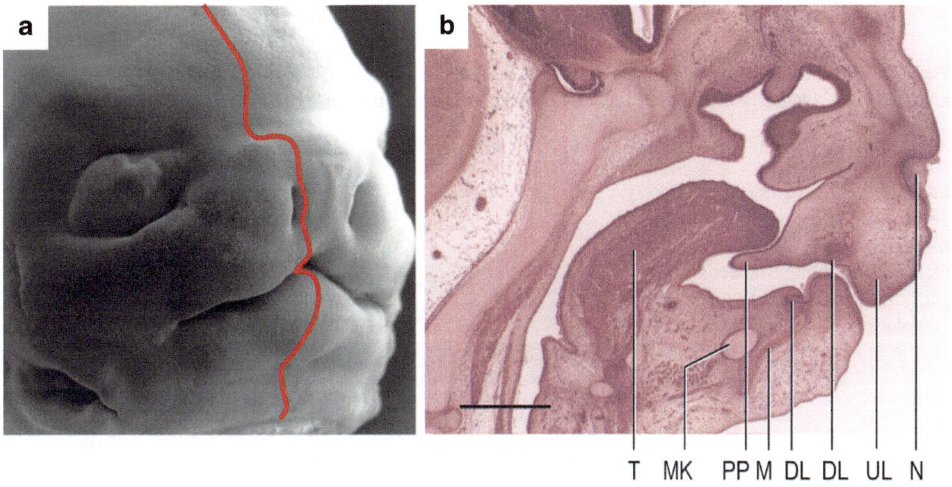

Fig. 5.1 Human embryo, 18 mm CRL, facial region. (a) Scanning electron microscopic image, the red line indicates the section plane of **b**. (**b**) Vertical section (staining: Hematoxylin-eosin), section plane indicated in **a**. *T* Tongue, *MK* Meckel's cartilage, *PP* Primary palate, *M* Mandible, *DL* Dental lamina (upper and lower), *UL* Upper lip, *N* Nasal placode; Bar: 250 mm. (From Radlanski [1]; with permission)

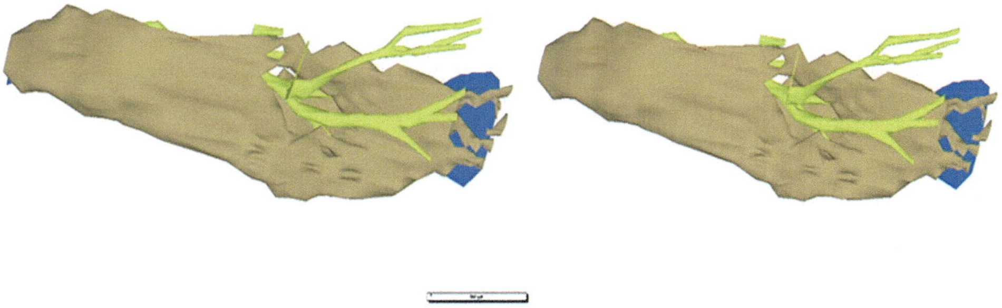

Fig. 5.2 Stereoscopical pair of images. Human embryo, 18 mm CRL. Right half of the mandible, anterior view. Ochre: bone; blue: Meckel's cartilage; yellow: inferior alveolar n. Bar: 500 µm

Mandible

As a flat, extended bone, the mandible runs laterally from Meckel's cartilage (Figs. 5.2 and 5.3).

R. J. Radlanski, *Atlas of Human Prenatal Craniofacial Development*, https://doi.org/10.1007/978-3-031-88812-0_5

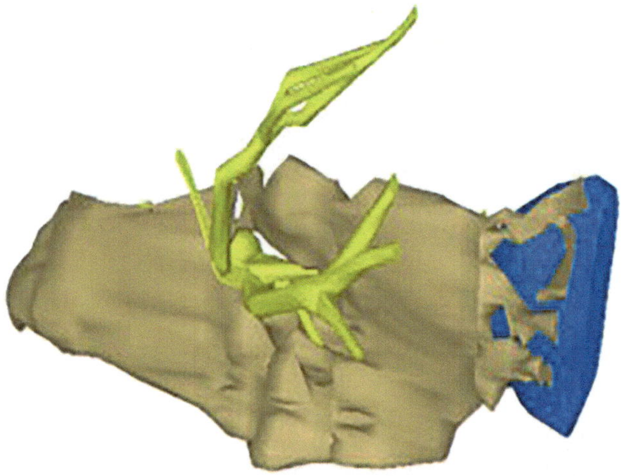

Fig. 5.3 Human embryo, 18 mm CRL. Right half of the mandible, anterior view. Ochre: bone; blue: Meckel's cartilage; yellow: inferior alveolar n

Cartilaginous Skeleton

Slightly curved, Meckel's cartilage is running medially of the mandibular bony formation. At this stage of development, the cartilage is not embraced by bone.

Nerves

The alveolar inferior nerve runs in the mesenchymal space between Meckel's cartilage and mandible. At the mental foramen, the mental nerve branches off laterally. At this point, the mandible forms a small trough whose medial wall is quite close to the cartilage. Past the branch point, the inferior alveolar nerve continues in anterior direction (Figs. 5.4 and 5.5).

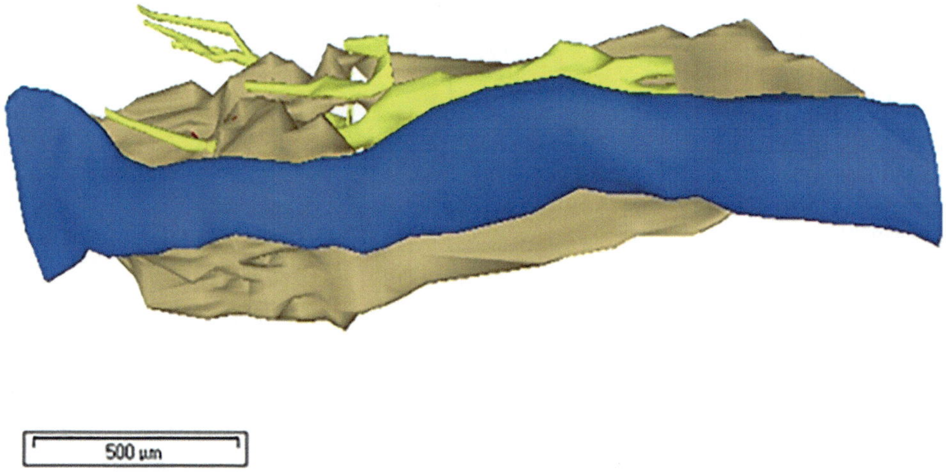

Fig. 5.4 Human embryo, 18 mm CRL. Right half of the mandible, medial view. Ochre: bone; blue: Meckel's cartilage; yellow: inferior alveolar n

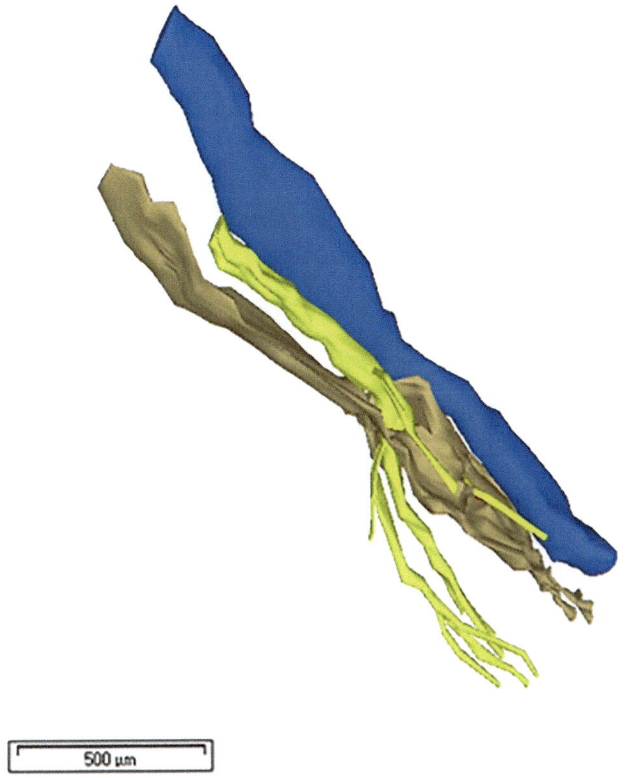

Fig. 5.5 Human embryo, 18 mm CRL. Right half of the mandible, cranial view. Ochre: bone; blue: Meckel's cartilage; yellow: inferior alveolar n

Mental Foramen

At the point where the mental nerve branches off from the alveolar nerve, it leaves free an opening, the mental foramen, which at this stage already encircles the nerve. The diameter of the foramen, however, is larger than the nerve and its accompanying vessels would require [2, 3].

Dental Primordia

At a certain "odontogenic area" [4] mediated by developmental mechanics [5–7] and by activity of specific growth factors [8–10], the epithelium of the oral cavity thickens. As it progresses, the epithelium sinks further and further into the underlying mesenchyme. At this stage of development, the separation into the dental lamina and the vestibular lamina just becomes visible (Fig. 5.6).

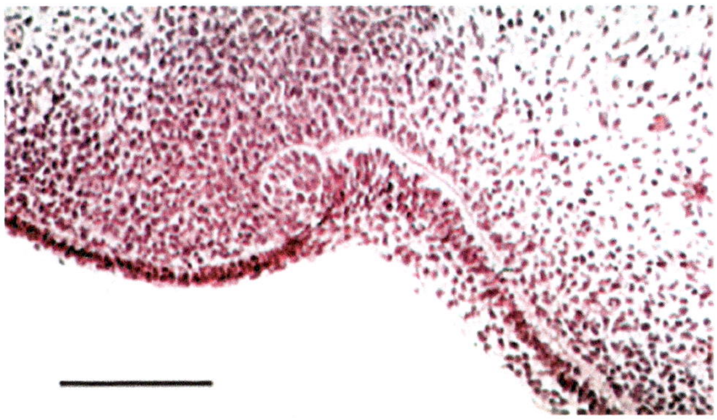

Fig. 5.6 Human embryo, 18 mm CRL. Thickening of the oral epithelium, formation of the dental (DL) and vestibular (VL) laminae, just slightly separated by the small wedge at the top center. Sagittal section, stained Hematoxylin-eosin, bar: 125 µm. (From Radlanski [1]; with permission)

References

1. Radlanski RJ. Oral structure and biology. Batavia: Quintessence Publishing; 2018.
2. Radlanski RJ, Renz H, Müller U, Schneider RS, Marcucio RS, Helms JA. Prenatal morphogenesis of the human mental foramen. Eur J Oral Sci. 2002;110:452–9.
3. Radlanski RJ, Renz H, Lajvardi S, Schneider R. A bone remodeling during prenatal morphogenesis of the human mental foramen. Eur J Oral Sci. 2004;112:301–10.
4. Nery EB, Croup M. Timing and topography of early human tooth development. Archs oral Biol. 1970;15:315–1326.
5. BlechschmidtE.MechanischeGenwirkungenFunktionsentwicklung I. Göttingen: Musterschmidt; 1948.
6. Blechschmidt E. Die Entwicklungsbewegungen der Zahnleiste. Funktionelle Faktoren bei der Frühentwicklung des menschlichen Kauapparates. Roux' Archiv für Entwicklungsmechanik. 1955;147:474–88.
7. Radlanski RJ, Renz H. Developmental movements of the inner enamel epithelium as derived from micromorphological features. Eur J Oral Sci. 2006;114(Suppl 1):343–8; discussion 349–50, 382. https://doi.org/10.1111/j.1600-0722.2006.00314.x.
8. Ruch JV, Karcher-Djuricic V. On odontogenic tissue interactions. In: Slavkin HC, Greulich RC, editors. Extracellular matrix influences on gene expression. New York: Academic Press; 1975. p. 549–54.
9. Slavkin HC. Molecular determinants during dental morphogenesis and cytodifferentiation: a review. J Craniofac Genet Dev Biol. 1991;11(4):338–49.
10. Thesleff I, Sharpe P. Signalling networks regulating dental development. Mech Dev. 1997;67(2):111–23. https://doi.org/10.1016/s0925-4773(97)00115-9.

6 Embryo 19 mm CRL (ca. 7th wk): CHR220687

Overall Remarks

In the seventh week, all essential anatomical structures of the face, or better the head region, are already well recognizable. The orofacial muscles are the first musculature to be formed in the human body. At least the attachments are already clearly defined, well recognizable, the final position and location will be reached only with about the 14th week. The origin and attachment are also not yet clearly defined. Because of the still clearly pronounced forward flexion of the head, the parts of the viscerocranium are still relatively little extended vertically. This becomes apparent in the overview image of the histological section (Fig. 6.1) and in the 3D reconstruction (see Fig. 6.2).

R. J. Radlanski, *Atlas of Human Prenatal Craniofacial Development*, https://doi.org/10.1007/978-3-031-88812-0_6

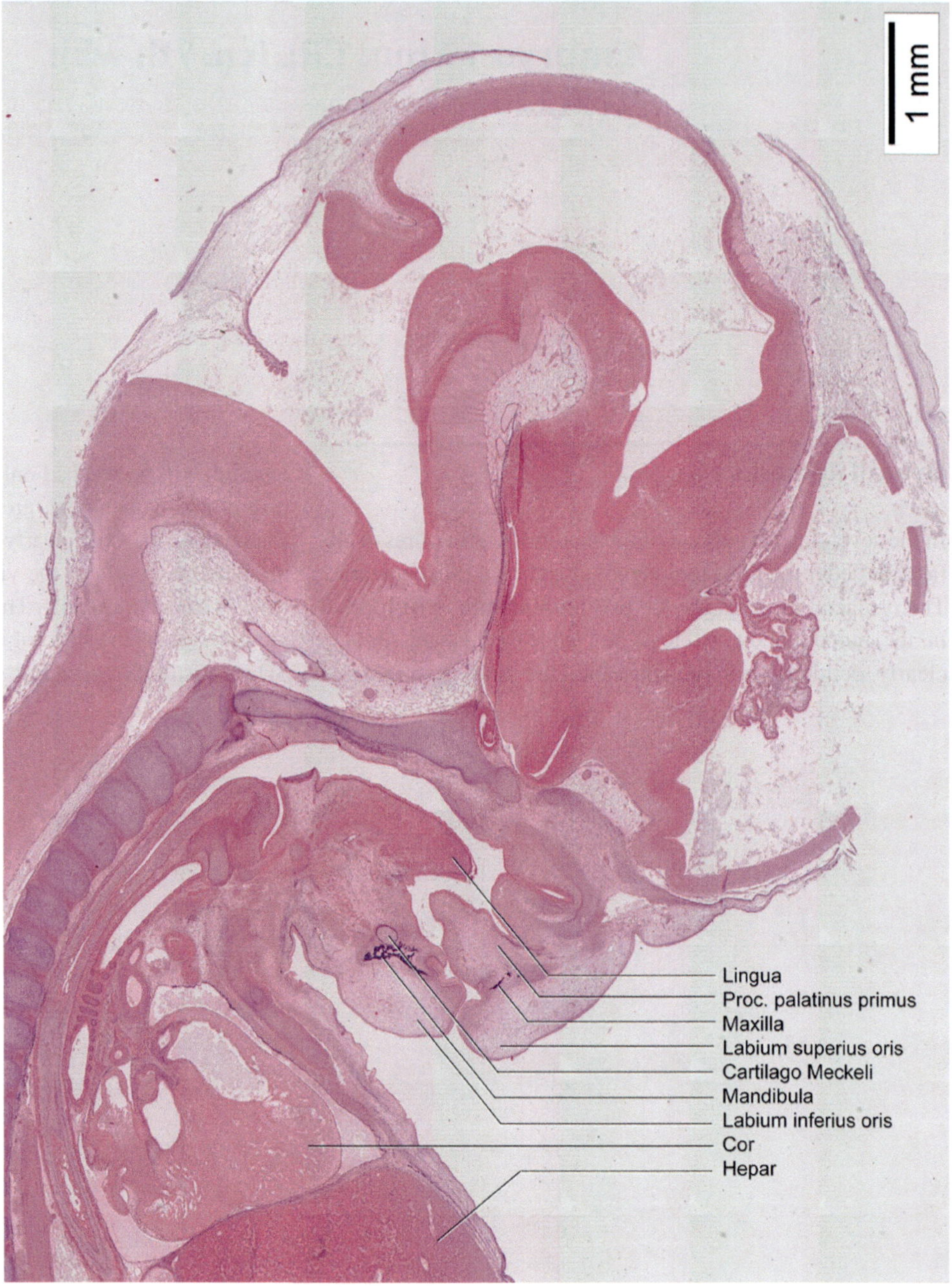

Fig. 6.1 Human embryo, 19 mm CRL (7th wk), mid-sagittal section through region of the head, staining: Hematoxylin-eosin

Muscles

The temporalis and masseter muscles were included in the reconstruction as masticatory muscles. The temporalis muscle appears as a compact muscle bulk, whereby at this stage of development only its rough position is recognizable. The known extensive origin at the side of the skull has not yet developed because no larger squamous part of the temporal is not present yet. The muscle already pulls into its insertion region, but because a coronoid process is not yet formed at the mandible, there is no insertion here. As far as histologically recognizable, individual muscle bundles pull into the tight, still barely differentiated connective tissue near the later coronoid process. The masseter muscle is also already present at this stage of development. However, its origin at the zygomatic arch, as we know it from the fetal anatomical situation, cannot yet be assumed because a zygomatic arch has not yet formed. Nevertheless, it begins its course in the corresponding region, starting from dense connective tissue, which, however, is difficult to recognize histologically. The muscle mass extends to the caudal margin of the early mandible, where some dense connective tissue is also present.

The medial pterygoid muscle follows a similar course on the medial side of the mandible. Here, too, there is still no bony origin and caudally it pulls into the dense connective tissue at the lingual margin of the mandible.

Already very well developed at this stage is the tongue and its internal musculature. In the region of the floor of the mouth, the mylohyoid, the geniohyoid, and the digastric muscles are nicely distinguishable from each other. The mylohyoid and the genioglossus are the first muscles masticatory muscles to be formed at their actual site, although interestingly, the mylohyoid has its insertion at Meckel's cartilage at this stage. Bone of the mandible is only found lateral to Meckel's cartilage, and only in later stages, there will be a detachment of the insertion from cartilage, a transition and a reattachment to its final bony insertion at the mylohyoid line, lingually at the mandible [1].

As soon as a muscle is present and can connect to the parts of the cartilaginous or bony skeleton (also via connective tissue if the fasciae are developed later), it can perform functional movements. So, an embryo can move its lips and jaws and swallow already at this age (Fig. 6.2) [2].

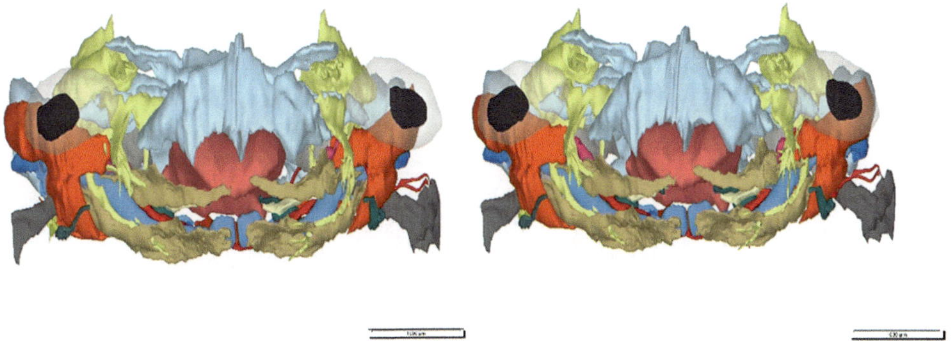

Fig. 6.2 Stereoscopical pair of images. Human embryo, 19 mm CRL (7th wk), head region, frontal view. Bar: 1 mm

Nerves

The ganglion gasseri is—compared to the rest of the face at this stage of development—impressively large and the three main branches of this fifth cranial nerve (trigeminal nerve) are recognizable as relatively thick strands.

Salivary Glands

The position and course of the excretory ducts of the major salivary glands now already correspond to the fetal situation. However, the gland bodies are only recognizable as small epithelial swellings (Fig. 6.3) [3, 4].

Maxillary Complex

In the midface, the voluminous nasal capsule is impressive. Paired and latero-caudal to it, even slightly underneath, the maxilla is following the curve of the contour of the midface as a flat bony band. Its dorsolateral extension at this stage reaches only about below the middle of the eye. The midfacial suture of the maxilla is still wide open at 250 μm. A frontal process is already recognizable; medially from the infraorbital nerve it projects about 200 μm cranially—but thus it just reaches the level of the lower edge of the eyeballs. There is no infraorbital foramen yet: The infraorbital nerve crosses the maxilla at its upper edge (Fig. 6.4).

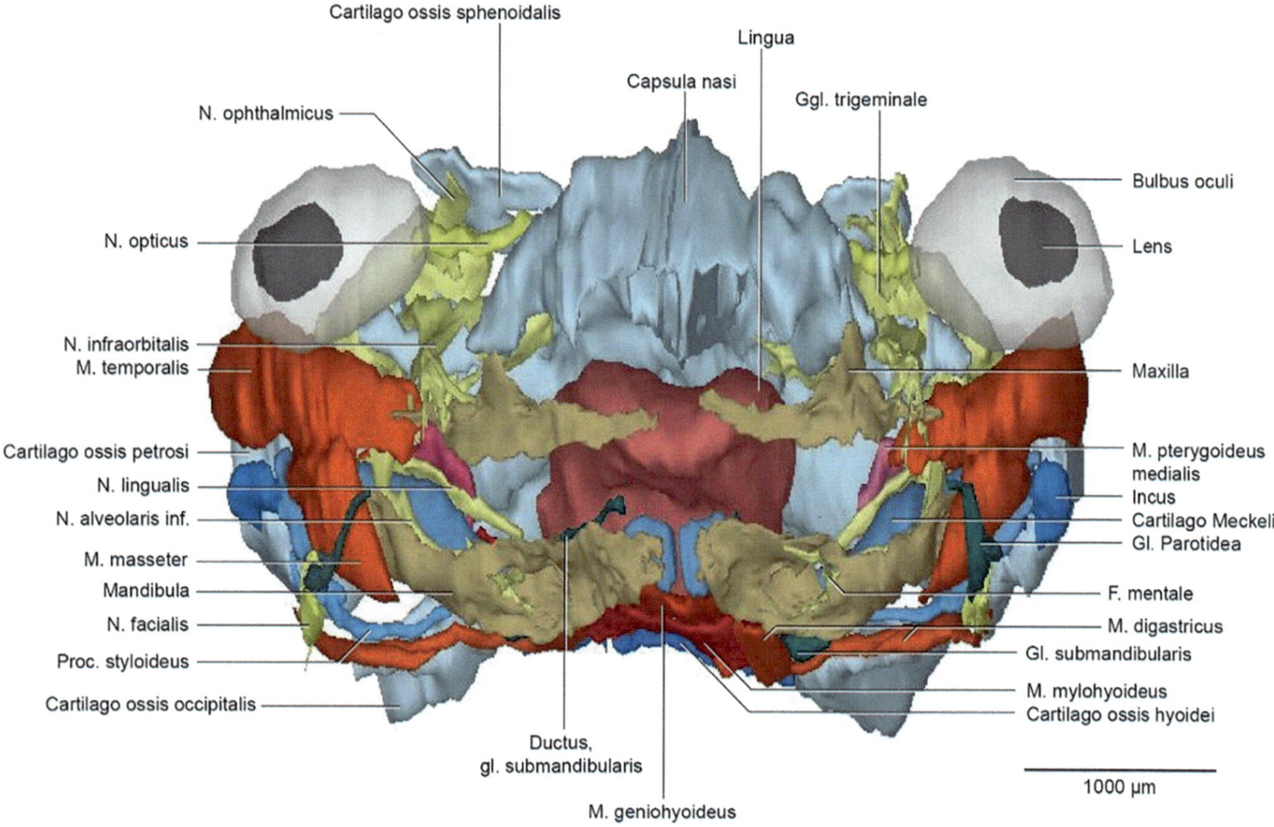

Fig. 6.3 Human embryo, 19 mm CRL (7th wk), head region, frontal view. Step 1 of preparation (*From* Radlanski and Renz [1]; *with permission*)

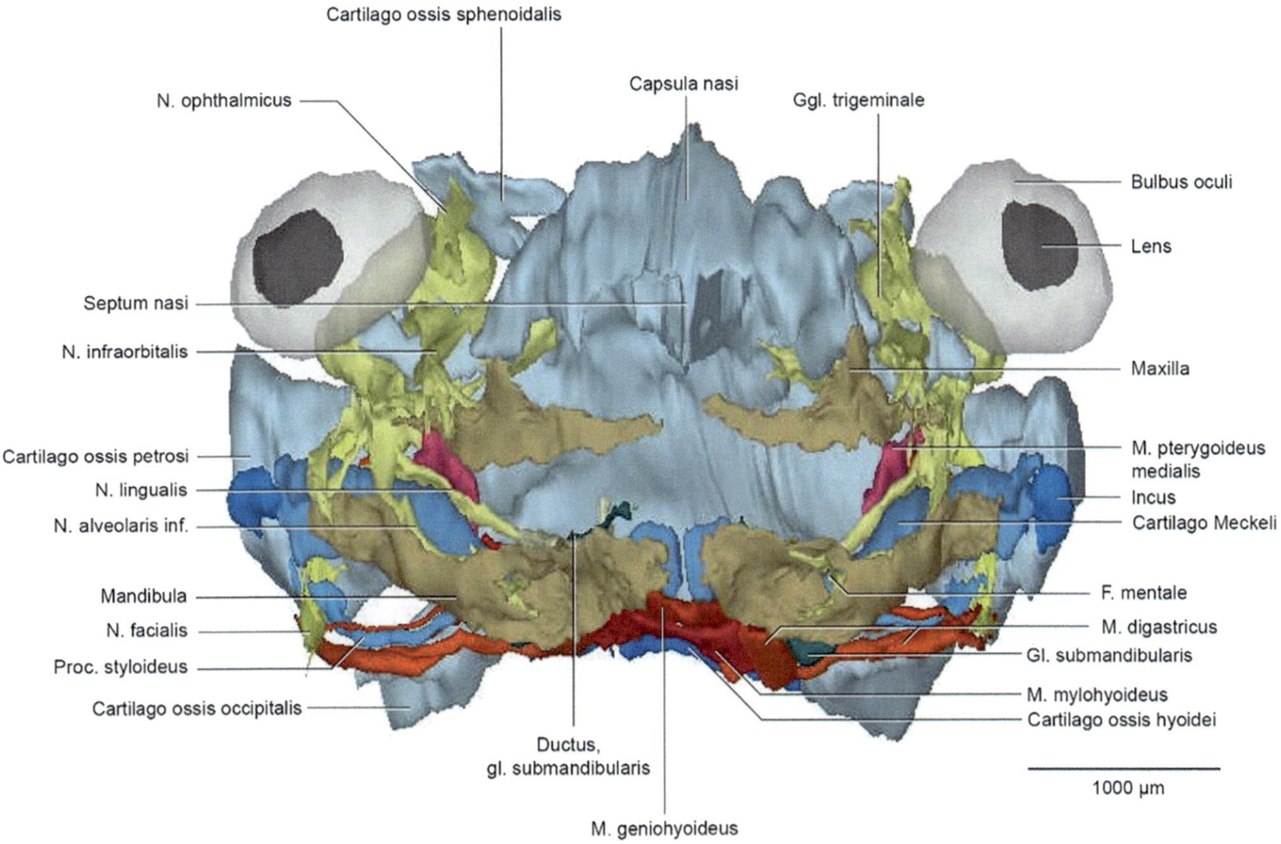

Fig. 6.4 Human embryo, 19 mm CRL (7th wk), head region, frontal view. Step 2 of preparation (*From* Radlanski and Renz [1]; *with permission*)

Mandible

The mandible lies lateral to Meckel's cartilage and follows its curved contour. In the area of the chin, the symphysis is still open by about 250 µm. The orifice of the mental foramen is closed as a ring around the passing mental nerve, whereby the diameter of the foramen is more than five times larger than that of the nerve. This leaves plenty of space for the vascular bundle and connective tissue that also passes through. At its dorsal end, the mandible noticeably swings laterally, where it also follows the sweep of Meckel's cartilage. Here, the buccal nerve and the masseteric nerve cross, but no signs of the formation of a condylar or coronoid process are yet visible (Fig. 6.5).

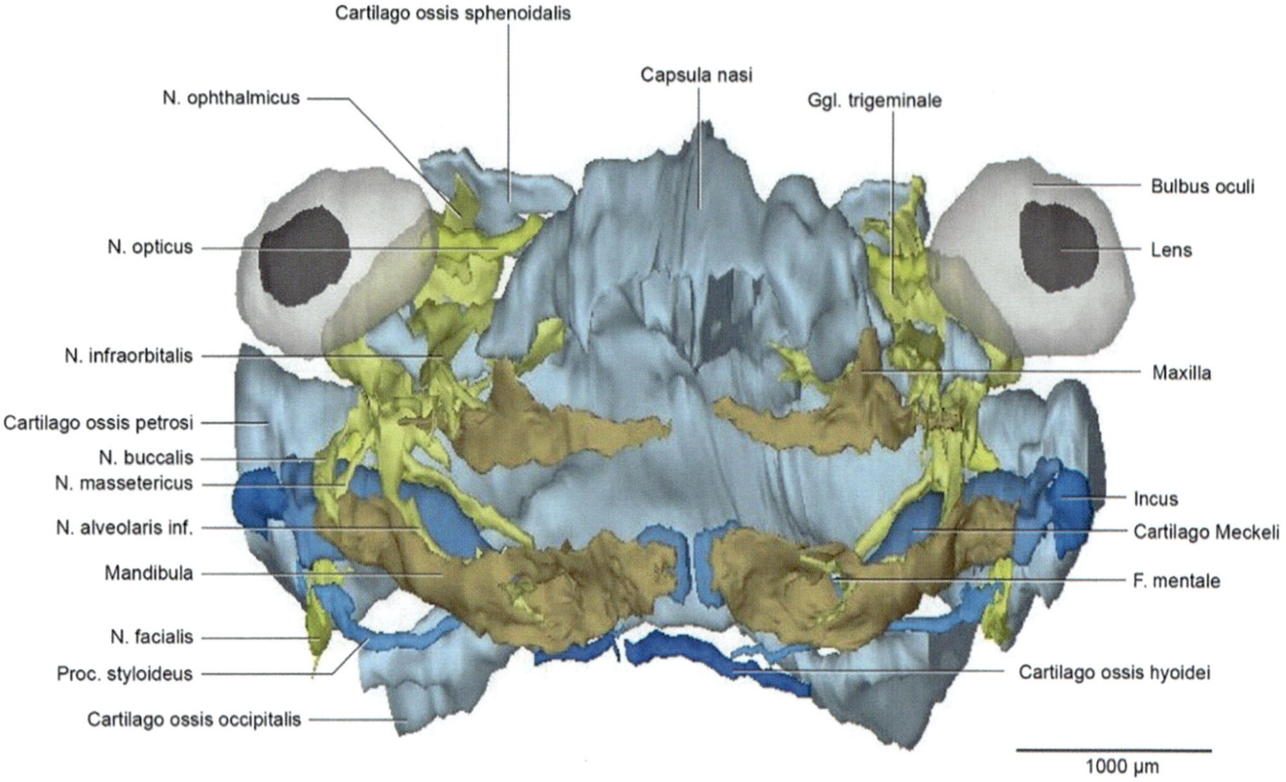

Fig. 6.5 Human embryo, 19 mm CRL (7th wk), head region, frontal view. Step 3 of preparation (*From* Radlanski and Renz [1]; *with permission*)

Cartilaginous Skeleton

The cartilaginous nasal capsule fills most of the volume of the midface. The two choanae are wide open, separated in the middle by the cartilaginous nasal septum. Lateral to the nasal capsule are the two eyeballs, with the cartilaginous primordium of the sphenoid bone already protruding dorso-medially behind the eye. At this stage, the skull base also still consists entirely of cartilage, as do the petrous bone and the occipital bone.

Meckel's cartilage runs as a paired structure in a curved arc from the mental symphysis, where the two halves do not make contact, laterally and dorsally, with its dorsal end articulating as manubrium mallei with the incus. The diameter of Meckel's cartilage is not the same at all points. In the middle of its course it is about 200 µm, in the area of the symphysis the two ends of the cartilage are thickened in a piston-shaped manner and dorsally the hammer projects in a hook-shaped manner about 750 µm caudal-medially; however, it is the upper end, which articulates with the incus.

The hyoid bone is present only as its cartilaginous precursor and the styloid process is also cartilaginous (Fig. 6.6).

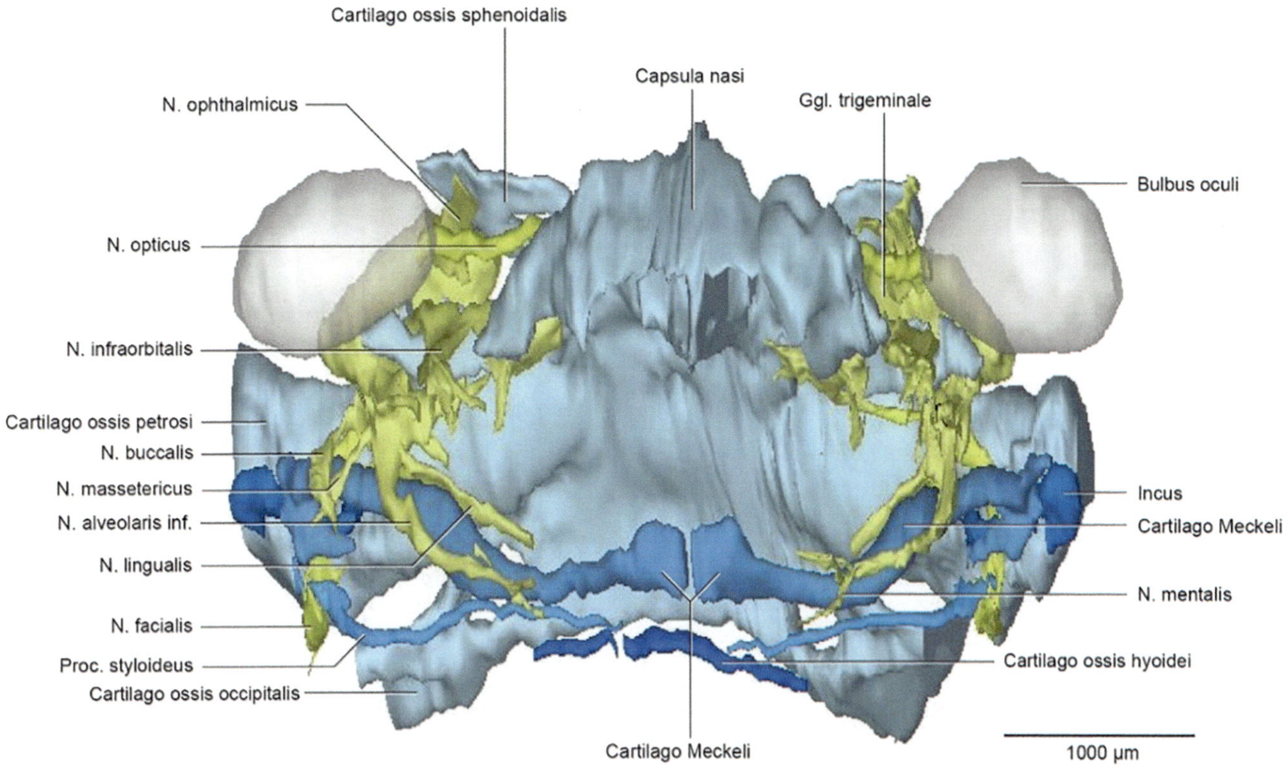

Fig. 6.6 Human embryo, 19 mm CRL (7th wk), head region, frontal view. Step 4 of preparation (*From* Radlanski and Renz [1]; *with permission*)

Temporomandibular Joint

In this stage of development, there is not yet any temporomandibular joint as would be found in the adult. At this time, Meckel's cartilage with its dorsal end, the malleus, enters into an articulated connection with the incus. When the embryo now opens its mouth, which it can do because of its already existing musculature, the connection between malleus and incus serves as a joint. Only later do these two cartilaginous structures, together with the stapes, enter the service of sound conduction, when they ossify to form the auditory ossicles. The definite temporomandibular joint develops later, when the coronoid processes will be formed and will articulate in the mandibular fossae. And it will not function as a sole temporomandibular joint until the connection between Meckel's cartilage and malleus has been fully severed. This is not the case until the 24th week of prenatal development (Figs. 6.7, 6.8, 6.9, 6.10, 6.11, 6.12, 6.13, 6.14, 6.15, and 6.16).

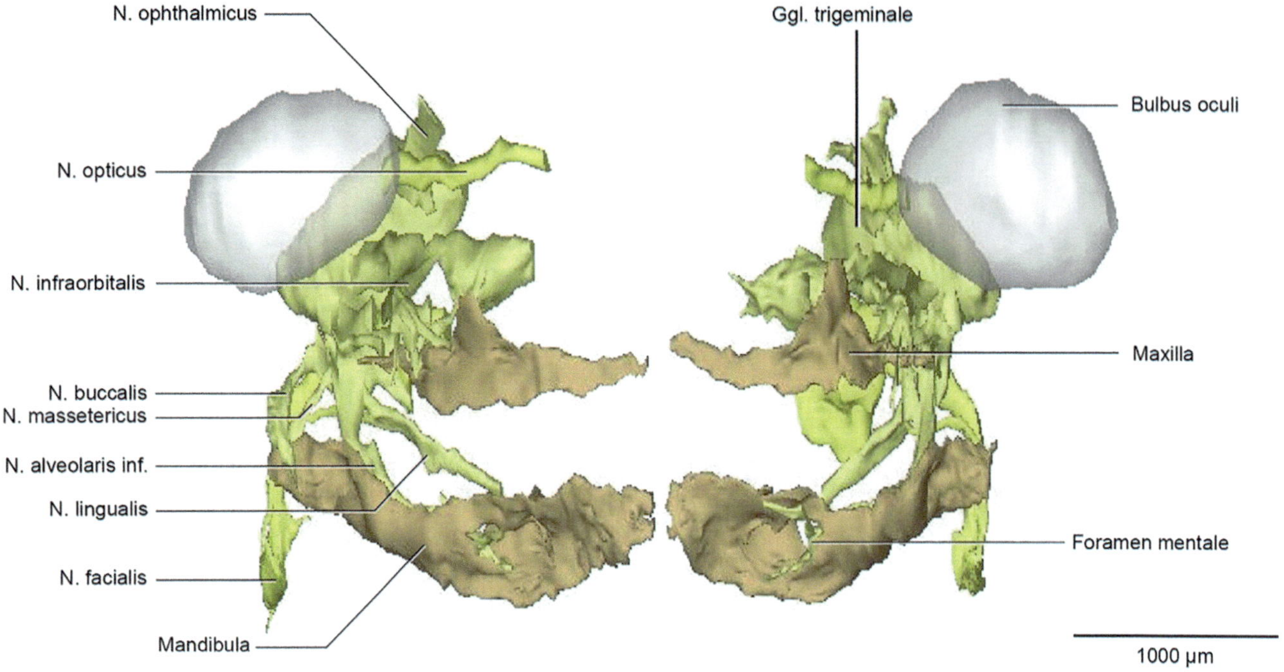

Fig. 6.7 Human embryo, 19 mm CRL (7th wk), head region, frontal view. Step 5 of preparation (*From* Radlanski and Renz [1]; *with permission*)

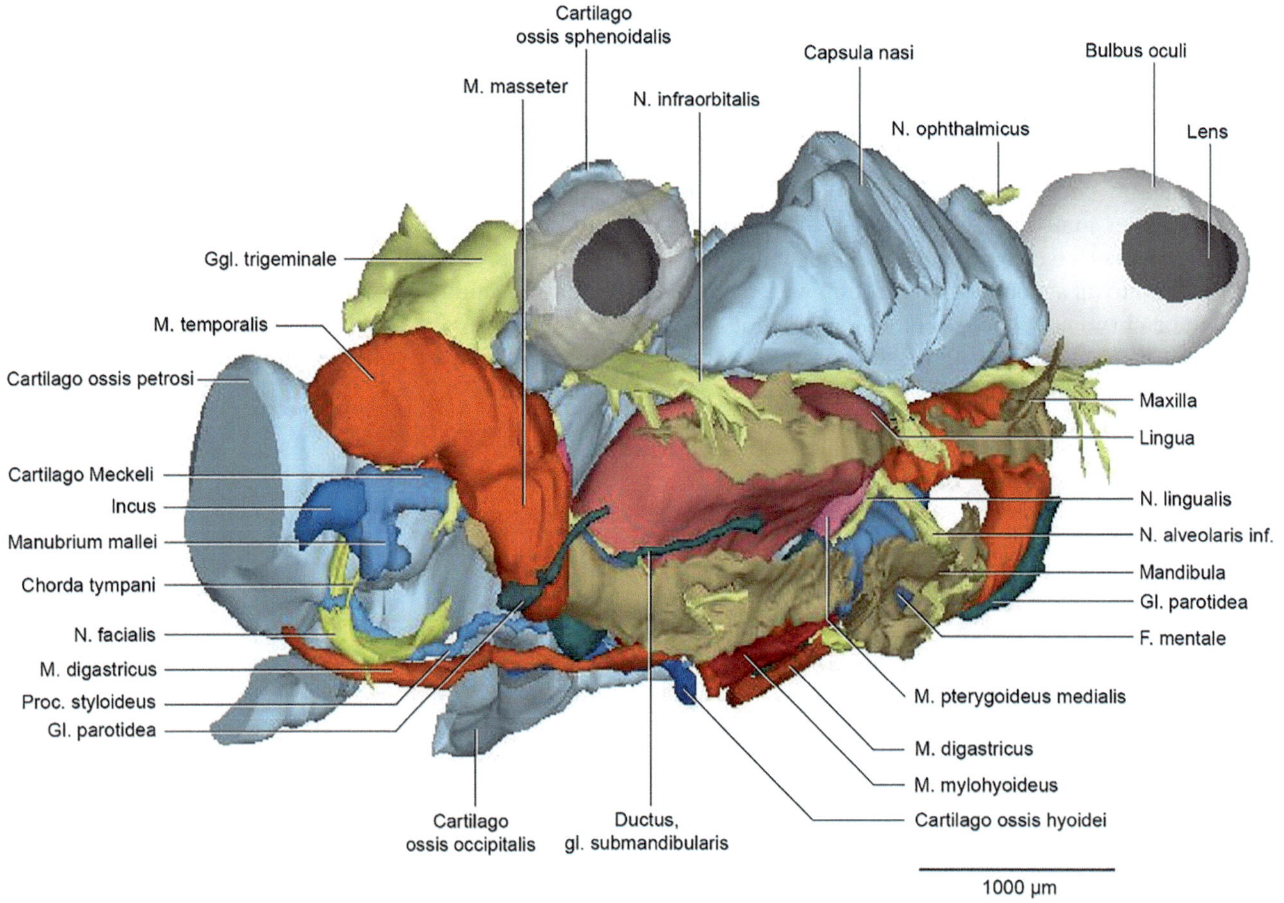

Fig. 6.8 Human embryo, 19 mm CRL (7th wk), head region, lateral and 45° frontal view. Step 1 of preparation (*From* Radlanski and Renz [1]; *with permission*)

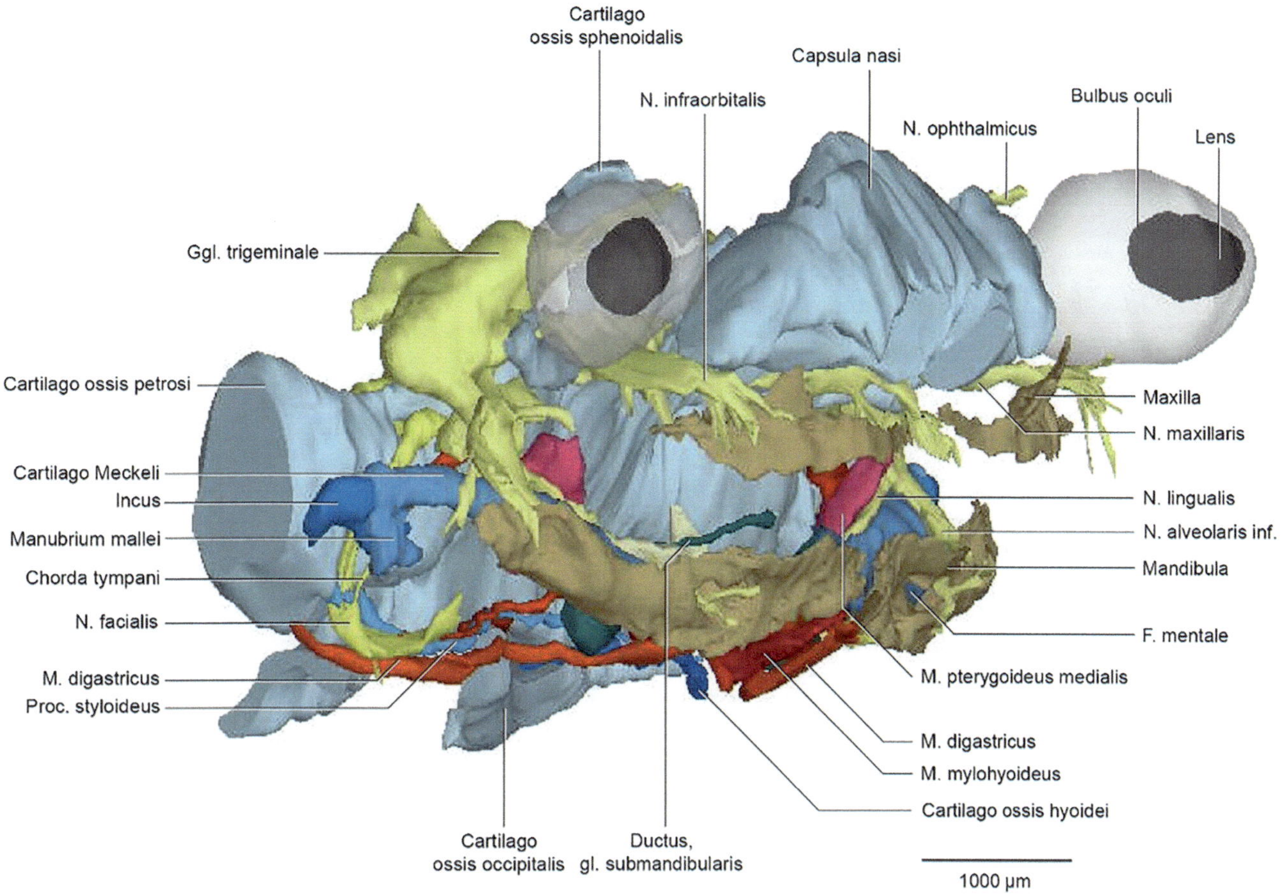

Fig. 6.9 Human embryo, 19 mm CRL (7th wk), head region, lateral and 45° frontal view. Step 2 of preparation (*From* Radlanski and Renz [1]; *with permission*)

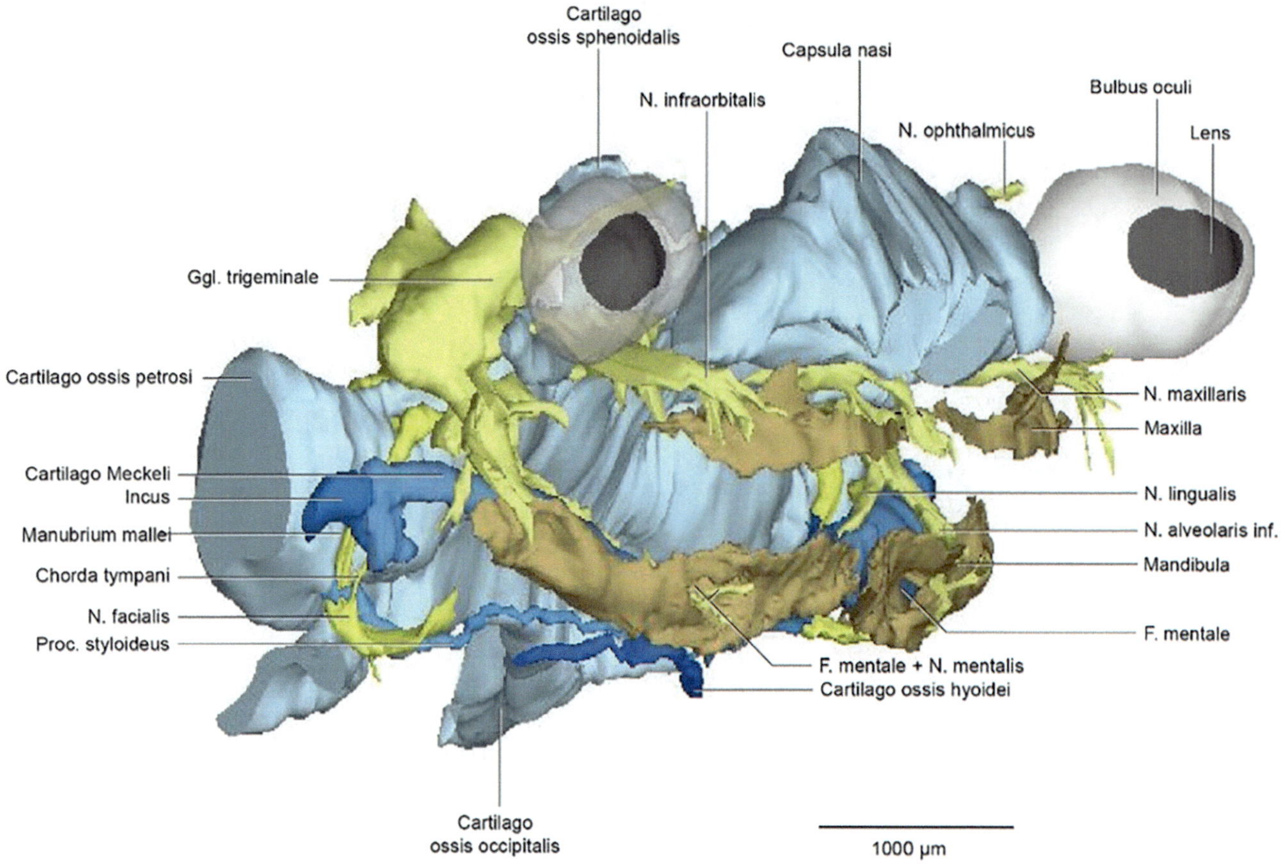

Fig. 6.10 Human embryo, 19 mm CRL (7th wk), head region, lateral and 45° frontal view. Step 3 of preparation (*From* Radlanski and Renz [1]; *with permission*)

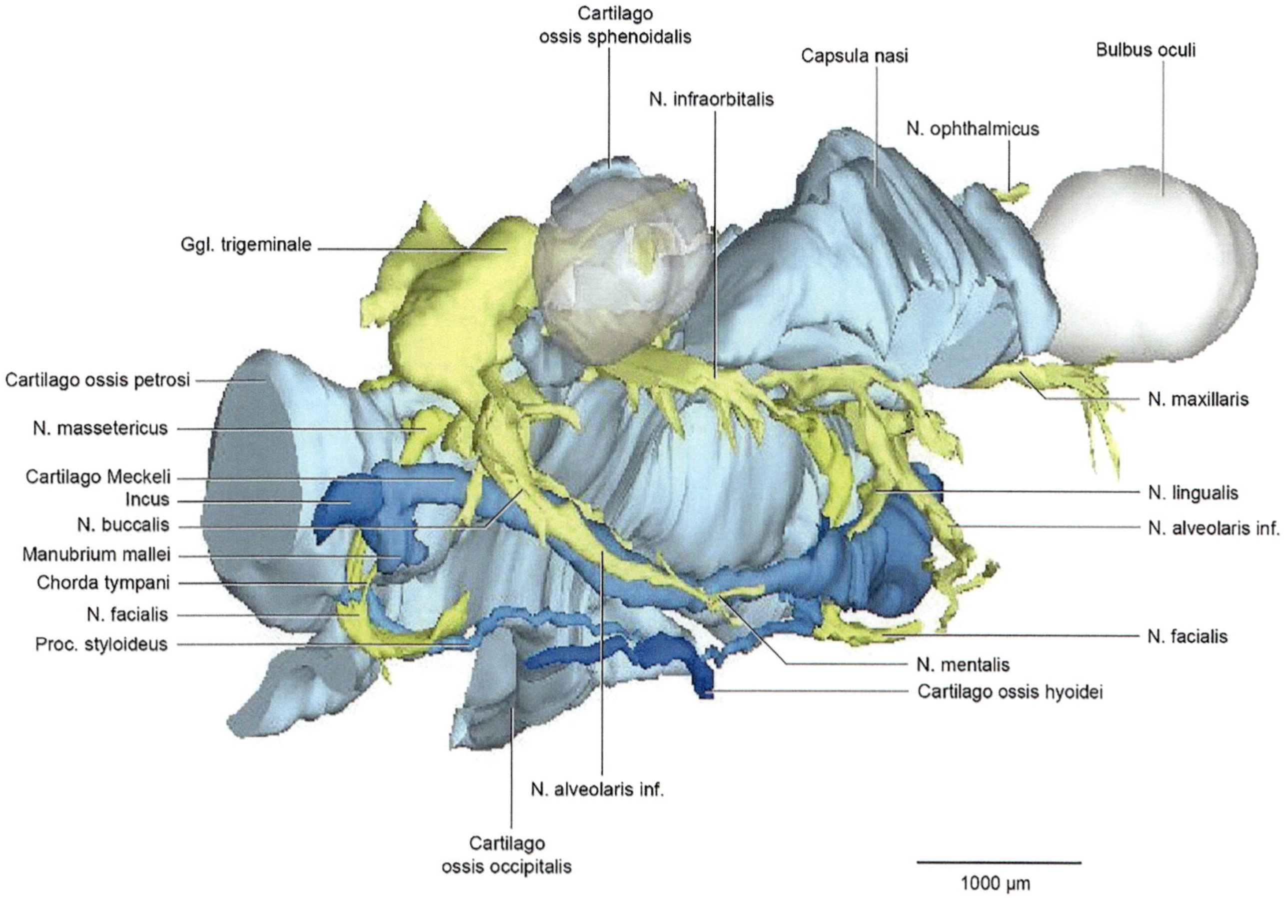

Fig. 6.11 Human embryo, 19 mm CRL (7th wk), head region, lateral and 45° frontal view. Step 4 of preparation (*From* Radlanski and Renz [1]; *with permission*)

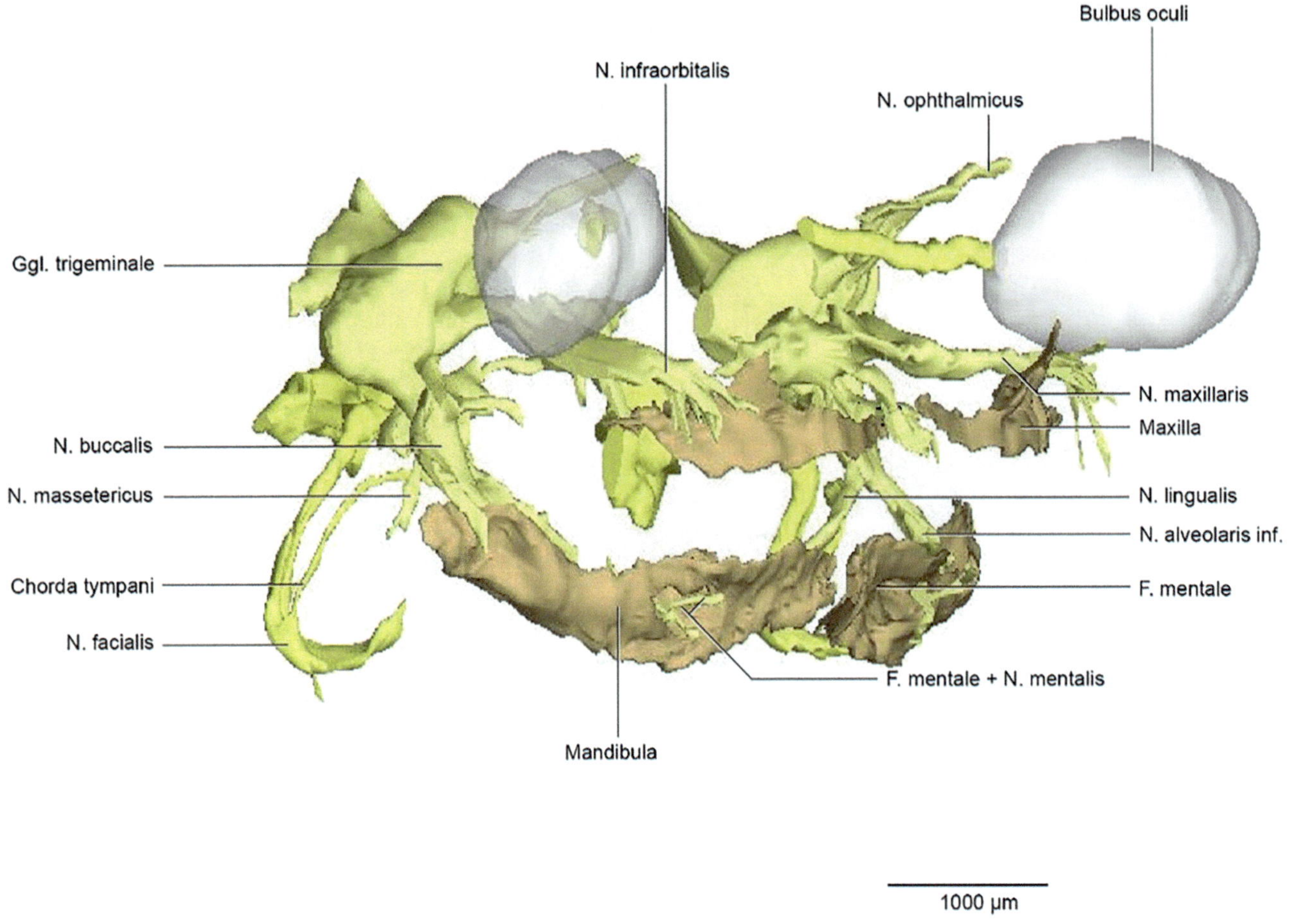

Fig. 6.12 Human embryo, 19 mm CRL (7th wk), head region, lateral and 45° frontal view. Step 5 of preparation (*From* Radlanski and Renz [1]; *with permission*)

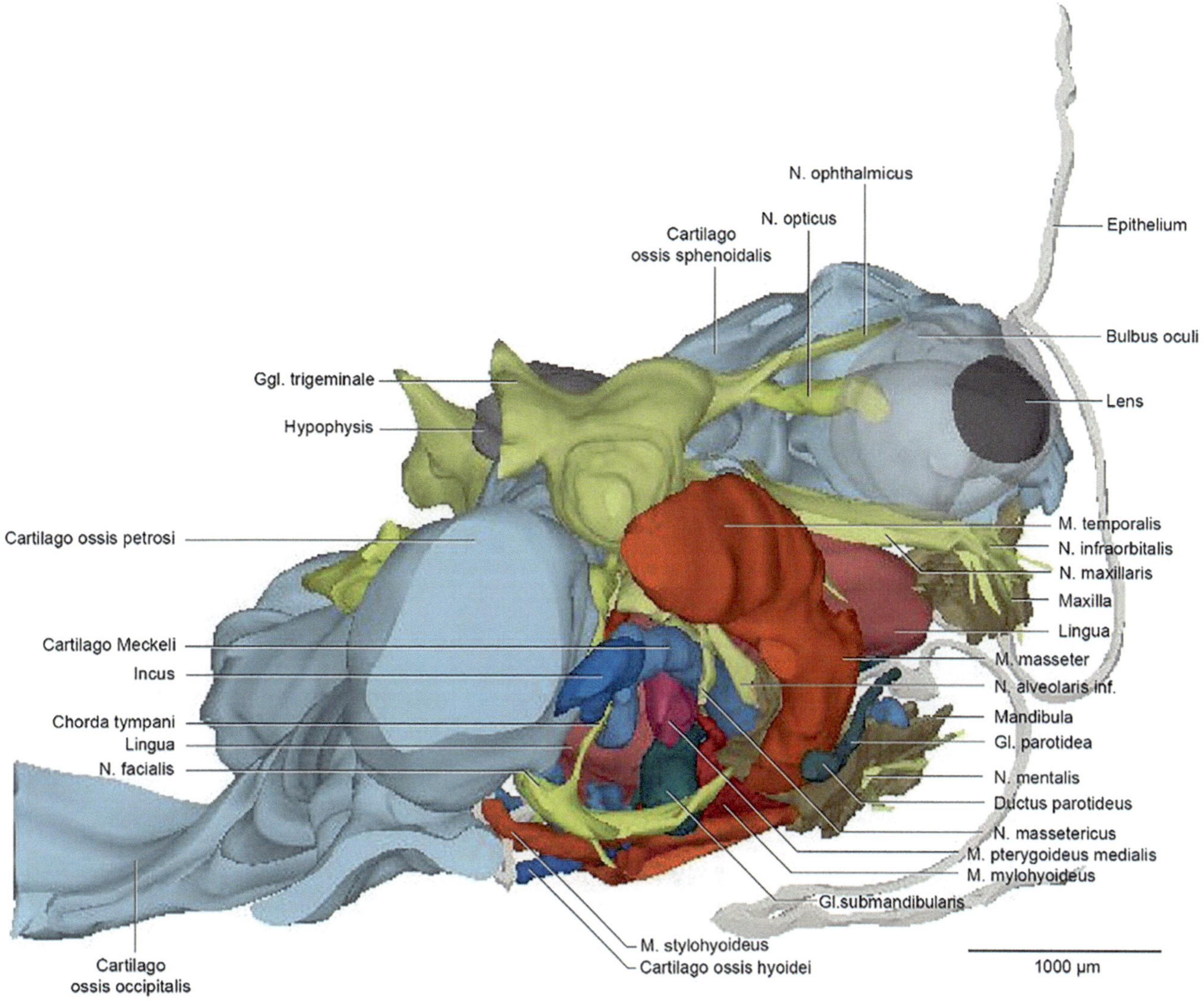

Fig. 6.13 Human embryo, 19 mm CRL (7th wk), head region, lateral view. Step 1 of preparation (*From* Radlanski and Renz [1]; *with permission*)

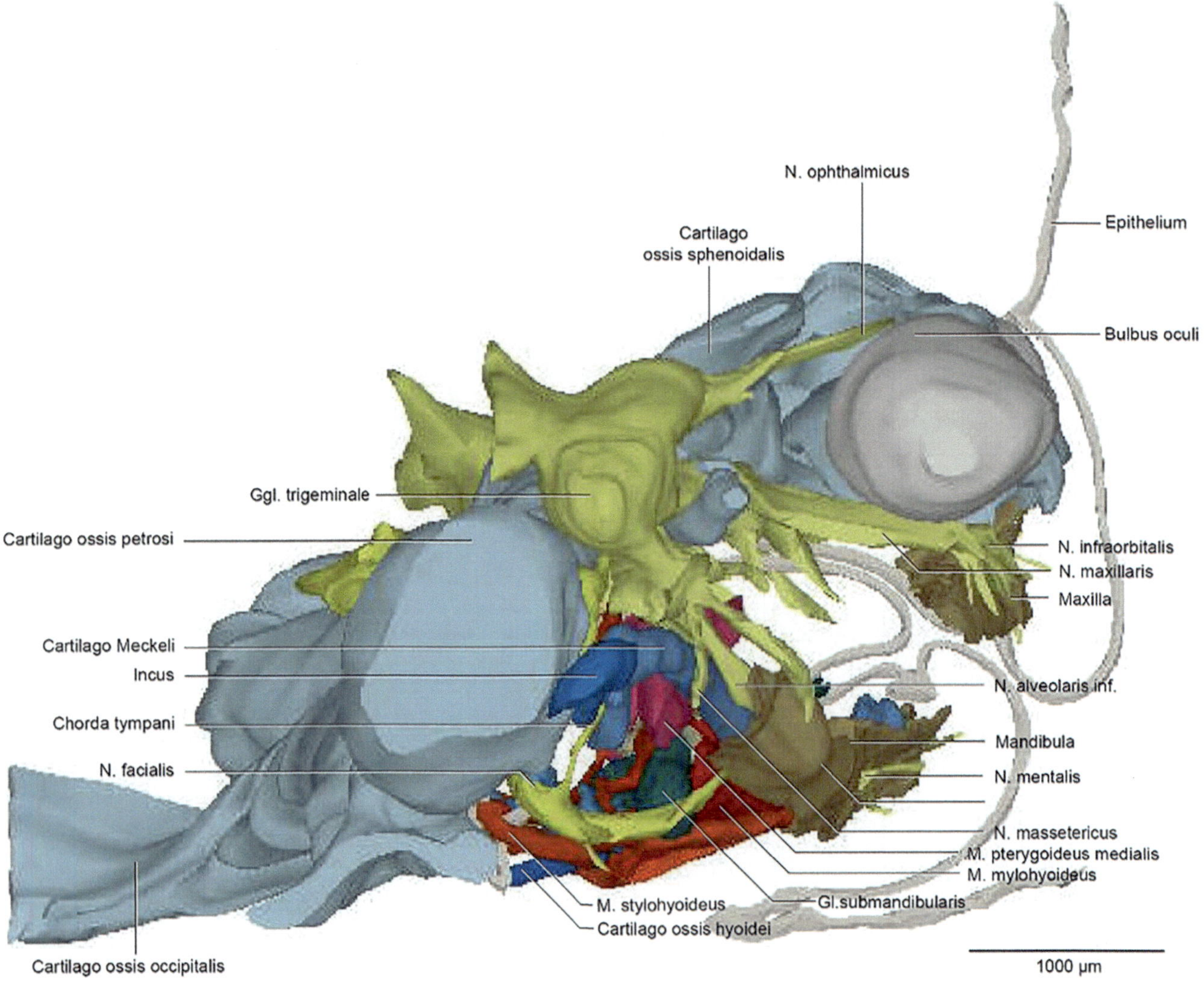

Fig. 6.14 Human embryo, 19 mm CRL (7th wk), head region, lateral view. Step 2 of preparation (*From* Radlanski and Renz [1]; *with permission*)

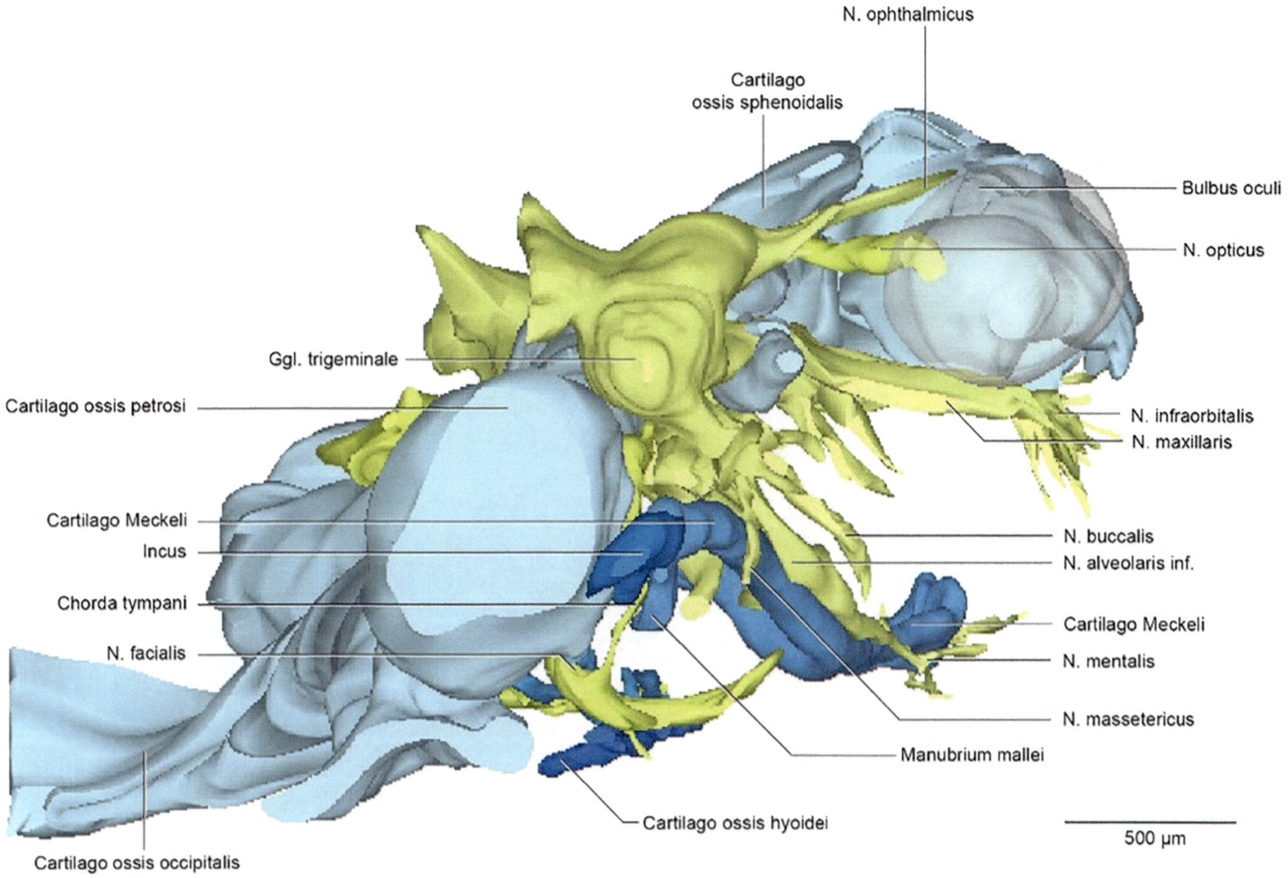

Fig. 6.15 Human embryo, 19 mm CRL (7th wk), head region, lateral view. Step 3 of preparation (*From* Radlanski and Renz [1]; *with permission*)

Fig. 6.16 Human embryo, 19 mm CRL (7th wk), head region, lateral view. Step 4 of preparation (*From* Radlanski and Renz [1]; *with permission*)

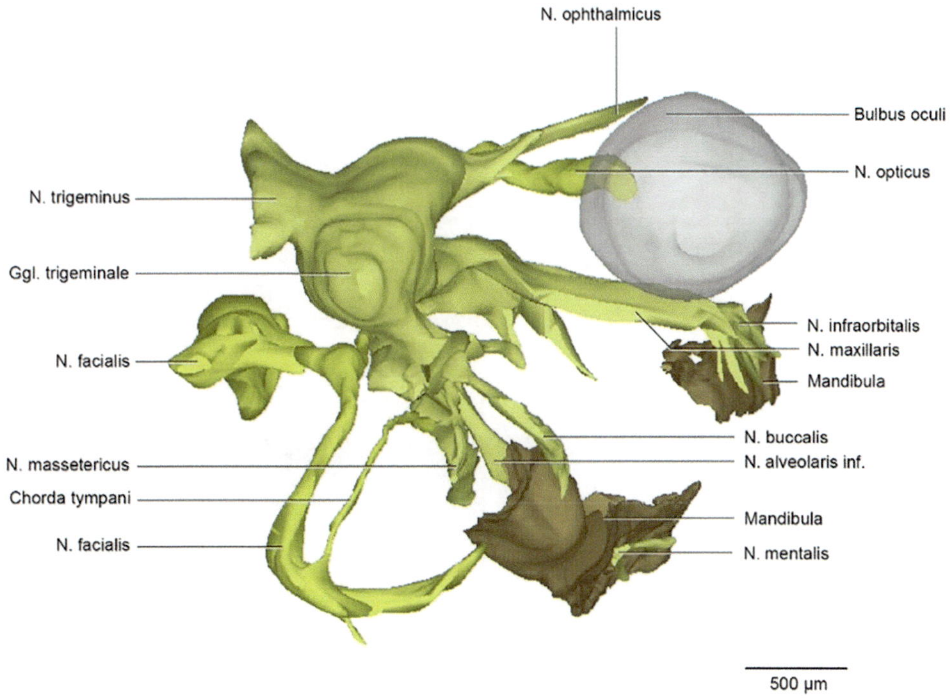

Dental Primordia

In this stage of development, all dental primordia (of the deciduous dentition) are in their lamina stage, and the site of later bud formation is already discernible in specific areas from the compaction of mesenchyme.

Formation of bone, which eventually will become the bony crypts for the dental primordia, is visible in this stage at distances of about 100–150 μm (Figs. 6.17, 6.18, 6.19, and 6.20).

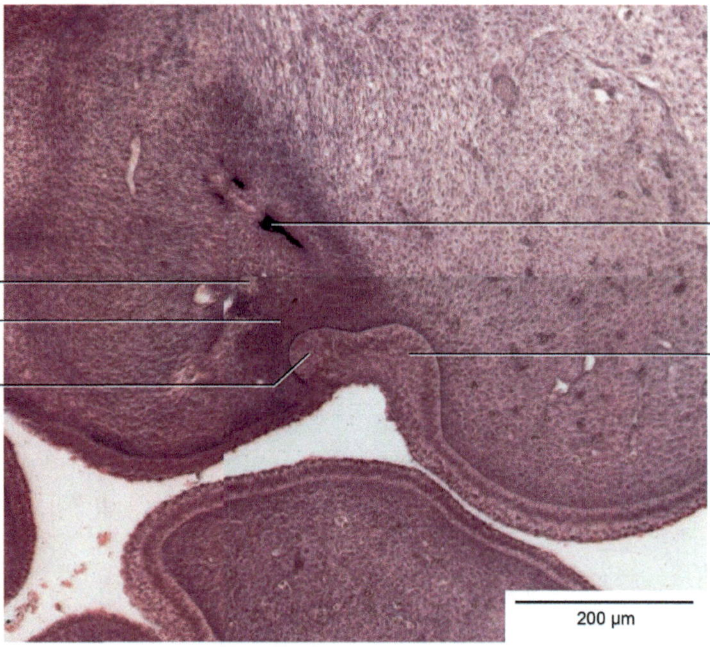

Fig. 6.17 Human embryo, 19 mm CRL (7th wk), sagittal section through region of the upper right dental primordium i^1, staining: Hematoxylin-eosin

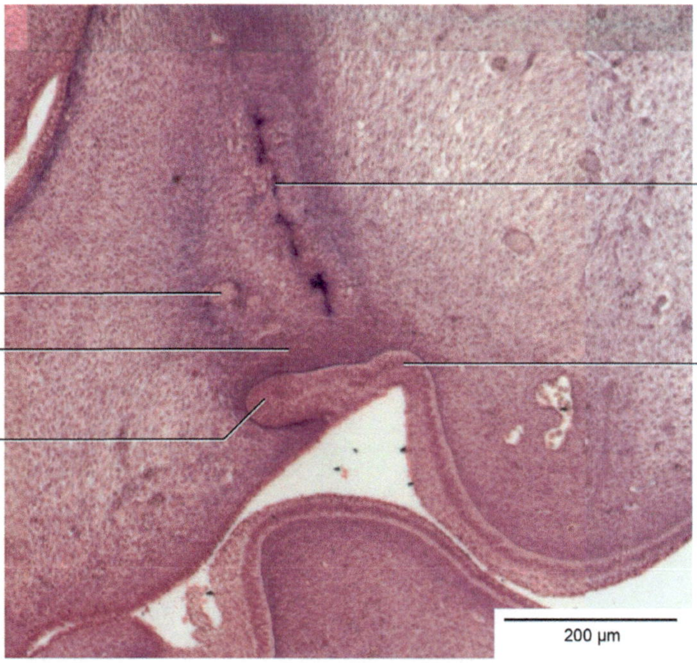

Fig. 6.18 Human embryo, 19 mm CRL (7th wk), sagittal section through region of the upper right dental primordium i^2, staining: Hematoxylin-eosin

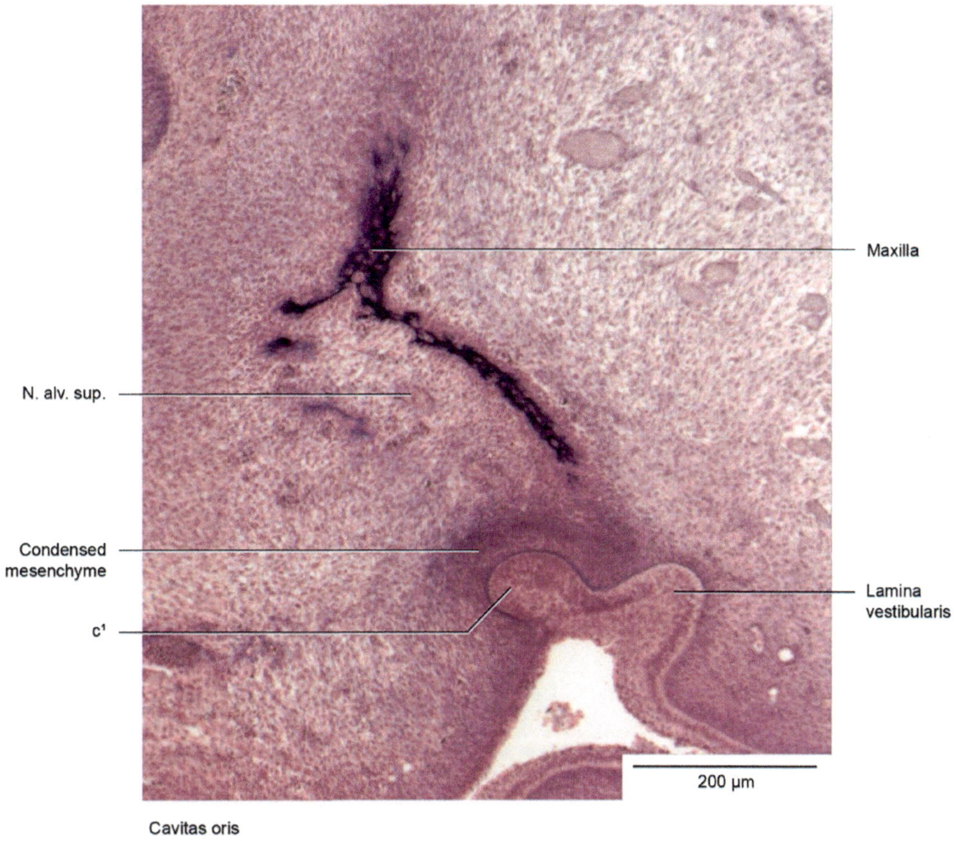

Fig. 6.19 Human embryo, 19 mm CRL (7th wk), sagittal section through region of the upper right deciduous canine primordium, staining: Hematoxylin-eosin

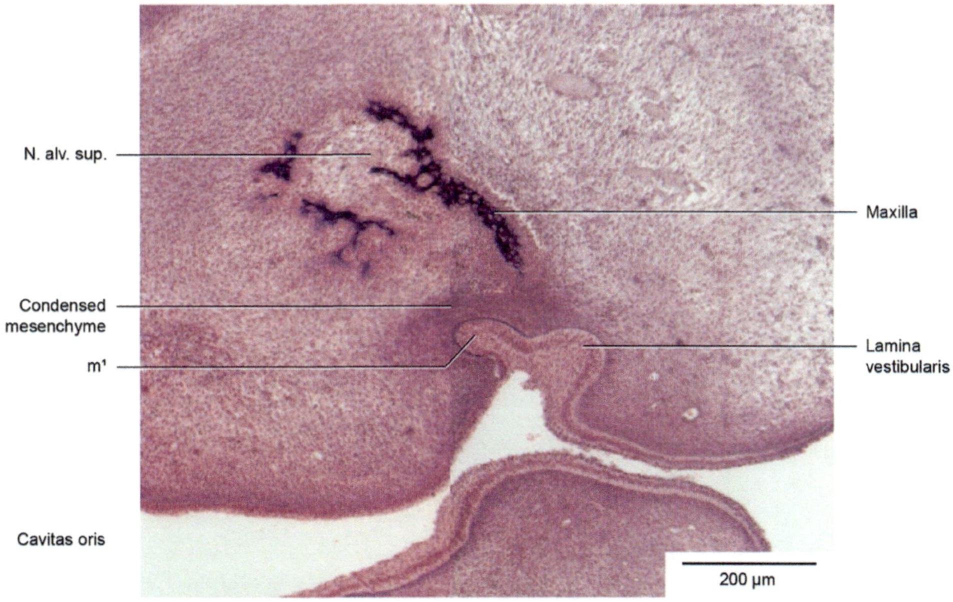

Fig. 6.20 Human embryo, 19 mm CRL (7th wk), sagittal section through region of the upper right dental primordium m^1, staining: Hematoxylin-eosin

References

1. Radlanski RJ, Renz H. An atlas of the human prenatal craniofacial development. Eur J Oral Sci. 2010;118:321–4.
2. Giambartolomei LA, Brunotto M, Gomez de Ferrais ME. Human temporomandibular joint disc: anatomy and measurements in prenatal development. Acta odontol Latino-americana. 2011;24:98–103.
3. Guizetti B, Radlanski RJ. Development of the parotid gland and its closer neighboring structures in human embryos and fetuses of 19–67 mm CRL. Ann Anat. 1996;178:503–8.
4. Guizetti B, Radlanski RJ. Development of the submandibular gland and its closer neighboring structures in human embryos and fetuses of 19–67 mm CRL. Ann Anat. 1996;178:509–14.

Further Reading

Fernández Rubio EM, Radlanski RJ. Development of the human primary and secondary jaw joint. Ann Anat. 2024;251:1–8.

7 Embryo 20 mm CRL (ca. 7th to 8th wk): ADE160787

Overall Remarks

At this time of development, there is still a shared oronasal cavity; separation does not occur until the palate will be formed between the 9th and 12th week. At least at this stage, the primary palate can already be seen, in the histological section (Fig. 7.1) showing well how the tip of the tongue lies cephal to it. It also still has contact with the nasal septum and the region of the later anterior skull base at this stage.

Furthermore, this histological section shows that the nostrils are still filled with massive epithelium, which is why they are now more correctly referred to as placodes (epithelial thickenings).

Nerves

The three branches of the trigeminal nerve branch off from the trigeminal ganglion: The supraorbital nerve emerges anteriorly between the cartilaginous primordium of the lesser wing of the sphenoid bone and the eyeball. The maxillary nerve runs anteriorly above the cartilaginous primordium of the greater wing of the sphenoid bone. Later, when the skull will ossify in this region, the foramen rotundum will be formed here. As the third branch of the trigeminal nerve, the mandibular nerve leaves the ganglion and continues below the cartilaginous primordium of the greater wing of the sphenoid bone in the direction of Meckel's cartilage. Here it is

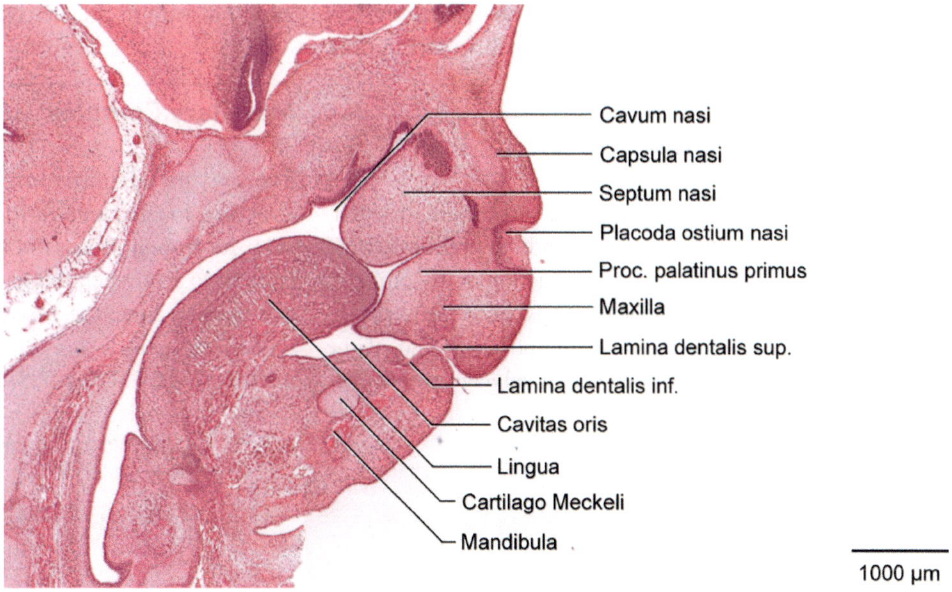

Fig. 7.1 Human embryo, 20 mm CRL (7th to 8th wk), head region, histological sagittal section (Staining: Hematoxylin-eosin)

R. J. Radlanski, *Atlas of Human Prenatal Craniofacial Development*, https://doi.org/10.1007/978-3-031-88812-0_7

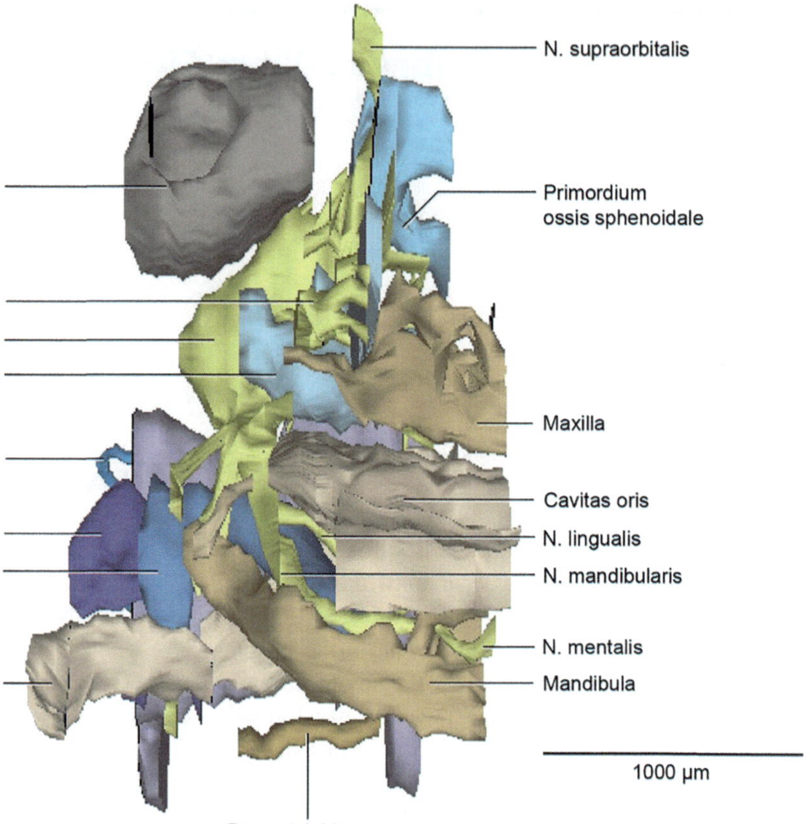

Fig. 7.2 Human embryo, 20 mm CRL (7th to 8th wk), head region, frontal view, right half. Step 1 of preparation

then called mandibular nerve and follows the course of the cartilaginous arch to the chin region. Along the way there are other branches (buccalis, masseteric, and lingual nerves) and at the mental foramen the mental nerve branches off (Fig. 7.2).

Cartilaginous Skeleton

Apart from the nasal capsule, which was given more attention in Chap. 5, the cartilaginous primordium of the sphenoid bone is well represented in the reconstruction of this embryo. The large and the small sphenoid wings are already recognizable in outline at this stage. At the posterior end of the Meckel's cartilage, it can be seen very nicely again how malleus, incus, and stapes are already spatially related to each other (see Figs. 7.4 and 7.5). In addition, the cartilaginous labyrinth with its arcades is shown in this reconstruction.

Maxillary Complex

The maxilla continues to be a paired, flat, slightly curved structure, located anteriorly and laterally from the nasal capsule. Its frontal process is well defined, but still only short enough to just reach the level of the lower edge of the eye. For the first time, the base of a zygomatic arch is now also recognizable, as a short stub at the dorsolateral ends of the maxilla (Figs. 7.3, 7.4, 7.5, and 7.6).

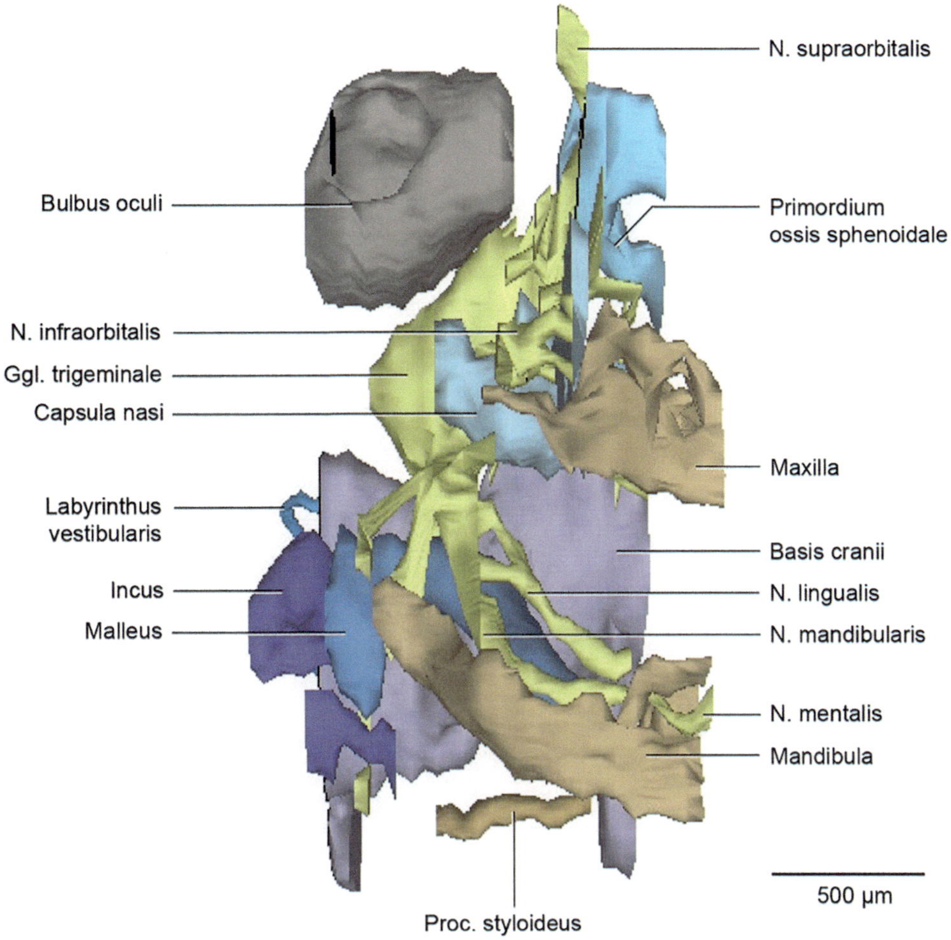

Fig. 7.3 Human embryo, 20 mm CRL (7th to 8th wk), head region, frontal view, right half. Step 2 of preparation

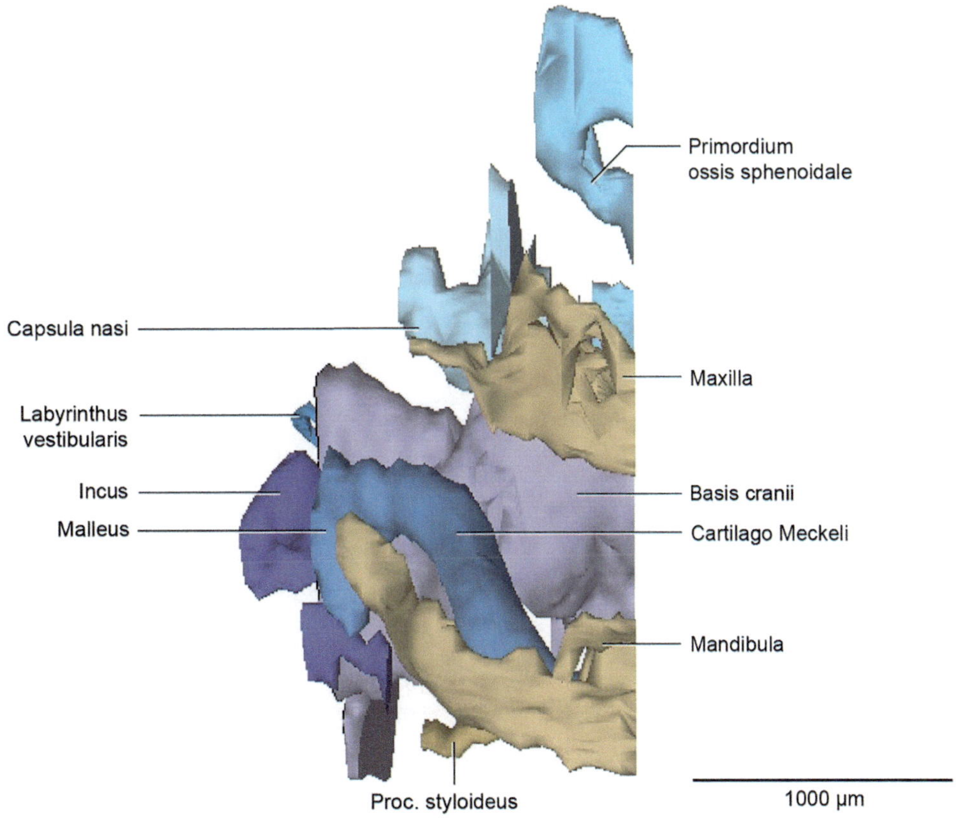

Fig. 7.4 Human embryo, 20 mm CRL (7th to 8th wk), head region, frontal view, right half. Step 3 of preparation

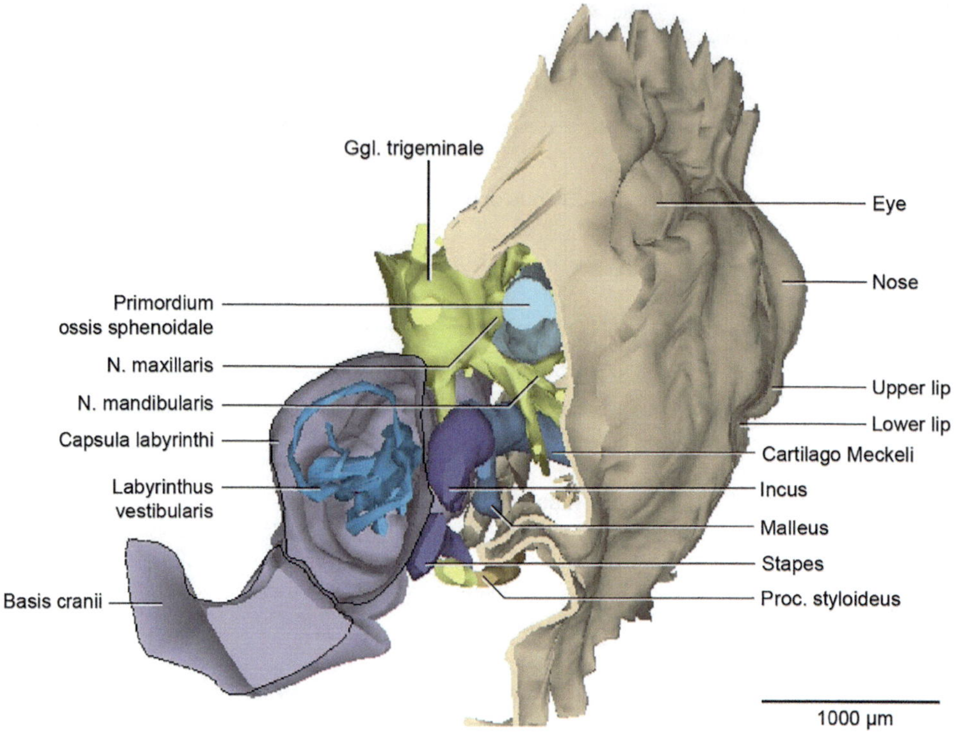

Fig. 7.5 Human embryo, 20 mm CRL (7th to 8th wk), head region, lateral right view. Step 1 of preparation (*From* Radlanski [1]; *with permission*)

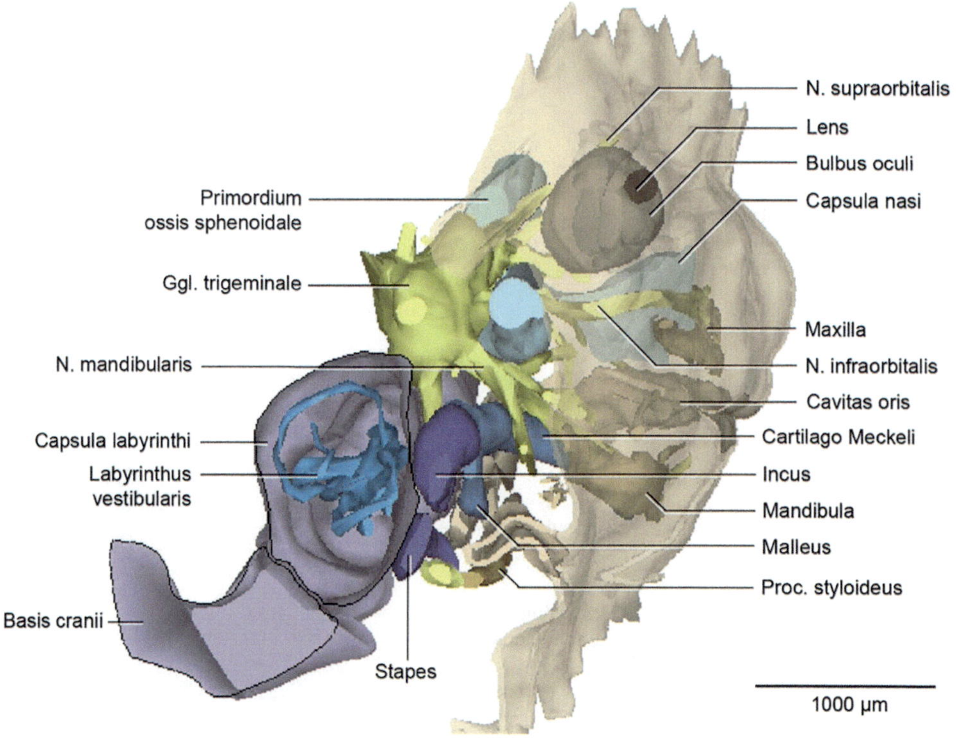

Fig. 7.6 Human embryo, 20 mm CRL (7th to 8th wk), head region, lateral right view. Step 1 of preparation, skin transparent (*From* Radlanski [1]; *with permission*)

Mandible

Here, where the facial skin has also been reconstructed, it becomes clear, that the mandible forms in exactly the space that remains available for it between Meckel's cartilage and the skin. Other than that, the mandible has not changed significantly, with the exception that the mental foramen is now closed circularly on all sides. And there is still no sign of the formation of coronoid or condylar processes (Figs. 7.7, 7.8, 7.9, 7.10, 7.11, 7.12, 7.13, 7.14, and 7.15).

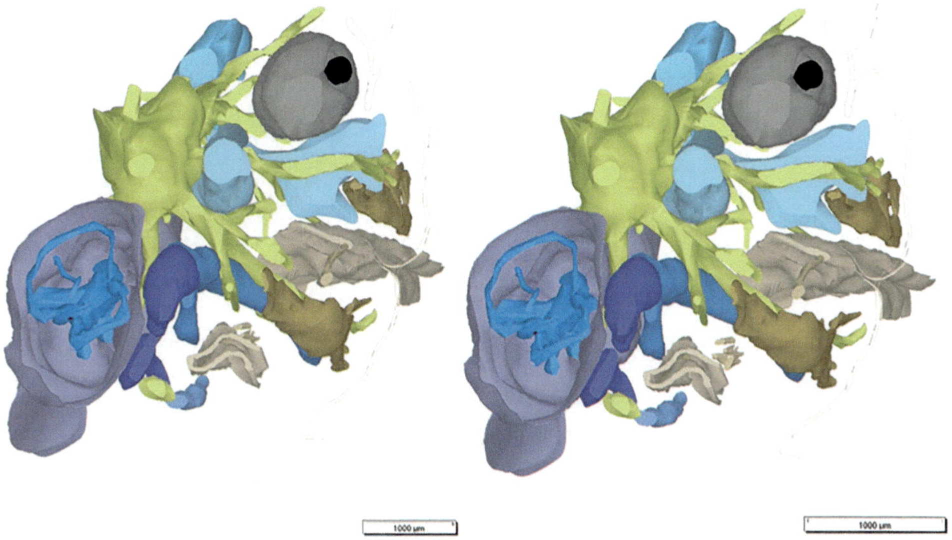

Fig. 7.7 Stereoscopical pair of images. Right view. Step 2 of preparation

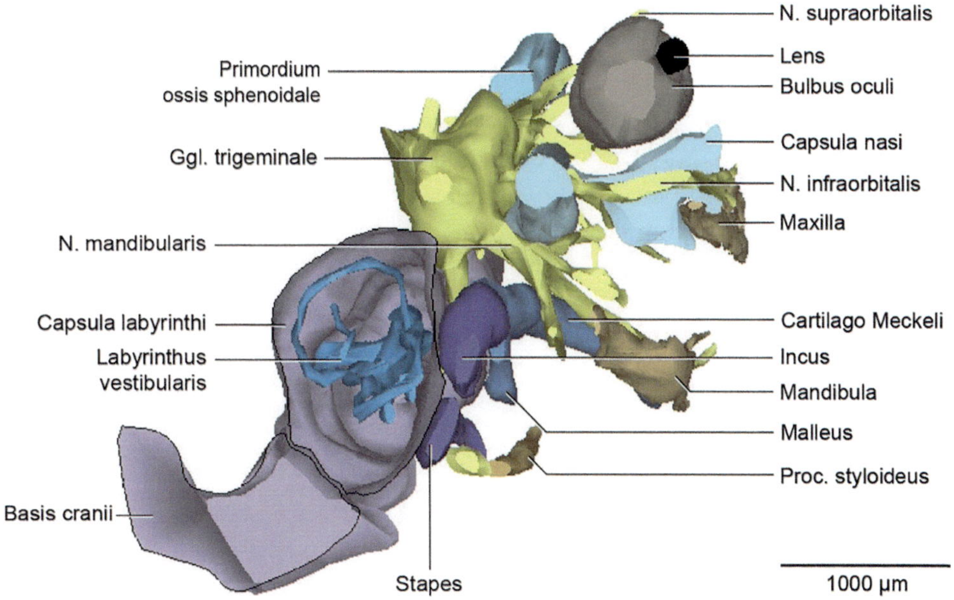

Fig. 7.8 Human embryo, 20 mm CRL (7th to 8th wk), head region, lateral right view. Step 2 of preparation (*From* Radlanski [1]; *with permission*)

Fig. 7.9 Human embryo, 20 mm CRL (7th to 8th wk), head region, right half, medial view. Step 1 of preparation

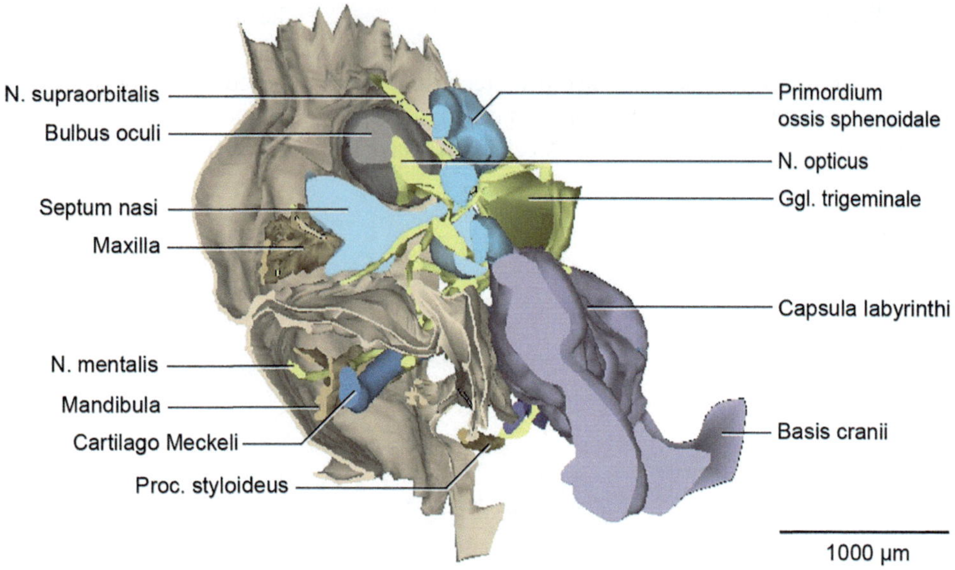

Fig. 7.10 Human embryo, 20 mm CRL (7th to 8th wk), head region, right half, medial view. Step 2 of preparation

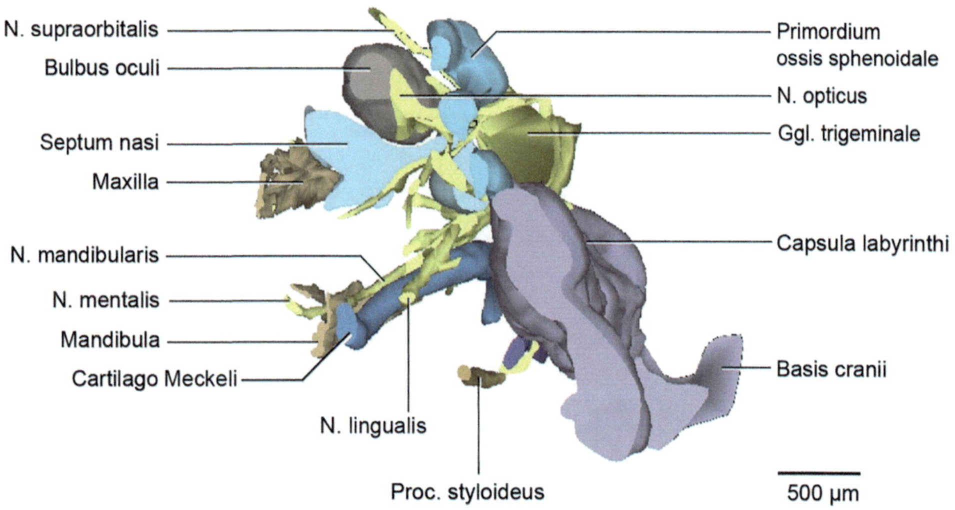

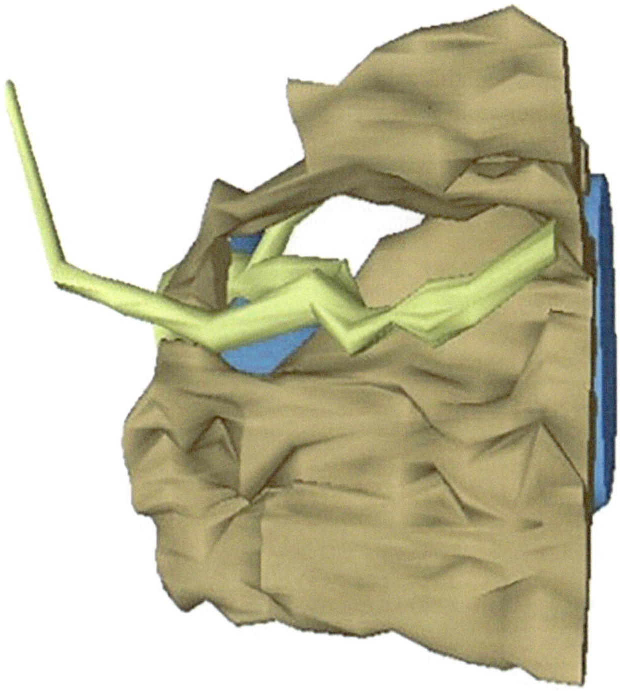

Fig. 7.11 Partial reconstruction of the mandible (ochre), right half with mental foramen, mental nerve (yellow), and Meckel's cartilage (blue). Anterior view

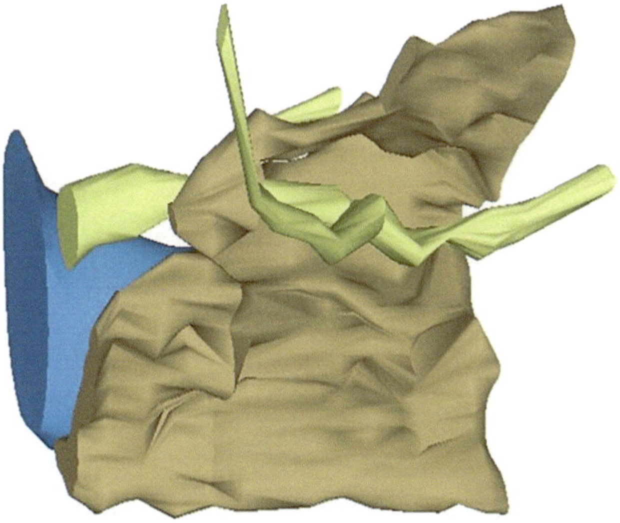

Fig. 7.12 Partial reconstruction of the mandible (ochre), right half with mental foramen, mental nerve (yellow), and Meckel's cartilage (blue). Lateral view

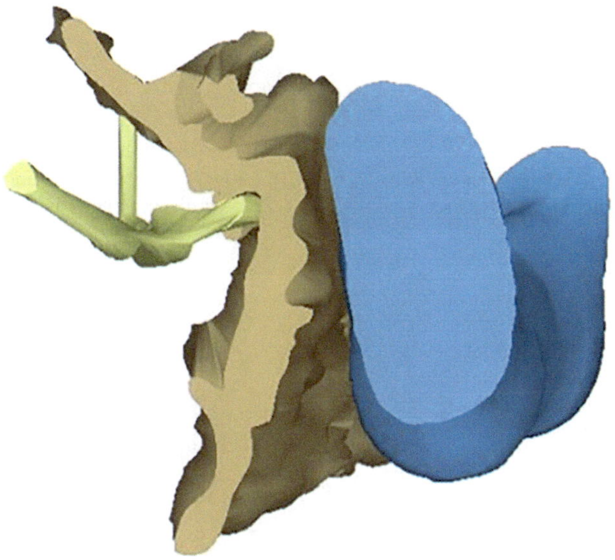

Fig. 7.13 Partial reconstruction of the mandible (ochre), right half with mental foramen, mental nerve (yellow), and Meckel's cartilage (blue). Dorsal view

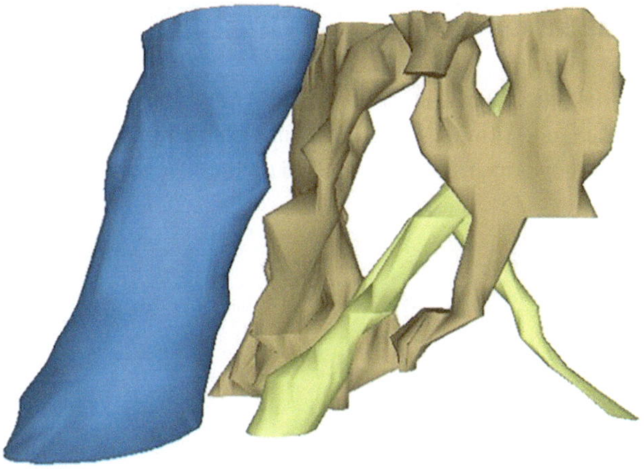

Fig. 7.14 Partial reconstruction of the mandible (ochre), right half with mental foramen, mental nerve (yellow), and Meckel's cartilage (blue). Cranial view

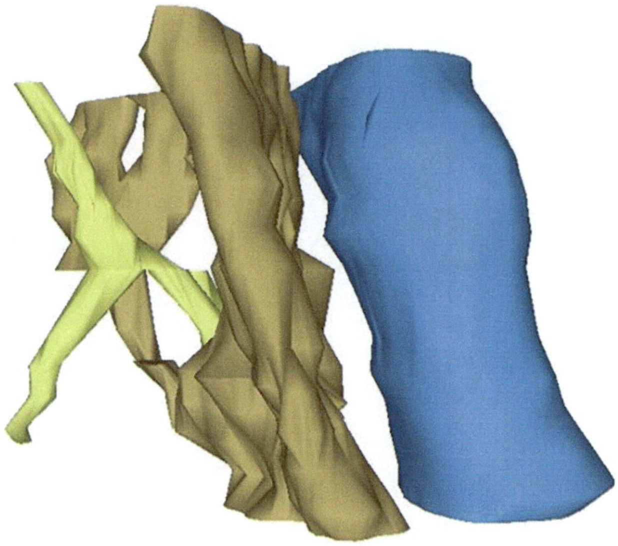

Fig. 7.15 Partial reconstruction of the mandible (ochre), right half with mental foramen, mental nerve (yellow), and Meckel's cartilage (blue). Caudal view

Oral Cavity

In this reconstruction, we have shown not only the contours of the facial skin (see Fig. 7.5), but also the lips and oral cavity as they appear as their external epithelial delineations. We can even see the formation of the parotid gland, of which only the compact excretory duct has now only been formed (see Fig. 7.7), and it is worthwhile to actually look at that stereo image pair in 3D.

Reference

1. Radlanski RJ. Oral structure and biology. Batavia: Quintessence Publishing; 2018.

Further Reading

Blechschmidt E. The stages of human development before birth. An introduction to human embryology. - Die vorgeburtlichen Entwicklungsstadien des Menschen. Eine Einführung in die Humanembryologie. London, New York, Basel, Freiburg i. B: Karger; 1960.

Radlanski RJ, Renz H, Müller U, Schneider RS, Marcucio RS, Helms JA. Prenatal morphogenesis of the human mental foramen. Eur J Oral Sci. 2002;110:452–9.

Radlanski RJ, Renz H, Lajvardi S, Schneider R. A bone remodeling during prenatal morphogenesis of the human mental foramen. Eur J Oral Sci. 2004;112:301–10.

Sperber GH. Craniofacial embryology. Hamilton: B.C. Decker; 2001.

Embryo 21 mm CRL (ca. 7th wk): NEL110989

8

Mandible: Meckel's Cartilage

This chapter now shows only the mandible as it has developed up to the seventh week. The cartilaginous framework is formed by Meckel's cartilage, which here has a very distinct V-shape (see Fig. 8.3). The medial bend at its right end may be an artifact: The embryo may have been lying on its right side during fixation, and the weight of the head may have led to this atypical deformation. However, the two single ends are obviously intact. Here the caudal bend of the cartilaginous bar is recognizable as the later malleus. In the area of the mental symphysis, the two halves are in contact with each other. And it is typical for Meckel's cartilage that it is thicker at its anterior end than along its course. Here, two individual processes are visible, and they project caudally to different extents (Fig. 8.1).

Mandible: Bone

The bone formation, which may well be called the "mandible" here, adapts to the course of the Meckel's cartilage as a flat band. The two ends swing out slightly laterally and anteriorly the two bone halves do not meet in the middle, but maintain a distance of about 300 μm from each other. Therefore, the mandible is now still a paired formation. The mental foramen, which is now already closed circularly, has already become a tunnel within the still thin mandibular bone, which already has a certain length because it passes obliquely through the bone. Its anterior opening accordingly also shows an oblique opening with a definite indentation on the mandibular bone (Figs. 8.2 and 8.3).

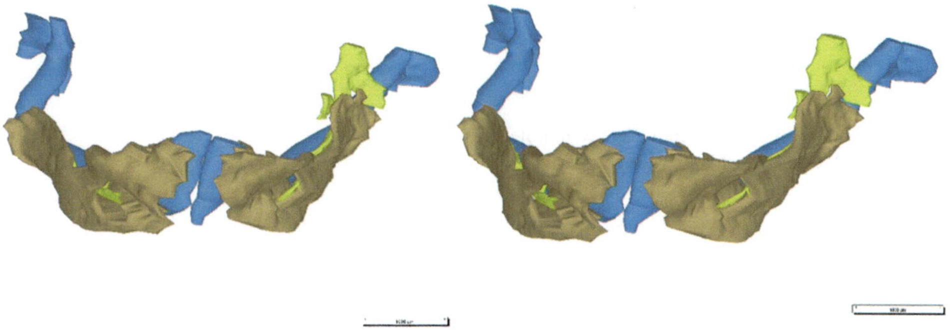

Fig. 8.1 Stereoscopical pair of images. Human embryo 21 mm CRL (ca. 7th wk), mandible, frontal view. Bar: 1000 μm

R. J. Radlanski, *Atlas of Human Prenatal Craniofacial Development*, https://doi.org/10.1007/978-3-031-88812-0_8

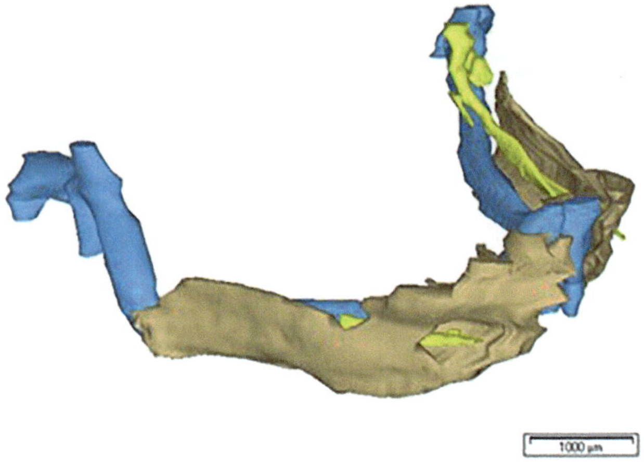

Fig. 8.2 Human embryo, 21 mm CRL. Mandible, right and 45°anterior view. Ochre: bone; blue: Meckel's cartilage; yellow: inferior alveolar n.

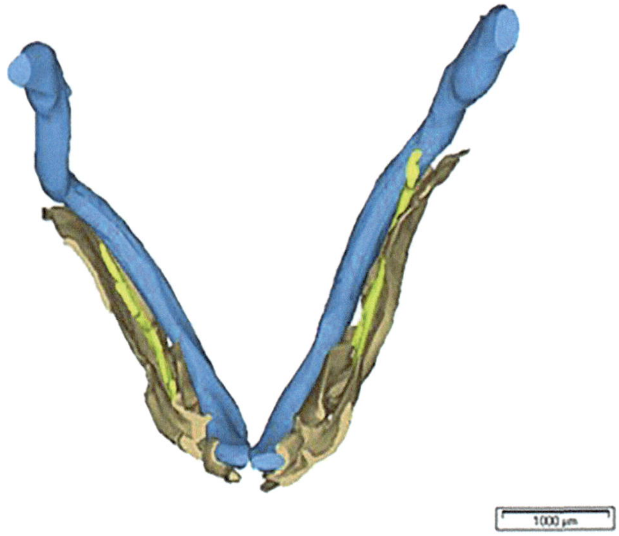

Fig. 8.3 Human embryo, 21 mm CRL. Mandible, cranial view. Ochre: bone; blue: Meckel's cartilage; yellow: inferior alveolar n.

Alveolar Inferior Nerve

The alveolar inferior nerve runs uncovered by bone for most of its course in the space between Meckel's cartilage and the mandible. Only in the anterior region, where the mental nerve branches off, does the nerve enter the bony tunnel, as can be clearly seen from a cranial view (see Fig. 8.3) and from an oblique lateral view (see Fig. 8.2).

Further Reading

Radlanski RJ. Oral structure and biology. Batavia: Quintessence Publishing; 2018.

Radlanski RJ, Renz H, Klarkowski MC. Prenatal development of the human mandible. 3D reconstructions, morphometry and bone remodelling pattern, sizes 12–117 mm CRL. Anat Embryol (Berl). 2003;207:221–32.

Sperber GH. Craniofacial embryology. Hamilton: B.C. Decker; 2001.

9 Embryo 22 mm CRL (ca. 7th wk): EMM150787

Overall Remarks

In this reconstruction, more attention is paid to the inframandibular anatomical structures. Thus, the clavicle is also included and the sternocleidomastoid muscle, as well as the larynx and the thyroid gland (Fig. 9.1). In the illustrations later in this chapter, some more of the neck muscles are shown (see Fig. 9.11).

Nerves

Removal of the maxilla and mandible reveals the branches of the trigeminal nerve. Quite laterally, the auriculo-temporalis nerve can be seen, which passes around anterior to the epithelium of the external auditory canal. Meckel's cartilage is accompanied by nerves on either side, the alveolaris inferior nerve runs laterally and the lingual nerve lingually.

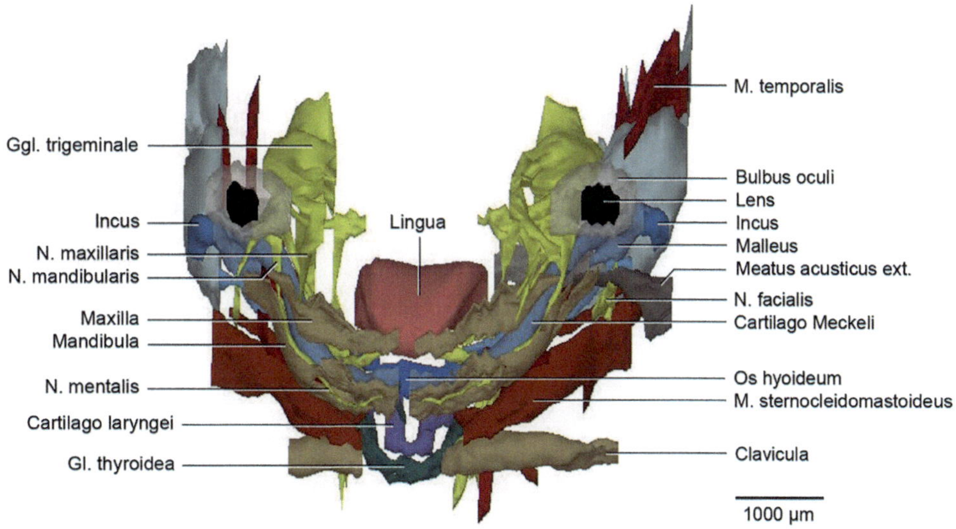

Fig. 9.1 Human embryo, 22 mm CRL, head, frontal view, skin removed, layer 1 of preparation

R. J. Radlanski, *Atlas of Human Prenatal Craniofacial Development*, https://doi.org/10.1007/978-3-031-88812-0_9

Cartilaginous Skeleton

In addition to the Meckel's cartilage, we have also depicted the stylohyoid process in this reconstruction, which is still fully cartilaginous at this stage. The hyoid bone is also still cartilaginous at this stage (Figs. 9.2, 9.3, and 9.4).

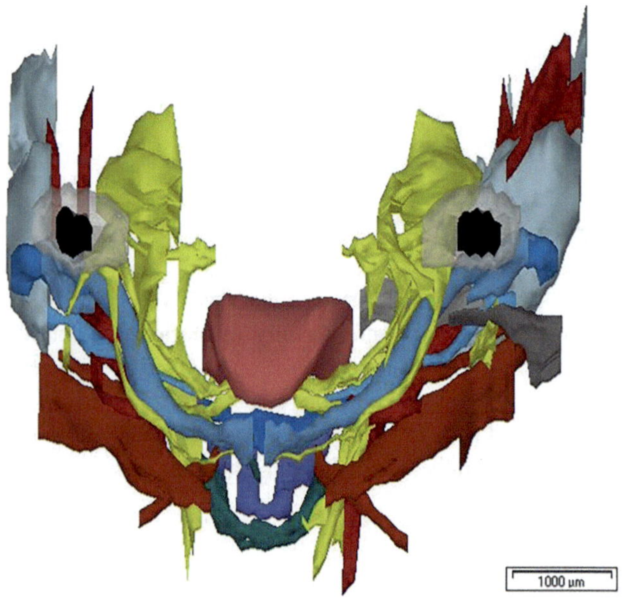

Fig. 9.2 Human embryo, 22 mm CRL, head, frontal view, layer 2 of preparation

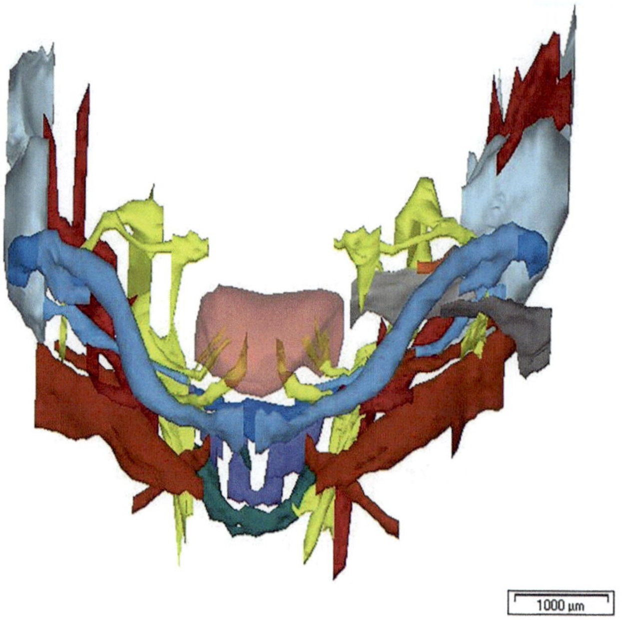

Fig. 9.3 Human embryo, 22 mm CRL, head, frontal view, layer 3 of preparation

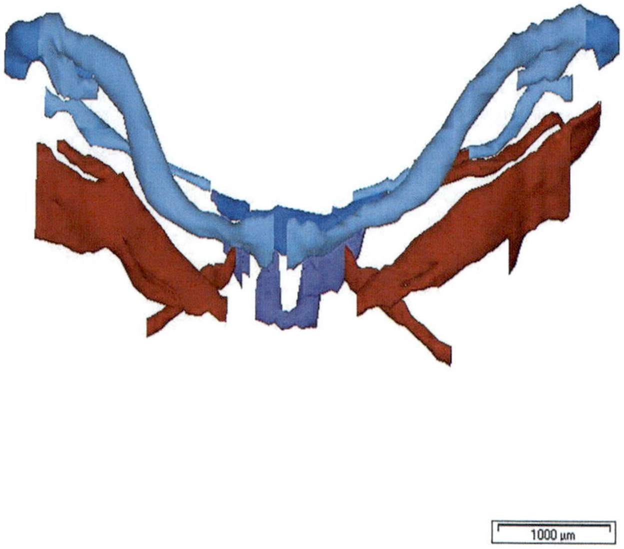

Fig. 9.4 Human embryo, 22 mm CRL, head, frontal view, muscles, cartilage, and nerves

Mandible

At this stage of development, or even in this individual alone, the mandible is no longer as pointedly V-shaped as it was in the embryo of Chap. 8 (also 22 mm CRL). Now or here it can be better described with a U-shape. In fact, a transition from a U-shape to a more pointed V-shape and then a U-shape again is described in the literature, also the double-S-form, as visible in Figs. 9.9 and 9.10 [1, 2]. However, one must also be prepared for individual variations (Figs. 9.5, 9.6, 9.7, 9.8, 9.9, 9.10, 9.11, 9.12, 9.13, 9.14, 9.15, and 9.16).

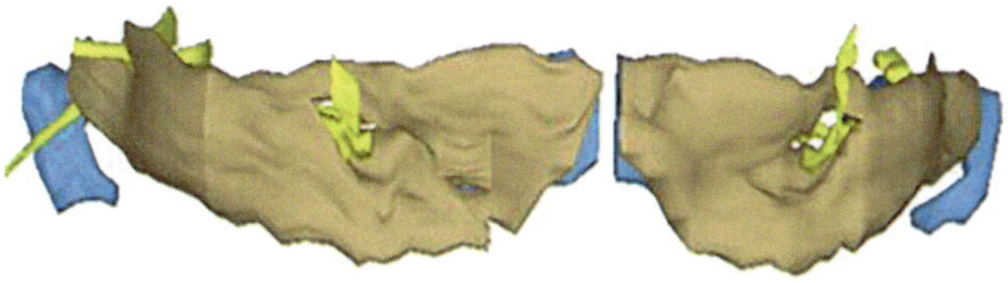

Fig. 9.5 Human embryo, 22 mm CRL, mandible, anterior view

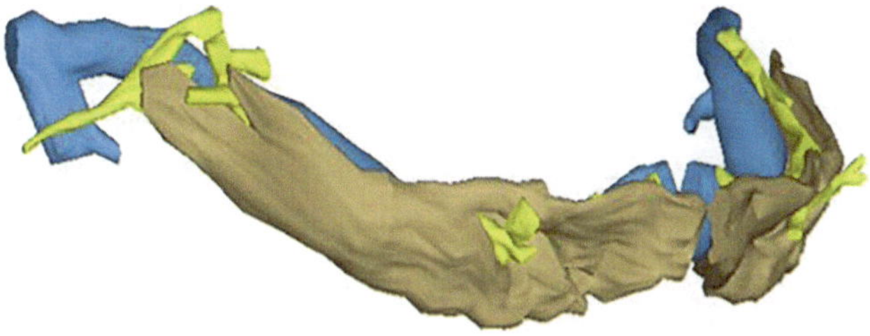

Fig. 9.6 Human embryo, 22 mm CRL, mandible, (45) lateral view

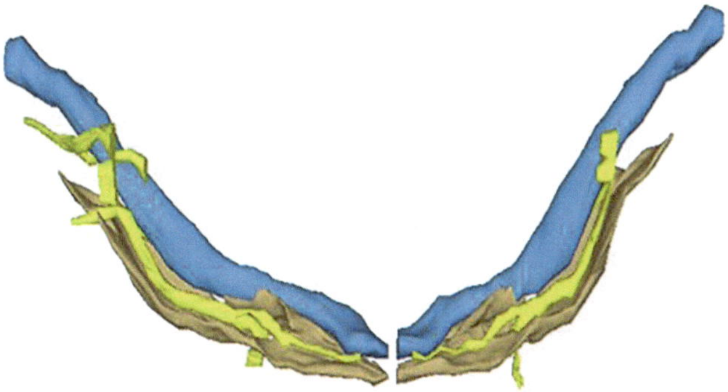

Fig. 9.7 Human embryo, 22 mm CRL, mandible, cranial view

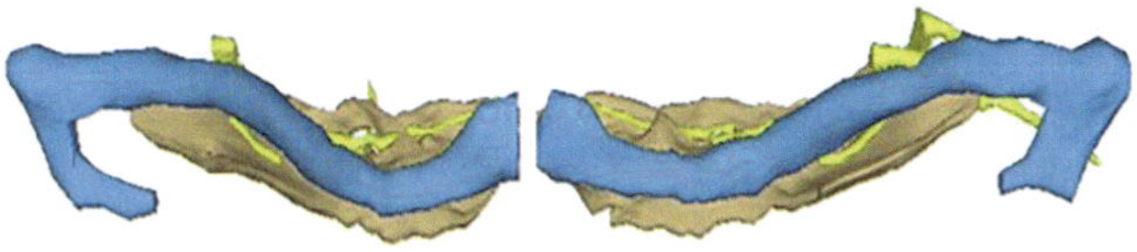

Fig. 9.8 Human embryo, 22 mm CRL, mandible, dorsal view

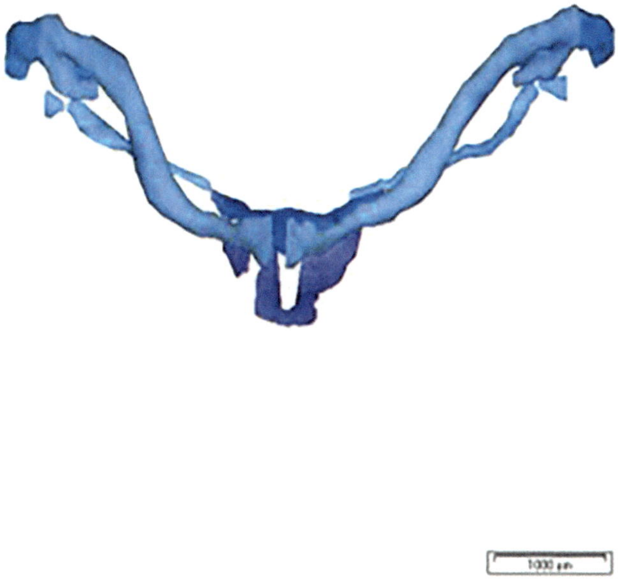

Fig. 9.9 Human embryo, 22 mm CRL, cartilaginous skeleton, mandibular region, anterior view, cartilage

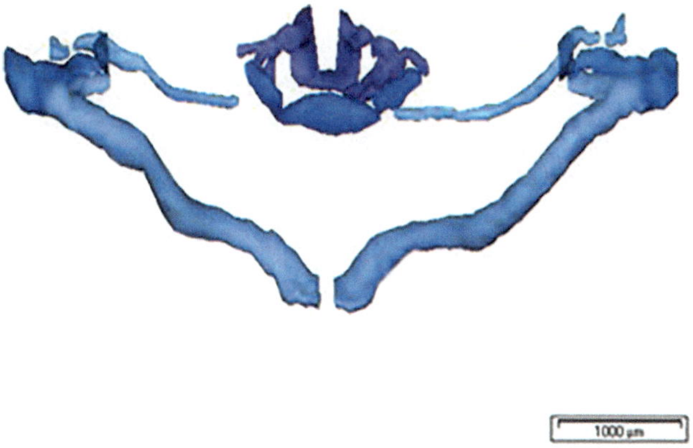

Fig. 9.10 Human embryo, 22 mm CRL, cartilaginous skeleton, mandibular region, cranial view, cartilage

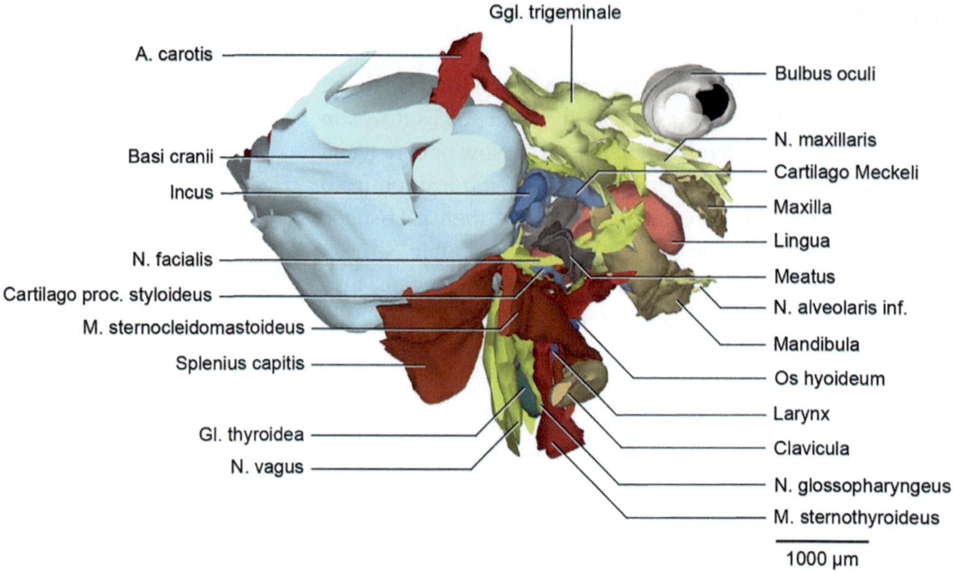

Fig. 9.11 Human embryo, 22 mm CRL, viscerocranium, lateral view. Step 1 of preparation

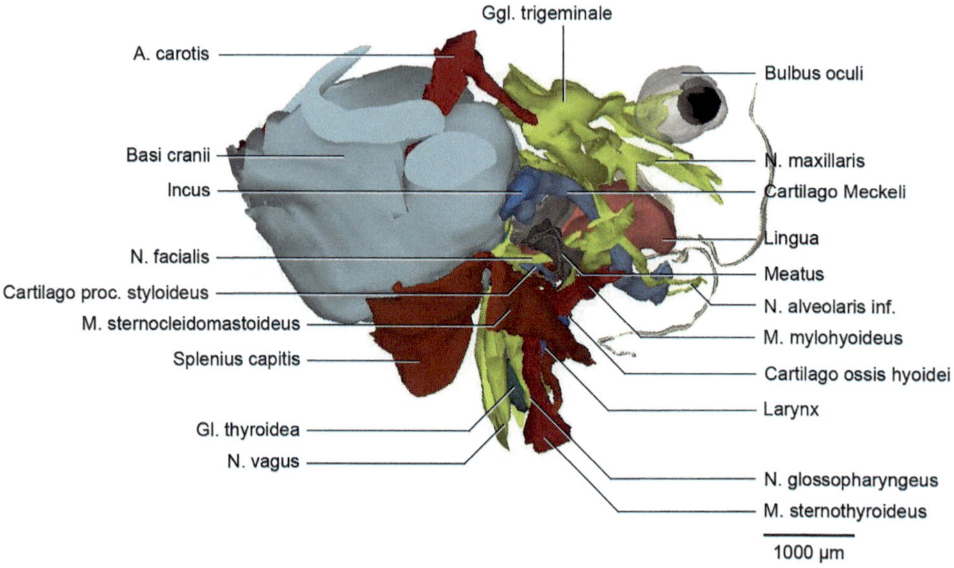

Fig. 9.12 Human embryo, 22 mm CRL, viscerocranium, lateral view, layer 2 of preparation

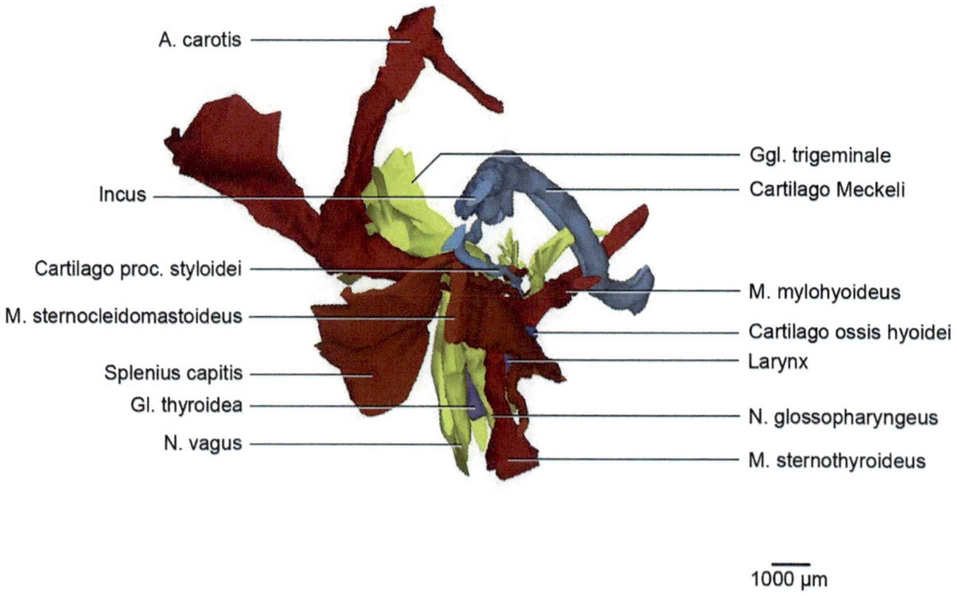

Fig. 9.13 Human embryo, 22 mm CRL, viscerocranium, lateral view, layer 3 of preparation

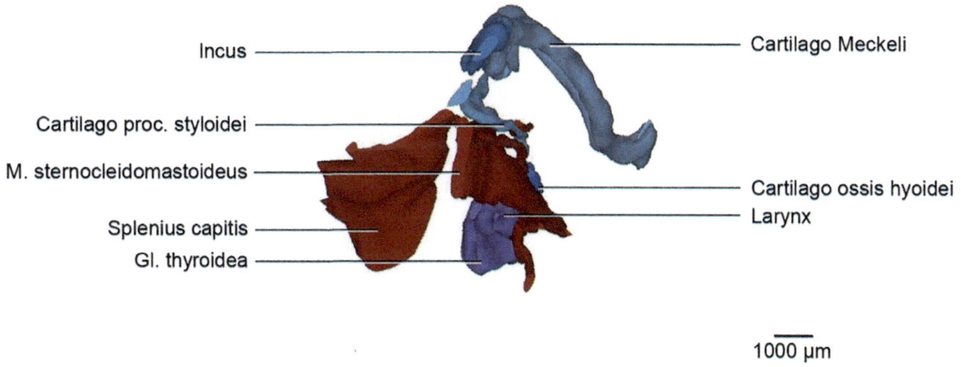

Fig. 9.14 Human embryo, 22 mm CRL, viscerocranium lateral view, layer 4 of preparation

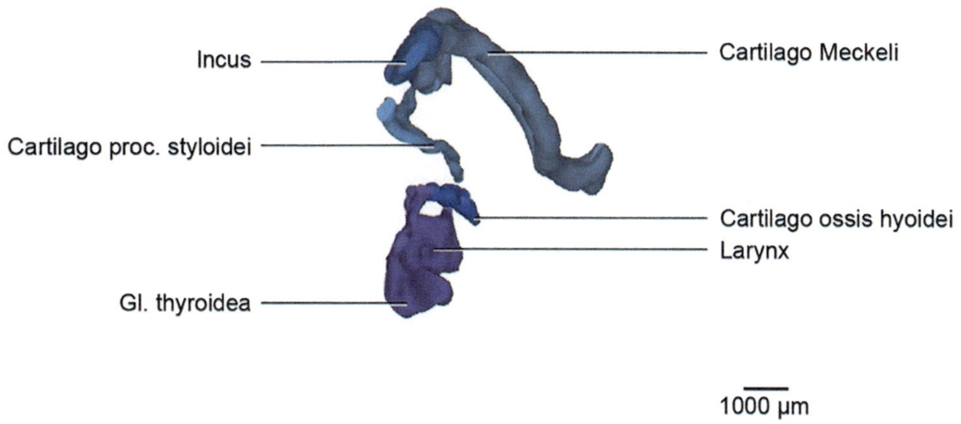

Fig. 9.15 Human embryo, 22 mm CRL, viscerocranium lateral view, layer 5 of preparation

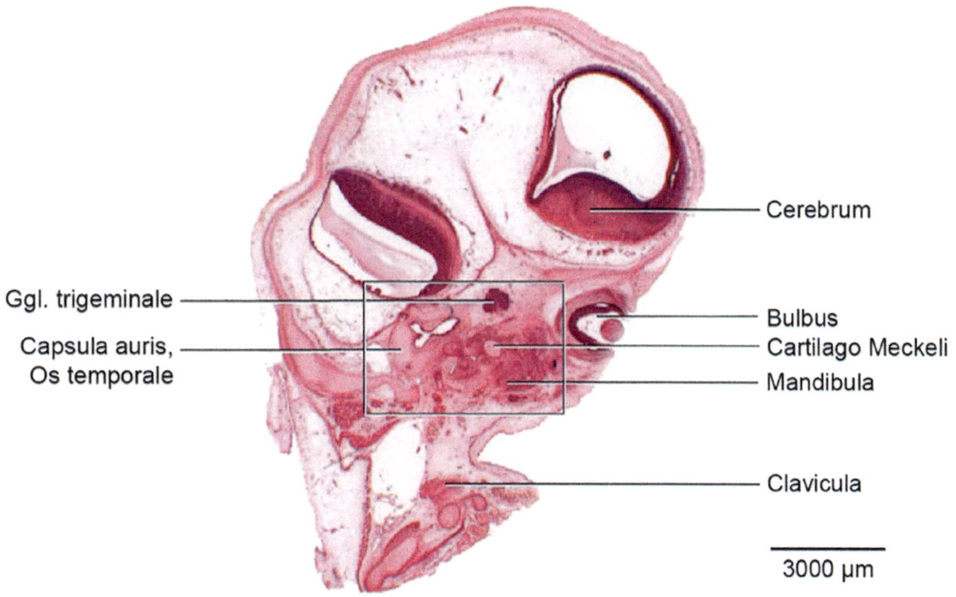

Fig. 9.16 Human embryo, 22 mm CRL, histological section, sagittal section at the level of the ear region, viscerocranium. Box: Fig. 9.17

Middle Ear

The anatomy of the auditory ossicles and their positional relationships to each other is shown in the reconstruction in Fig. 9.18. The malleus is a part of Meckel's cartilage, its posterior end. It articulates with the incus and medial to it lies the stapes. The tympanic cavity does not yet provide the room for these ossicles, but they are already close to the external acoustic meatus. The facial nerve and a branch of it, the tympanic chord, which will be seen later running across the eardrum, are also included in this representation (Figs. 9.17 and 9.18).

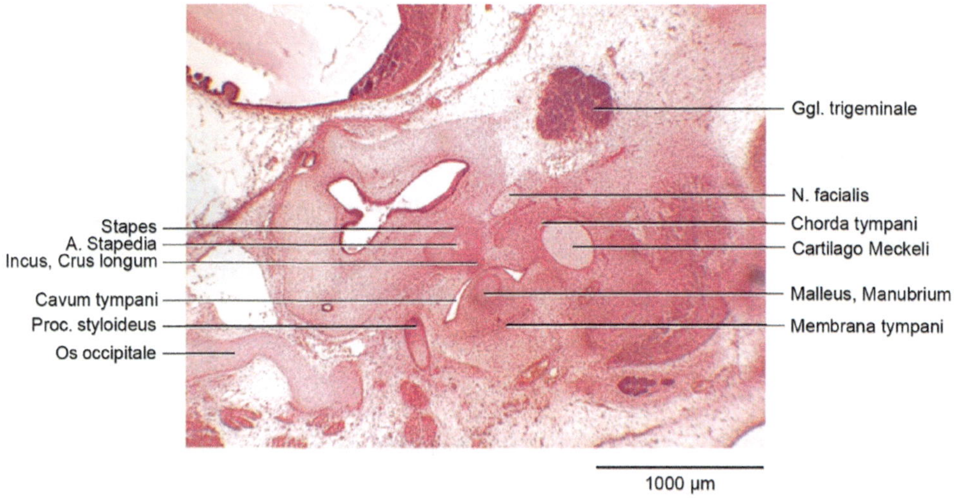

Fig. 9.17 Human embryo, 22 mm CRL, serial section, sagittal section, detailed view of Fig. 9.16

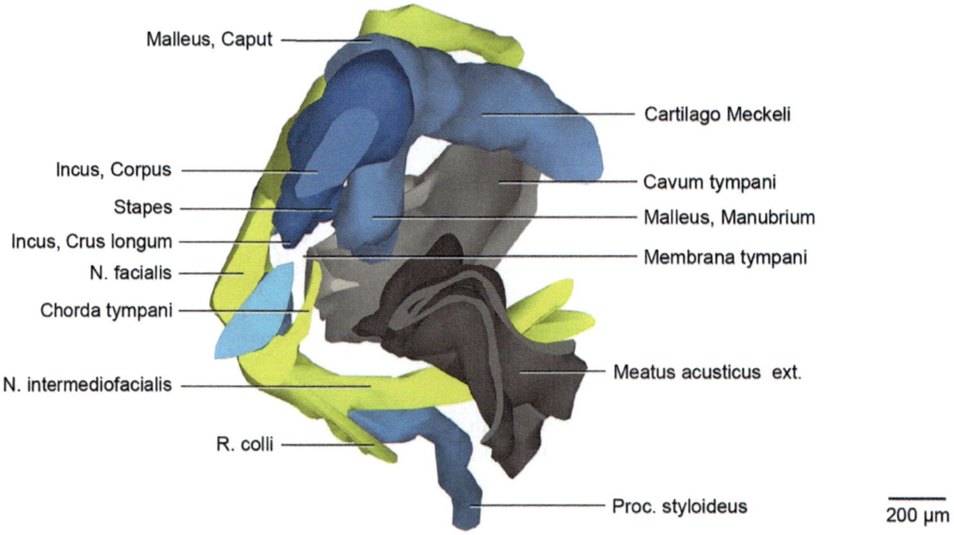

Fig. 9.18 Human embryo, 22 mm CRL, right auditory system, lateral view, presentation of the right auditory system

References

1. Kjaer I, Keeling JW, Fischer-Hansen B. The prenatal human cranium - normal and pathologic development. Copenhagen: Munksgaard; 1999.
2. Radlanski RJ, Renz H, Klarkowski MC. Prenatal development of the human mandible. 3D reconstructions, morphometry and bone remodelling pattern, sizes 12–117 mm CRL. Anat Embryol (Berl). 2003;207:221–32.

Further Reading

Germelmann AR. (née Than) Beitrag zur pränatalen Morphogenese des Os hyoideum des Menschen – Computergestützte 3D-Rekonstruktionen und histologische Untersuchungen menschlicher Embryonen/Feten von 19 – 150 mm CRL (7. bis 18. Woche) [Med. dissertation]. Charité-Universitätsmedizin Berlin; 2008.

Radlanski RJ, Renz H, Tabatabai A. Prenatal development of the muscles in the floor of the mouth in human embryos and fetuses from 6.9 to 76 mm CRL. Ann Anat. 2001;183:511–8.

Radlanski RJ. Oral structure and biology. Batavia: Quintessence Publishing; 2018.

Sperber GH. Craniofacial embryology. Hamilton: B.C. Decker; 2001.

Embryo 22 mm CRL (ca. 7th wk): ALI070785

10

Mandible

The mandible is also broadly U-shaped in this embryo, which is the same size as the previous one. The anterior ends of Meckel's cartilage continue to be typically bulb-shaped upward and well in contact, but still well demarcated from each other by a perichondrium. The bony symphysis remains open by about 300 µm. The bony groove, in which the inferior alveolar nerve runs, is already somewhat more markedly developed in this embryo (Figs. 10.1, 10.2, and 10.3).

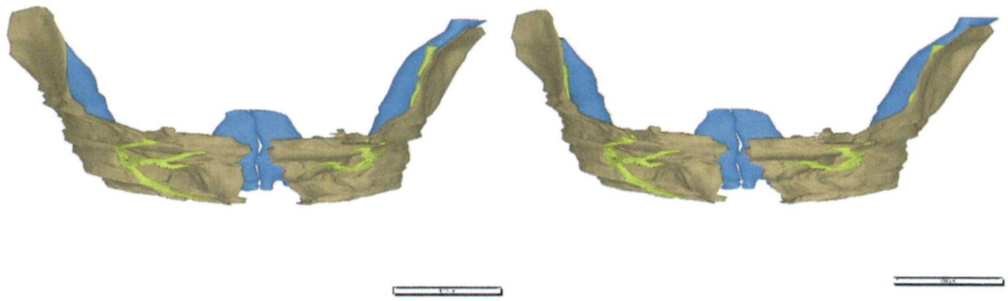

Fig. 10.1 Human embryo, 22 mm CRL, mandible (ochre) with Meckel's cartilage (blue), and inferior alveolar nerve (yellow), frontal view. Bar: 1000 µm

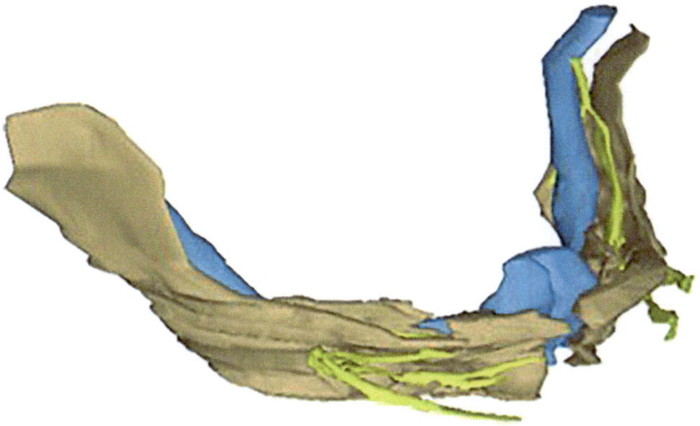

Fig. 10.2 Human embryo, 22 mm CRL. Mandible, right and 45°anterior view. Ochre: bone blue: Meckel's cartilage; yellow: inferior alveolar n.

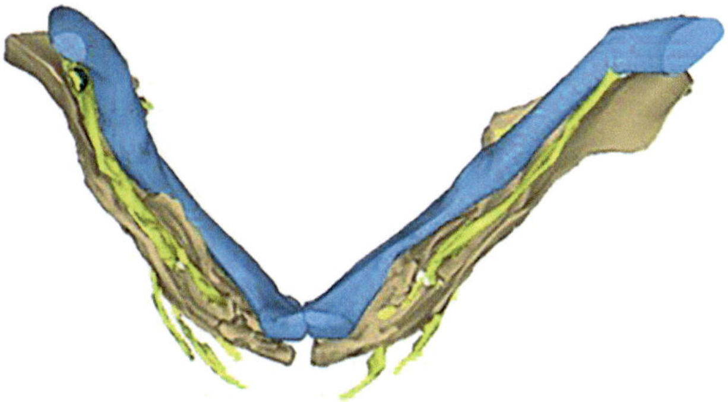

Fig. 10.3 Human embryo, 22 mm CRL. Mandible, cranial view. Ochre: bone; blue: Meckel's cartilage; yellow: inferior alveolar n.

R. J. Radlanski, *Atlas of Human Prenatal Craniofacial Development*, https://doi.org/10.1007/978-3-031-88812-0_10

Dental Primordia

In the horizontal section passing through the dental lamina of the maxilla, it can be seen, that it possesses condensations of its epithelium at regular intervals (see Fig. 10.6). The spacing of these condensations is about 425 μm (measured from edge to edge). These knot-like swellings are the dental primordia or also called tooth buds. These buds are almost round in this developmental stage and are approximately 200 μm in diameter.

In the tooth buds, the marginal epithelial cells assume a wedge-shape outline of its bulgings (see Fig. 10.7). In the areas between the tooth buds, the cells of the dental lamina obtain a more cuboid outline. The mesenchymal tissue is condensed around each tooth bud (Figs. 10.4, 10.5, 10.6, and 10.7).

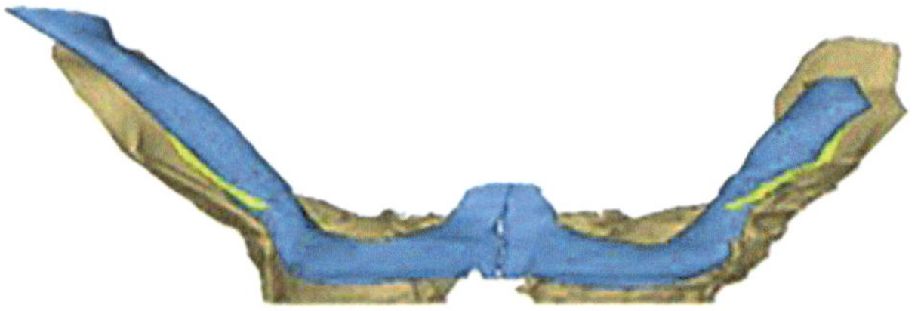

Fig. 10.4 Human embryo, 22 mm CRL. Mandible, dorsal view. Ochre: bone; blue: Meckel's cartilage; yellow: inferior alveolar n.

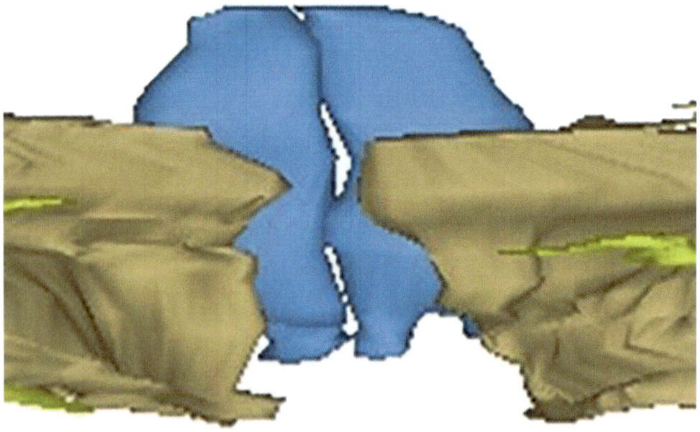

Fig. 10.5 Human embryo, 22 mm CRL. Mandible, frontal view. Enlarged symphyseal region from Fig. 10.1. Ochre: bone; blue: Meckel's cartilage; yellow: inferior alveolar n.

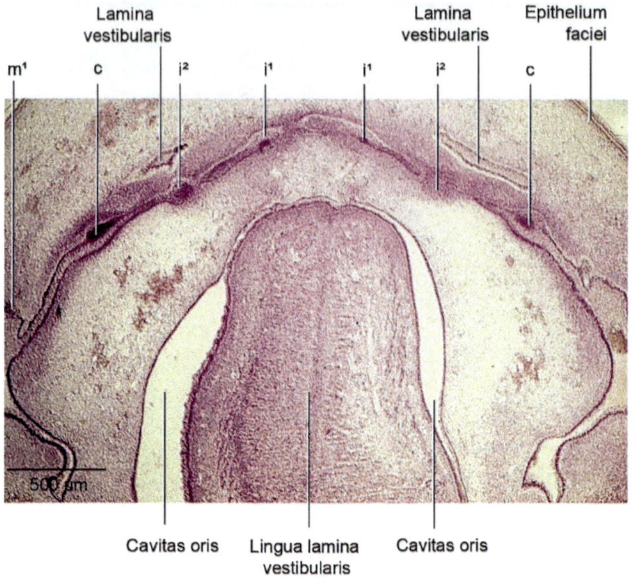

Fig. 10.6 Human embryo, 22 mm CRL. Histological horizontal section, maxillary region, staining: Hematoxylin-eosin. (From Radlanski [1] and Radlanski [2]; with permission)

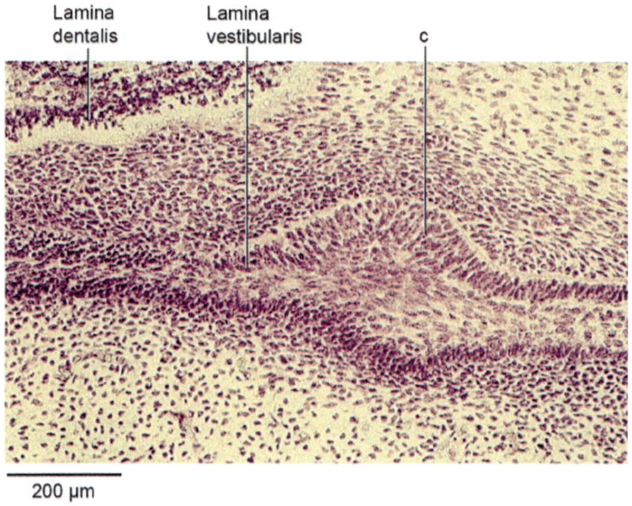

Fig. 10.7 Human embryo, 22 mm CRL. Histological horizontal section, maxillary region, primordium c, staining: Hematoxylin-eosin (From Radlanski [1] and Radlanski [2]; with permission)

References

1. Radlanski RJ. Contributions to the development of human deciduous tooth primordia. Chicago: Quintessence Publishing; 1993.
2. Radlanski RJ. Oral structure and biology. Batavia, NY: Quintessence Publishing; 2018.

Further Reading

Radlanski RJ, Renz H, Klarkowski MC. Prenatal development of the human mandible. 3D reconstructions, morphometry and bone remodelling pattern, sizes 12–117 mm CRL. Anat Embryol (Berl). 2003;207:221–32.

Embryo 22 mm CRL (ca. 7th wk): DID110688

11

Mandible

Although this embryo is listed in the data sheet with a crown-rump length of 22 mm and thus has an estimated age of about 7 weeks. However, the morphological findings show a less mature mandible. The two cartilage braces of Meckel's cartilage show their malleus-forming extensions dorsally, do not meet at the middle of the chin, but maintain a distance of about 500 µm. Also, the bone formation of the mandible is less developed and represents only a flat bony plate, even in front still far apart. Both, cartilage and bone run almost straight, show no curves, but stand in a not very acutely pointed V to each other (see Figs. 11.3 and 11.4). The inferior alveolar nerve runs in the space between cartilage and bone—a groove, as already developed in the previous embryos, does not exist here yet. The mental foramen is still open obliquely anteriorly at the top. This finding should not be surprising, but rather shows the variability of size, age, and state of maturity (Figs. 11.1, 11.2, 11.3, 11.4 and 11.5).

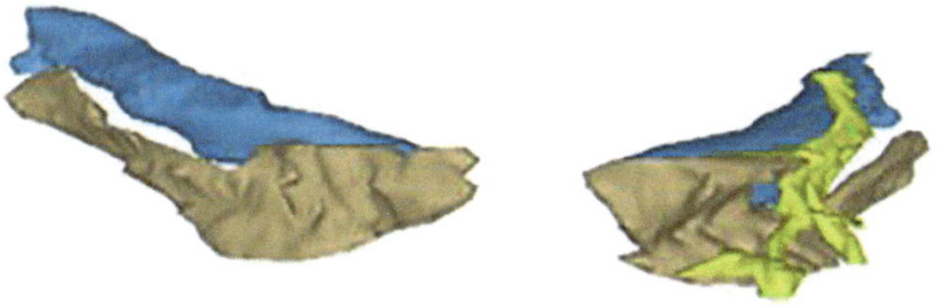

Fig. 11.1 Human embryo, 22 mm CRL, mandible (ochre) with Meckel's cartilage (blue), and mental n. (yellow) on left side, frontal view

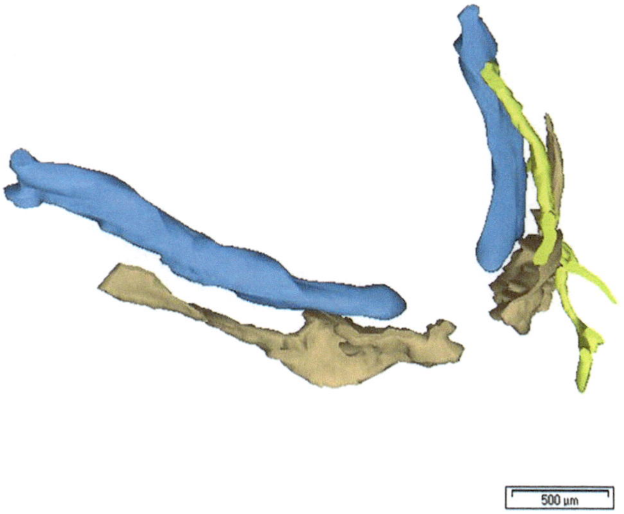

Fig. 11.2 Human embryo, 22 mm CRL, mandible with mental n. on left side, right lateral, 45° anterior, and 45° cranial view. Ochre: bone; blue: Meckel's cartilage; yellow: inferior alveolar n

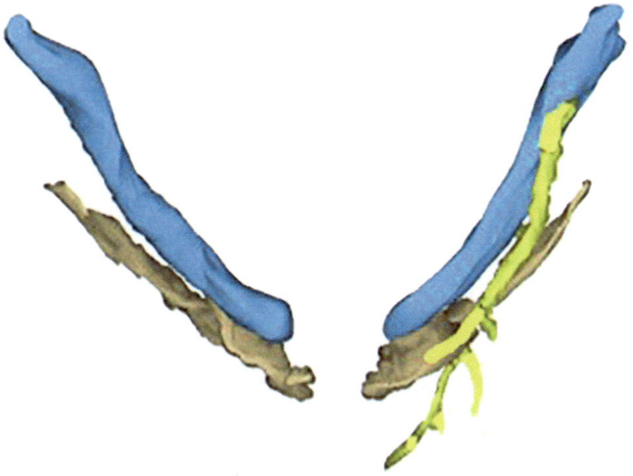

Fig. 11.3 Human embryo, 22 mm CRL, mandible with mental n. on left side, cranial view

R. J. Radlanski, *Atlas of Human Prenatal Craniofacial Development*, https://doi.org/10.1007/978-3-031-88812-0_11

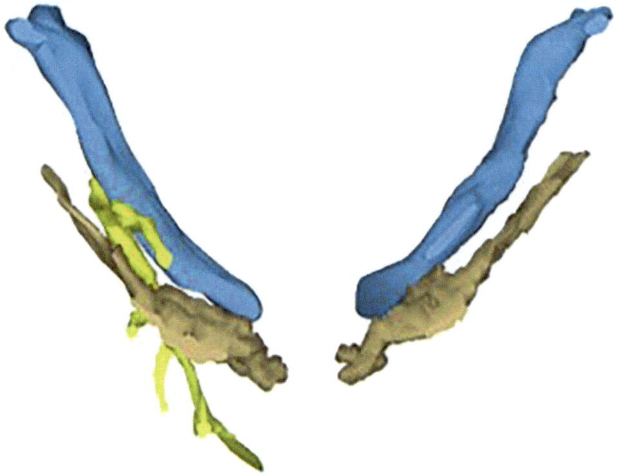

Fig. 11.4 Human embryo, 22 mm CRL, mandible with mental n. on left side, caudal view Ochre: bone; blue: Meckel's cartilage; yellow: inferior alveolar n

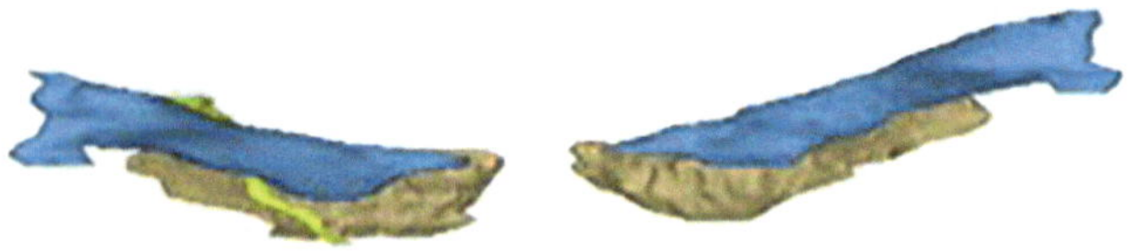

Fig. 11.5 Human embryo, 22 mm CRL, mandible with mental n. on left side, dorsal view. Ochre: bone; blue: Meckel's cartilage; yellow: inferior alveolar n

Further Reading

Radlanski RJ, Renz H, Müller U, Schneider RS, Marcucio RS, Helms JA. Prenatal morphogenesis of the human mental foramen. Eur J Oral Sci. 2002;110:452–9.

Radlanski RJ, Renz H, Klarkowski MC. Prenatal development of the human mandible. 3D reconstructions, morphometry and bone remodelling pattern, sizes 12-117 mm CRL. Anat Embryol (Berl). 2003;207:221–32.

Radlanski RJ, Renz H, Lajvardi S, Schneider R. A Bone remodeling during prenatal morphogenesis of the human mental foramen. Eur J Oral Sci. 2004;112:301–10.

Radlanski RJ. Oral structure and biology. Batavia: Quintessence Publishing; 2018.

Sperber GH. Craniofacial embryology. Hamilton: B.C. Decker; 2001.

12 Embryo 24 mm CRL (ca. 8th wk): BAL110787

Mandible

In this embryo, Meckel's cartilage and the mandible converge anteriorly in a well-proportioned V; there is even a slight curve in both bone and cartilage at the transition to the dorsal third of its course (see Figs. 12.3 and 12.4). The bone of the mandible is so well developed vertically that the piston-shaped distensions of Meckel's cartilage at its anterior ends are almost covered by bone. Both ends of the cartilage are touching in the symphysis and the bone formations come close to each other anteriorly to within 100 μm. The alveolar inferior nerve runs in a well-developed bone groove, the mental foramen is properly circularly closed (Figs. 12.1, 12.2, 12.3, and 12.4).

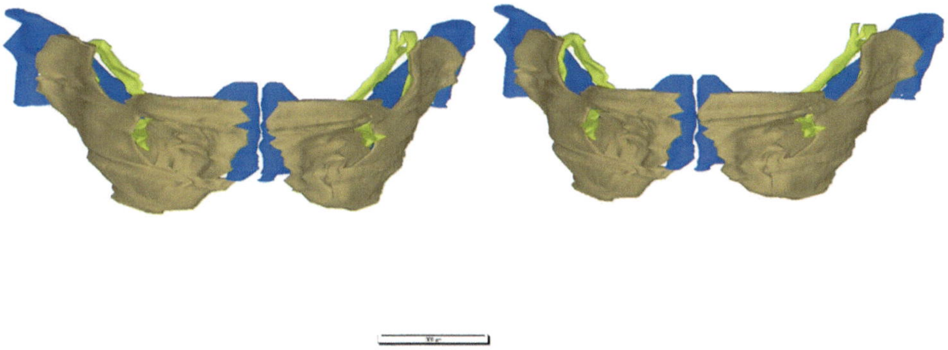

Fig. 12.1 Stereoscopical pair of images. Human embryo 24 mm CRL (ca. 8th week), mandible (ochre) and Meckel's cartilage (blue), and inferior alveolar nerve (yellow), frontal view. Bar: 500 μm

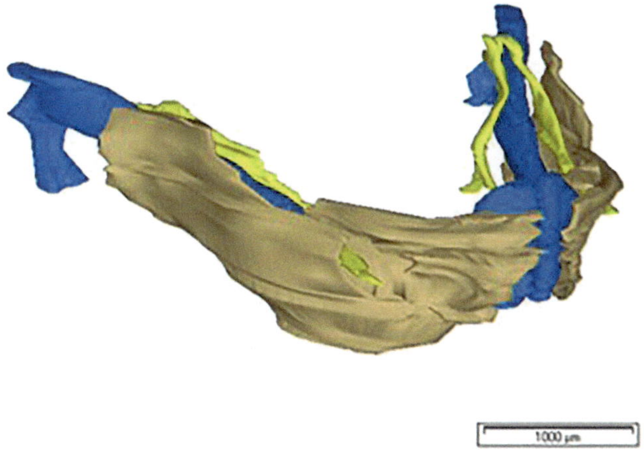

Fig. 12.2 Human embryo, 24 mm CRL, mandible, lateral view. Mandible: ochre; Meckel's cartilage: blue; lingual and alveolar inf. nerves: yellow

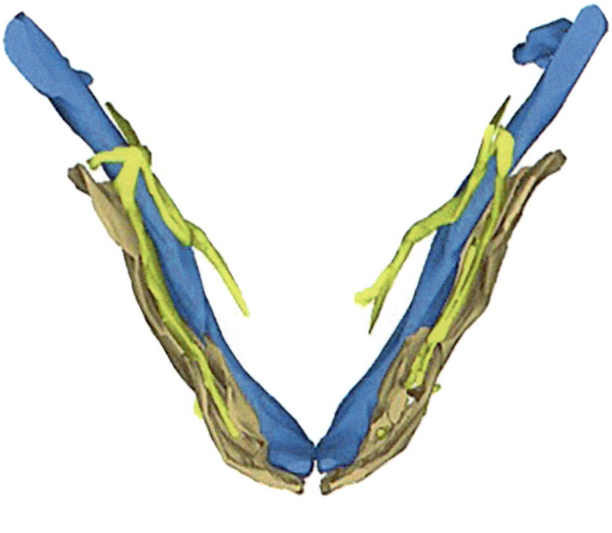

Fig. 12.3 Human embryo, 24 mm CRL, mandible, cranial view. Mandible: ochre; Meckel's cartilage: blue; lingual and alveolar inf. nerves: yellow

R. J. Radlanski, *Atlas of Human Prenatal Craniofacial Development*, https://doi.org/10.1007/978-3-031-88812-0_12

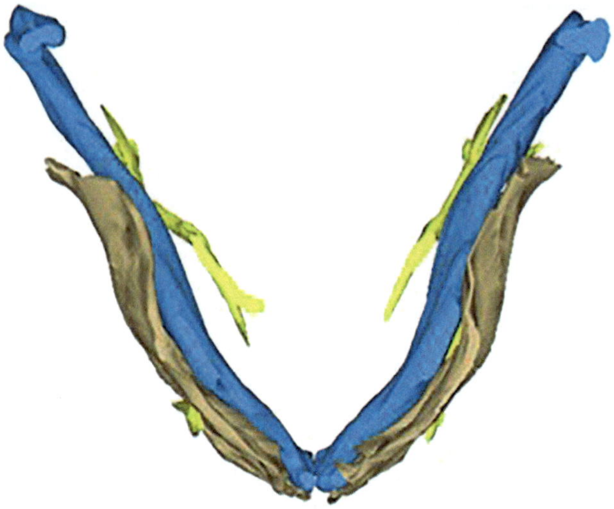

Fig. 12.4 Human embryo, 24 mm CRL, mandible, cranial view. Mandible: ochre; Meckel's cartilage: blue; lingual and alveolar inf. nerves: yellow

Further Reading

Radlanski RJ, Renz H, Müller U, Schneider RS, Marcucio RS, Helms JA. Prenatal morphogenesis of the human mental foramen. Eur J Oral Sci. 2002;110:452–9.

Radlanski RJ, Renz H, Klarkowski MC. Prenatal development of the human mandible. 3D reconstructions, morphometry and bone remodelling pattern, sizes 12-117 mm CRL. Anat Embryol (Berl). 2003;207:221–32.

Radlanski RJ, Renz H, Lajvardi S, Schneider R. A Bone remodeling during prenatal morphogenesis of the human mental foramen. Eur J Oral Sci. 2004;112:301–10.

Radlanski RJ. Oral structure and biology. Batavia: Quintessence Publishing; 2018.

Sperber GH. Craniofacial embryology. Hamilton: B.C. Decker; 2001.

Embryo 25 mm CRL (ca. 8th wk): JOS080289

13

Overall Remarks

This reconstruction again shows the most important structures of the face in their spatial relationships. In the midface, the nasal capsule occupies the largest space, and both maxilla and mandible have increased significantly in size and height (Figs. 13.1 and 13.2).

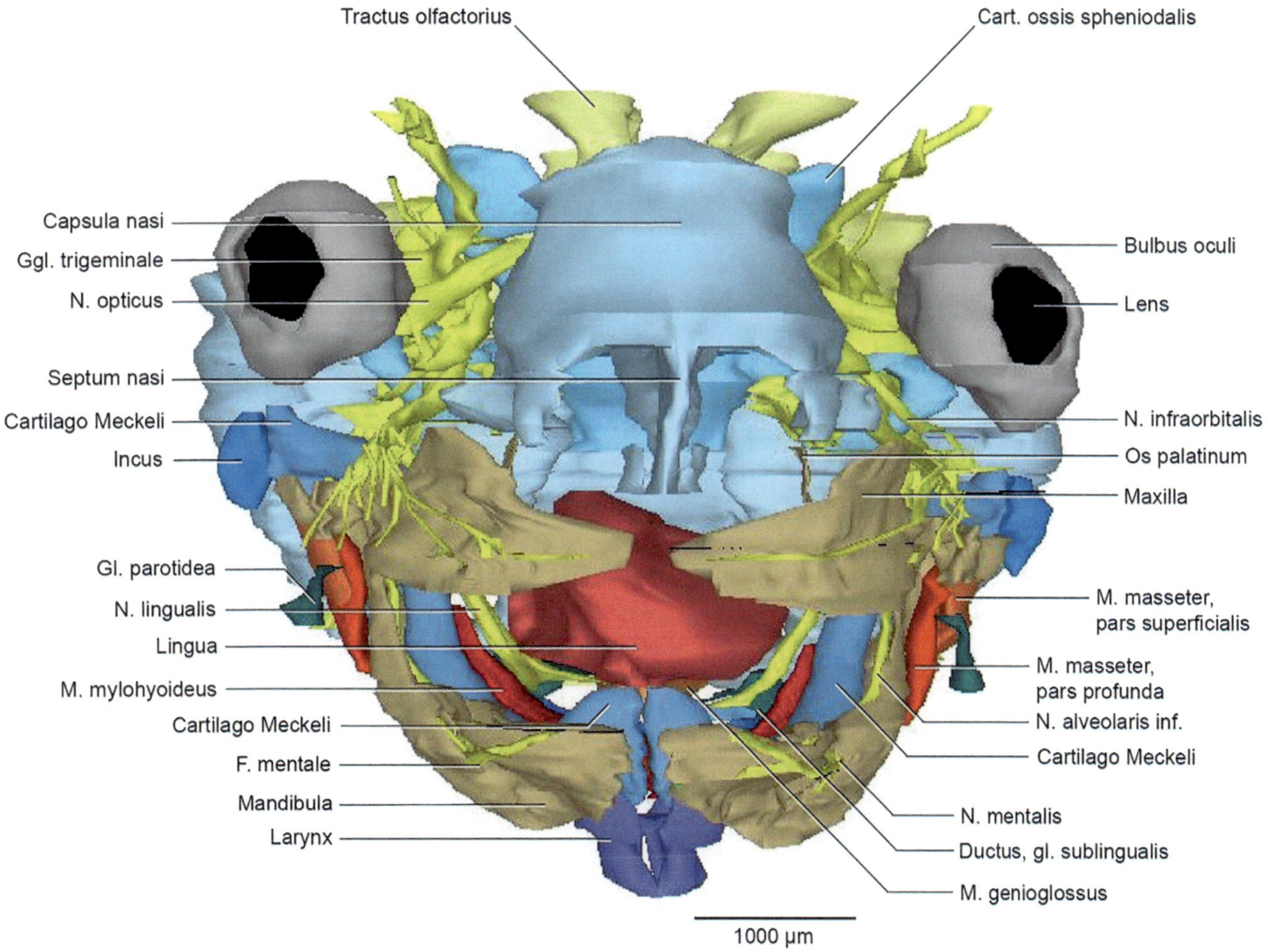

Fig. 13.1 Human embryo, 25 mm CRL, head, frontal view, skin removed, layer 1 of preparation

R. J. Radlanski, *Atlas of Human Prenatal Craniofacial Development*, https://doi.org/10.1007/978-3-031-88812-0_13

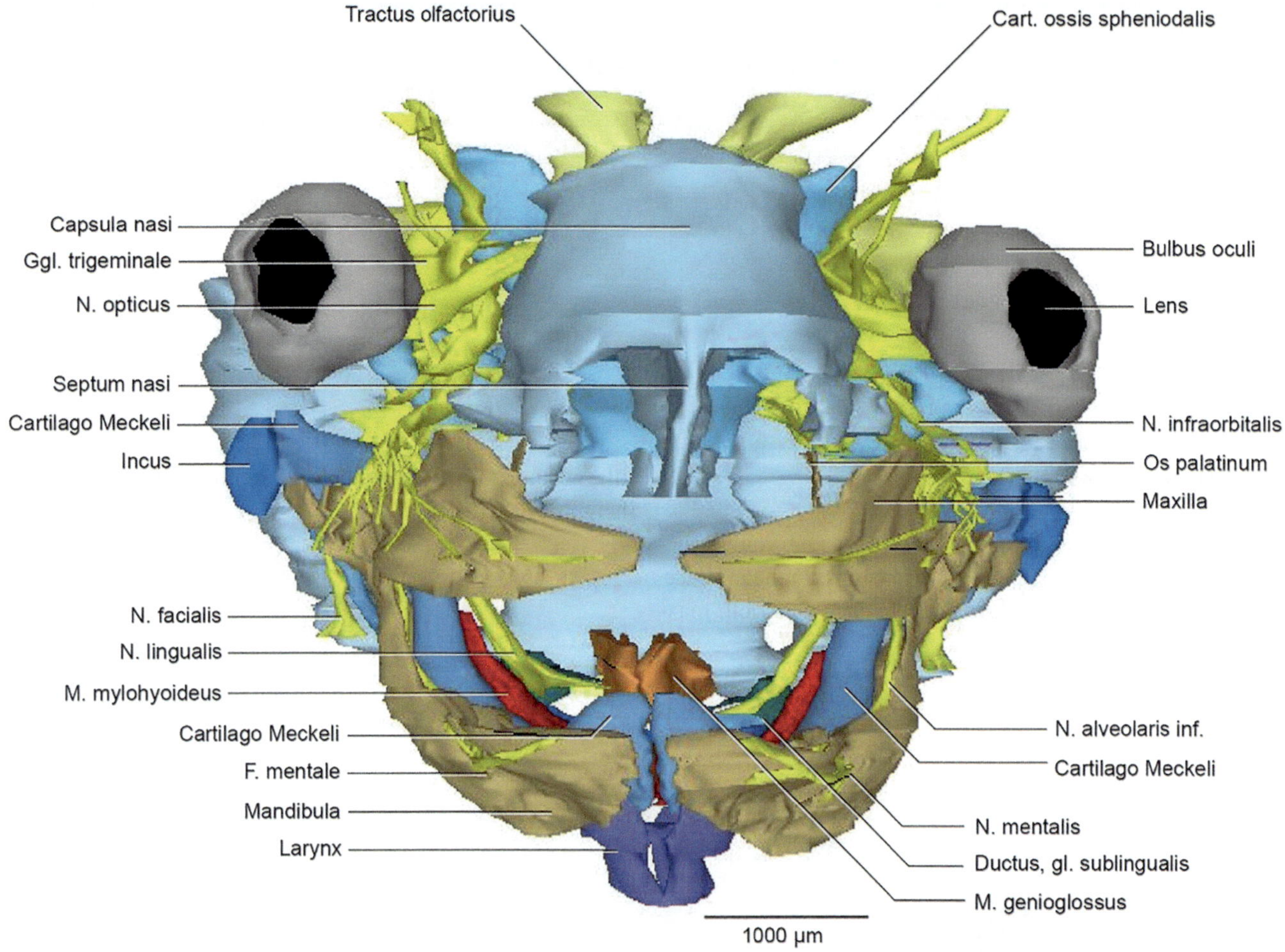

Fig. 13.2 Human embryo, 25 mm CRL, head, frontal view, layer 2 of preparation

Muscles

The tongue is well developed and fills a large part of the joint oronasal cavity. The masseter muscle can be distinguished as a superficial and a deep portion. Since the zygomatic arch is already recognizably well developed as a bony formation (see Fig. 13.6), its area of origin is by now clearly defined. The mylohyoid muscle extends transversely between the two halves of the mandible. At this stage, its insertion is still at Meckel's cartilage and in its course, it sags downward like a hammock (Fig. 13.3).

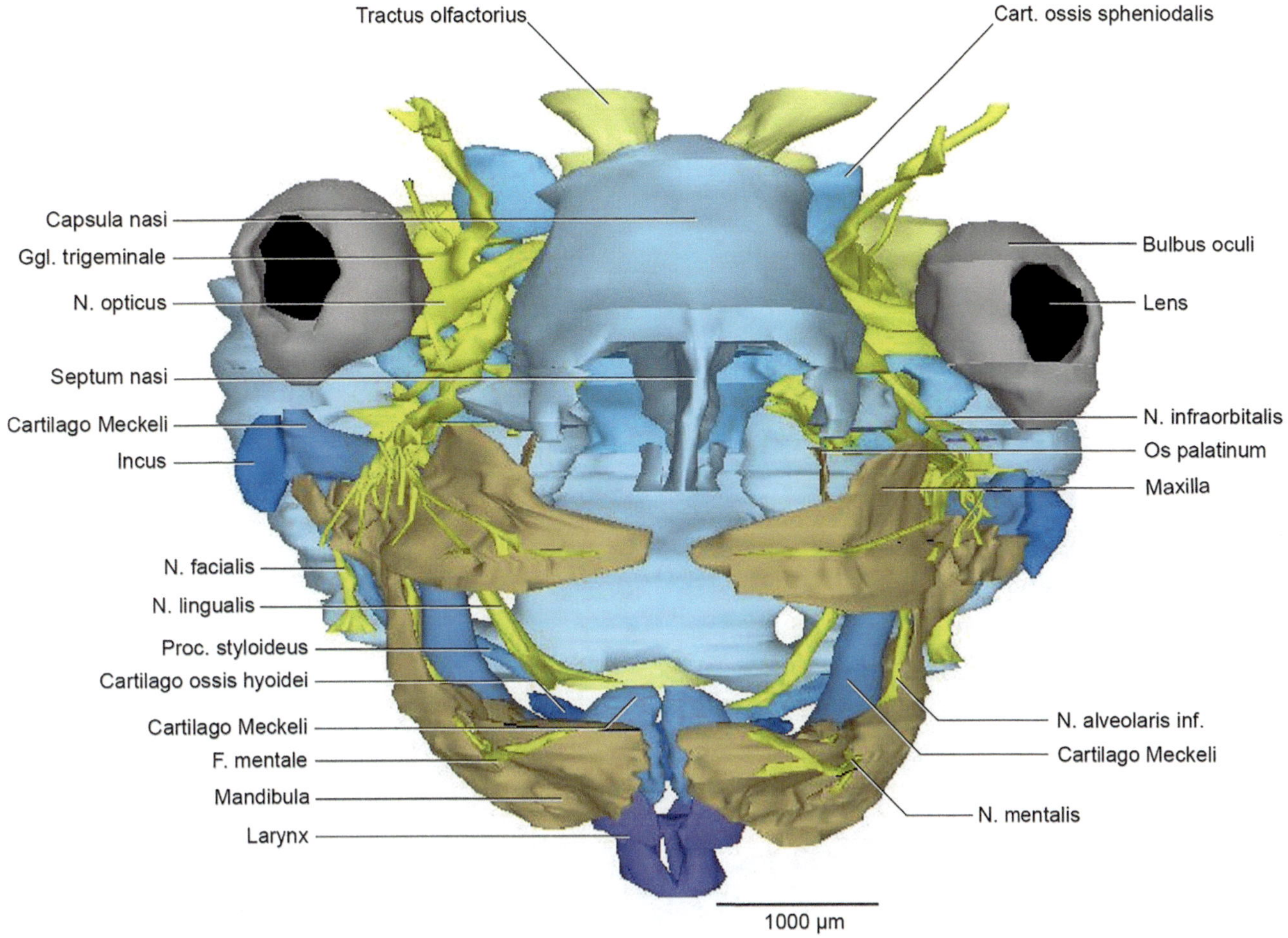

Fig. 13.3 Human embryo, 25 mm CRL, head, frontal view, layer 3 of preparation

Nerves

Cranially, the paired extensions of the olfactory tract are visible (see Figs. 13.1 and 13.6). The optical nerve enters the eyeballs directly from behind, slightly medially. The branches of the trigeminal nerve take their typical courses. The supraorbital nerve finds its way anteriorly between the cartilaginous primordium of the lesser wing of the sphenoid bone and the eyeball. The infraorbital nerve passes into anterior direction below the eye. Thereby it already gives off some small ramifications into the region of the forming maxilla, and with many single branches it also crosses the upper edge of the maxilla. The later infraorbital foramen is now only indicated as a notch. The most important branches of the mandibular nerve are the alveolar inferior nerve and the lingual nerve, which run along laterally and medially to Meckel's cartilage, respectively. The mental nerve exits through the mental foramen anterior-laterally through the mandible into the chin region. Portions of the facial nerve with the tympanic chord are shown (see Fig. 13.7; Fig. 13.4).

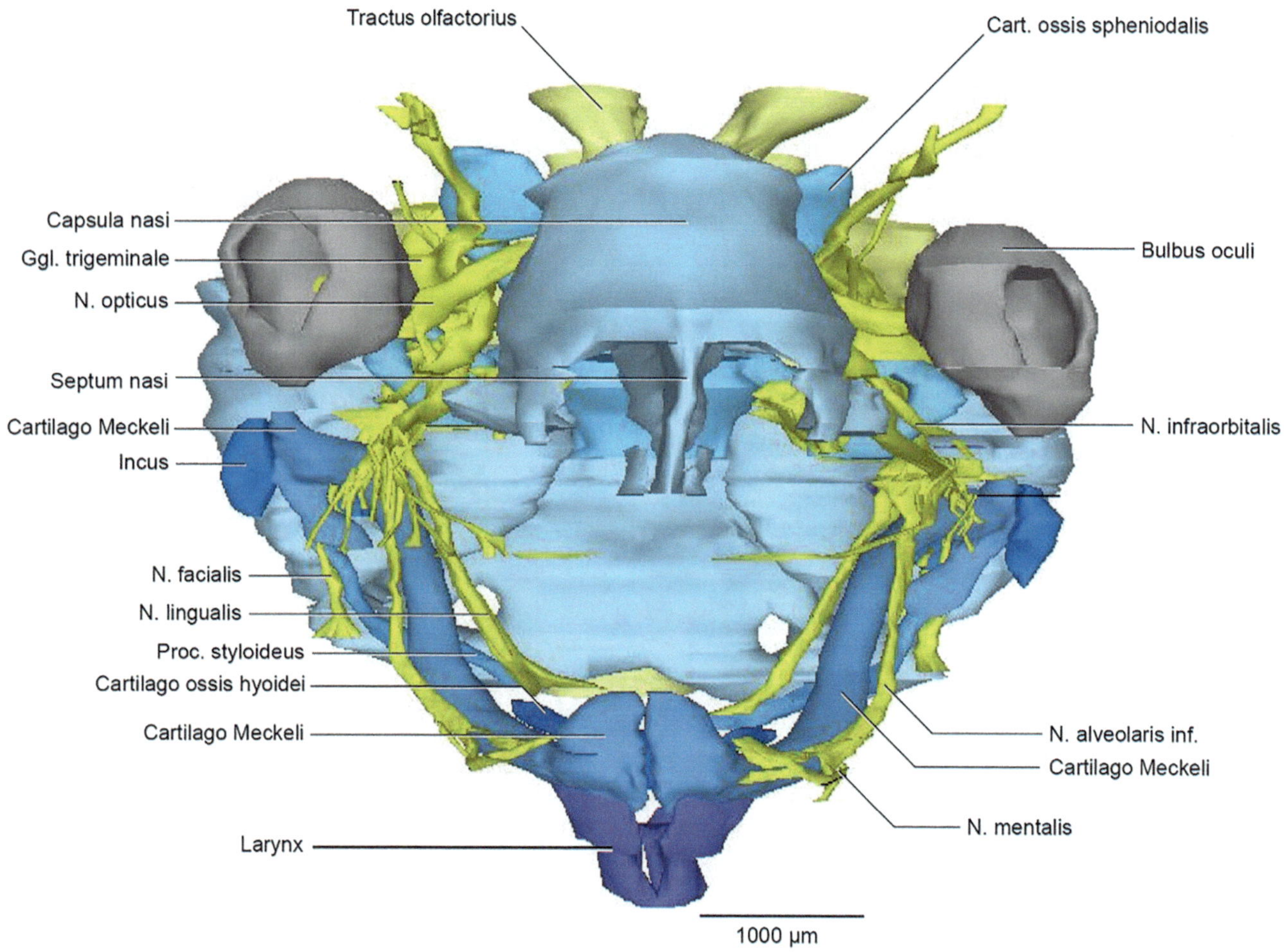

Fig. 13.4 Human embryo, 25 mm CRL, head, frontal view, cartilage, nerves, and eyeballs

Salivary Glands

In this reconstruction, the anlage of the parotid gland is included (Fig. 13.5). Its excretory duct arises from the buccal epithelium of the oral cavity and extends laterally across the mandible to lie on the outer surface of the masseter muscle. A cob-shaped extension can be seen here at this stage. At this time, the canal lumen is still compactly filled with epithelium, and the gland is not yet formed at the end (see Figs. 13.1, 13.6, and 13.11).

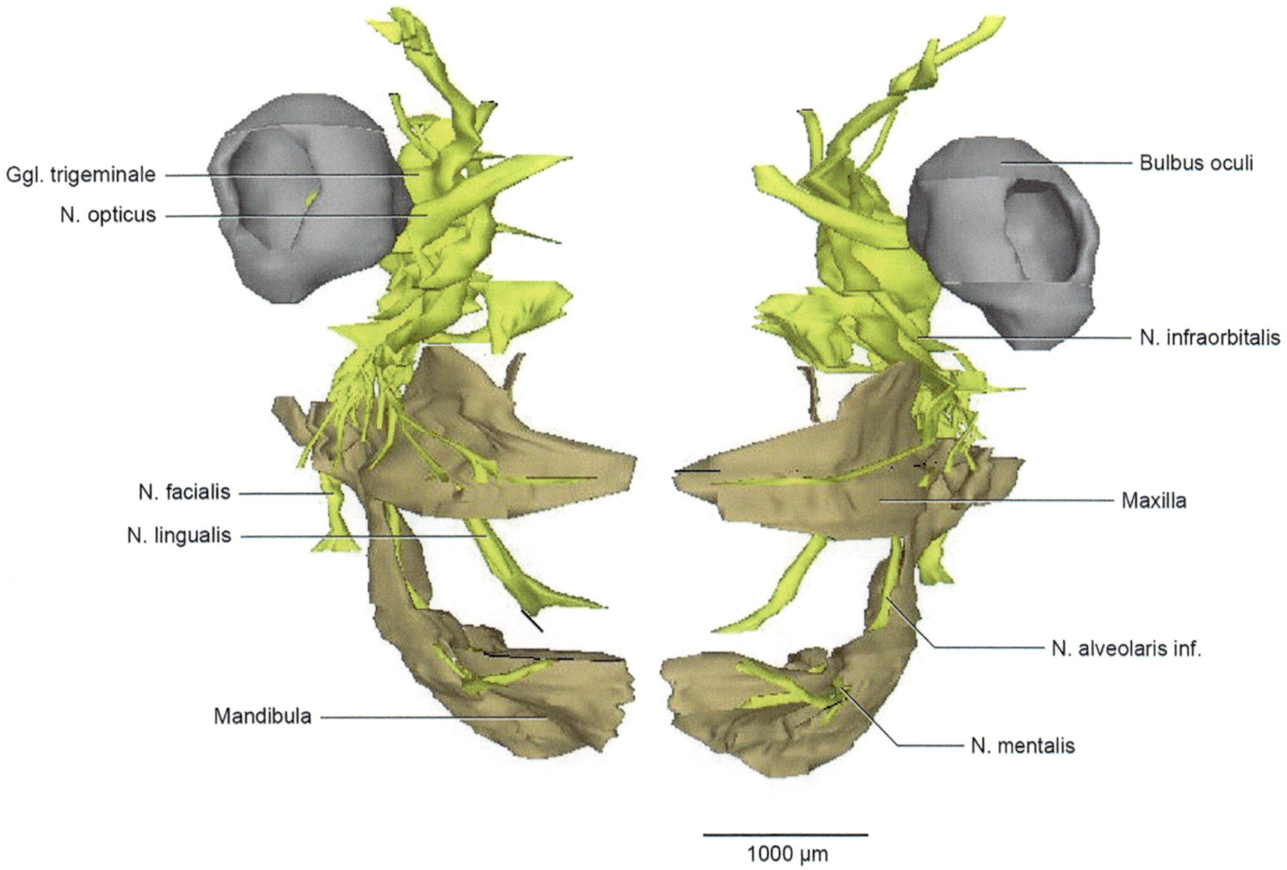

Fig. 13.5 Human embryo, 25 mm CRL, head, frontal view, bone, nerves, and eyeballs

Maxillary Complex

The maxilla lies in pairs below and next to the nasal capsule (Fig. 13.6). Overall, the maxilla has the shape of a flattened triangle, the tip of which can be identified as a frontal process. Its height, however, is far from reaching the lower edge of the eyeball. The entire bone mass develops from a multitude of individual bony islands, which gradually become confluent with each other.

In contrast to embryo ADE160787 (20 mm CRL, Chap. 7), which showed an already well-developed infraorbital foramen, only a notch is visible here in this reconstruction where the infraorbital nerve passes over the maxilla. At the lateral-posterior ends of the maxilla, bone formations of the zygomatic arch are already visible.

The nasal capsule occupies a large volume in the midfacial region. It represents a bilaterally closed formation separated in its center by a nasal septum, creating two nasal cavities. Dorsally, the still cartilaginous primordia of the sphenoid bone are attached.

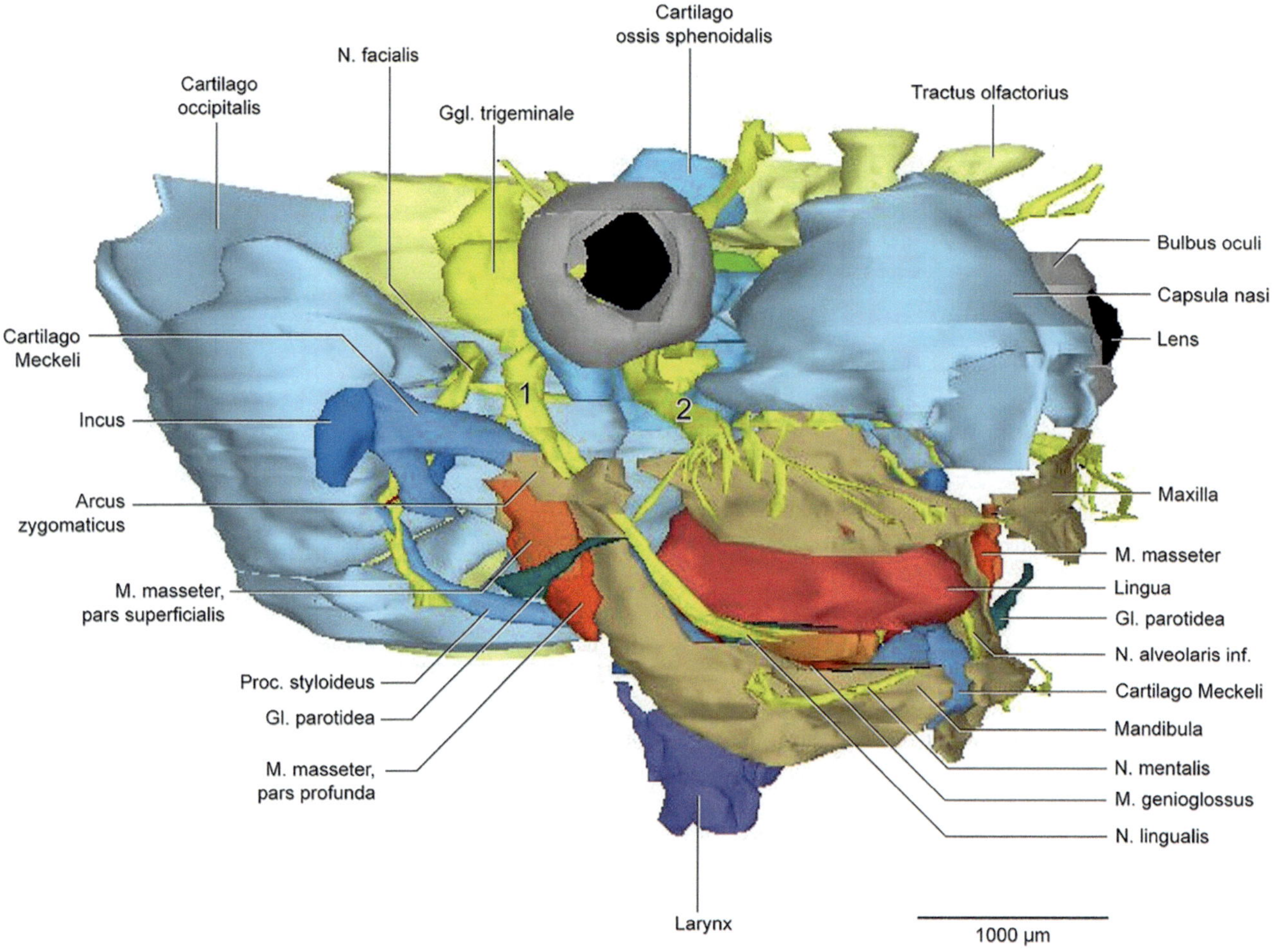

Fig. 13.6 Human embryo, 25 mm CRL, head, 45° lateral view, skin removed, layer 1 of preparation. 1: mandibular n. 2: infraorbital n. (*From* Fernández Rubio and Radlanski [2]; *with permission.*)

Mandible

The mandible reveals a transition from a V to a U shape, with the curves now in transition from anterior third to middle third of their course (see Fig. 13.22). The curvilinear protrusions are more accentuated at this stage than in the preceding ones and the posterior end of Meckel's cartilage swings quite far outward beyond the dorsal edge of the mandible. The anterior ends of Meckel's cartilage are in contact, but the bony portions of the mandible are still about 300 μm apart in the chin region (Fig. 13.7)

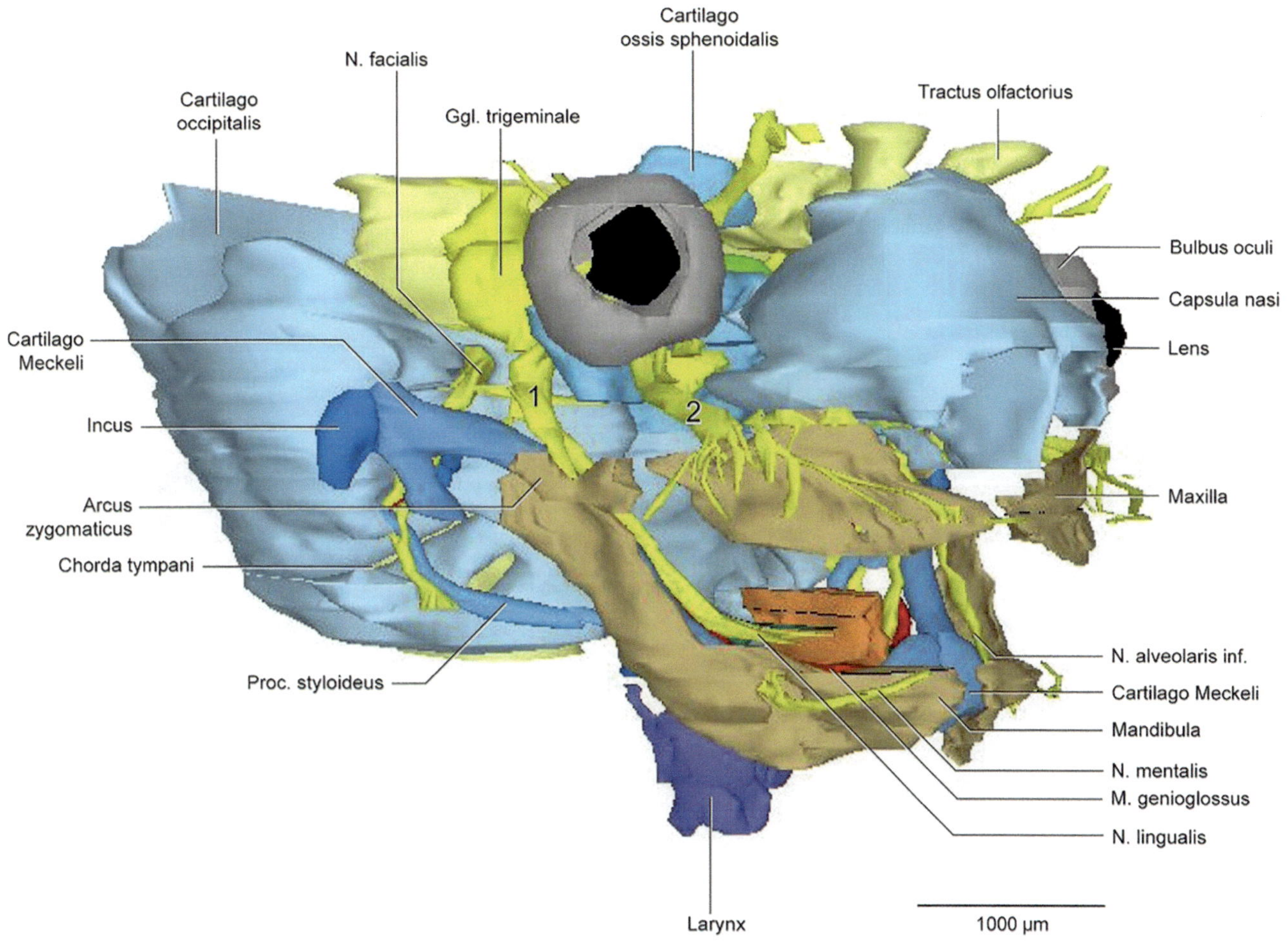

Fig. 13.7 Human embryo, 25 mm CRL, head, 45° lateral view, layer 2 of preparation. 1: mandibular n. 2: infraorbital n.

Cartilaginous Skeleton and Calvaria

While most of the bones of the calvaria show desmal ossification, there is a cartilaginous primordium for the occipital bone (lateral and basal part) (see Fig. 13.6). The petrous bone and the labyrinth capsule are also performed cartilaginously. The body and the smaller wing of the sphenoid bone also have a cartilaginous preformation. The styloid process also has a cartilaginous preform that ossifies later. Of the auditory ossicles, the incus is always clearly visible in these reconstructions, the stapes lies medial to it and is usually hidden in standard views, therefore. From more favorable viewing angles, however, it is better visible, as shown in Figs. 13.28. And the malleus is represented by the dorsal end of Meckel's cartilage, which will separate from it only during the course of later development. All the primordia of the auditory ossicles are still present as cartilaginous formations at this stage. The hyoid bone is created from cartilaginous primordia of the second (lesser horn) and third (greater horn) visceral arches, which are fused together and only later ossify. The larynx remains cartilaginous even after birth (Figs. 13.8, 13.9, 13.10, 13.11, 13.12, 13.13, 13.14, 13.15 and 13.16).

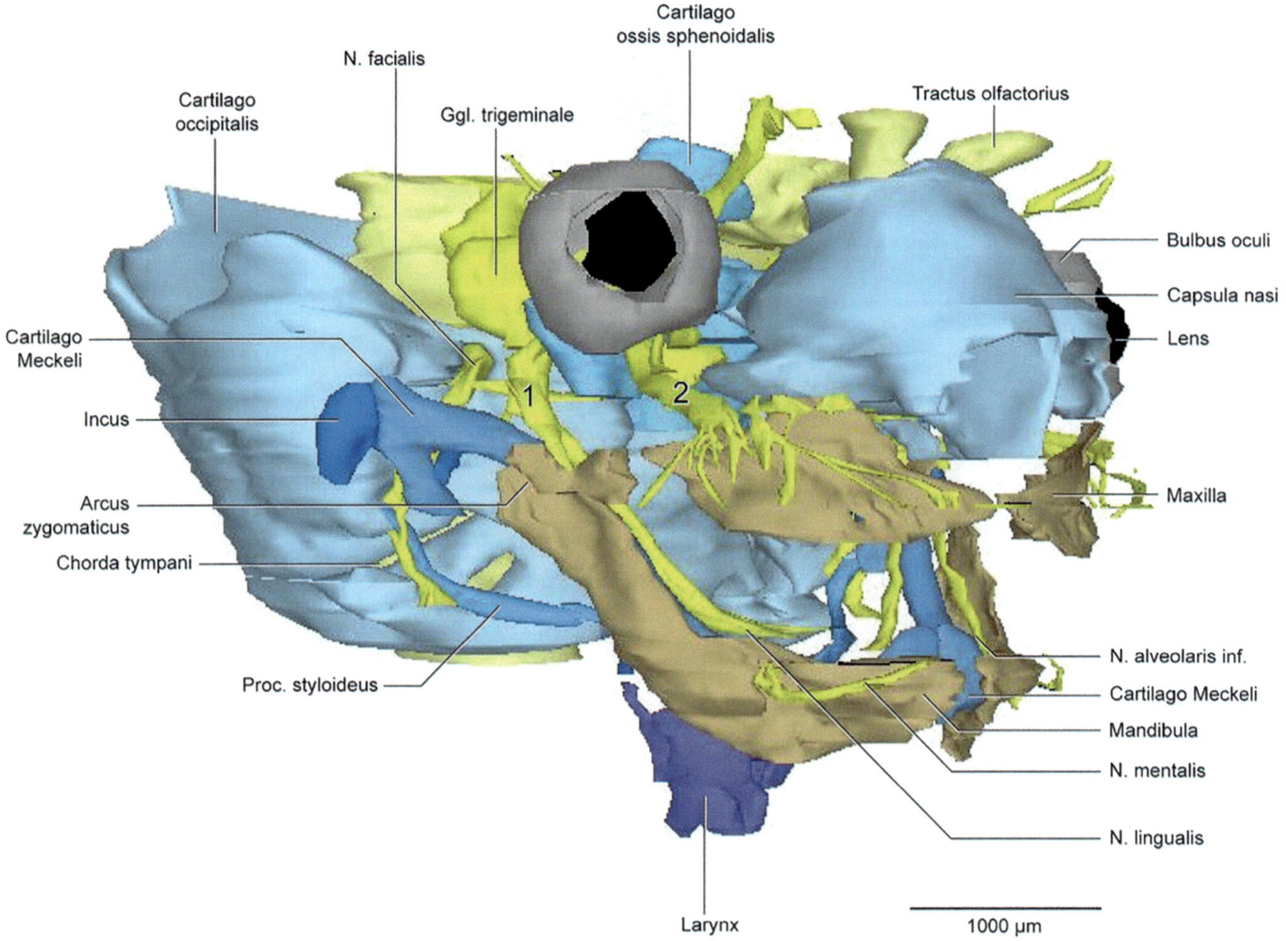

Fig. 13.8 Human embryo, 25 mm CRL, head, 45° lateral view, layer 3 of preparation. 1: mandibular n. 2: infraorbital n.

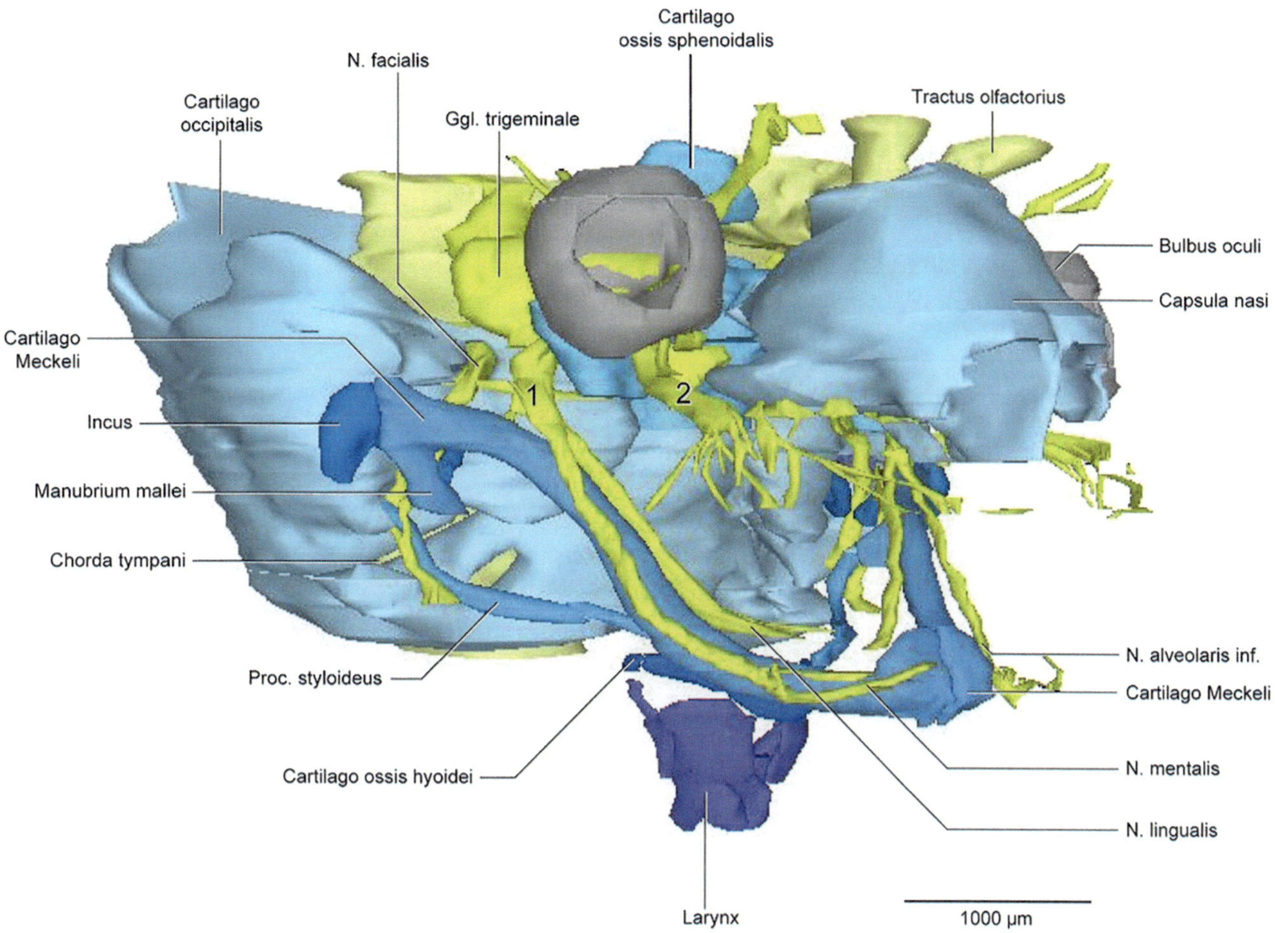

Fig. 13.9 Human embryo, 25 mm CRL, head, 45° lateral view, cartilage, nerves, and eyeballs1: mandibular n. 2: infraorbital n.

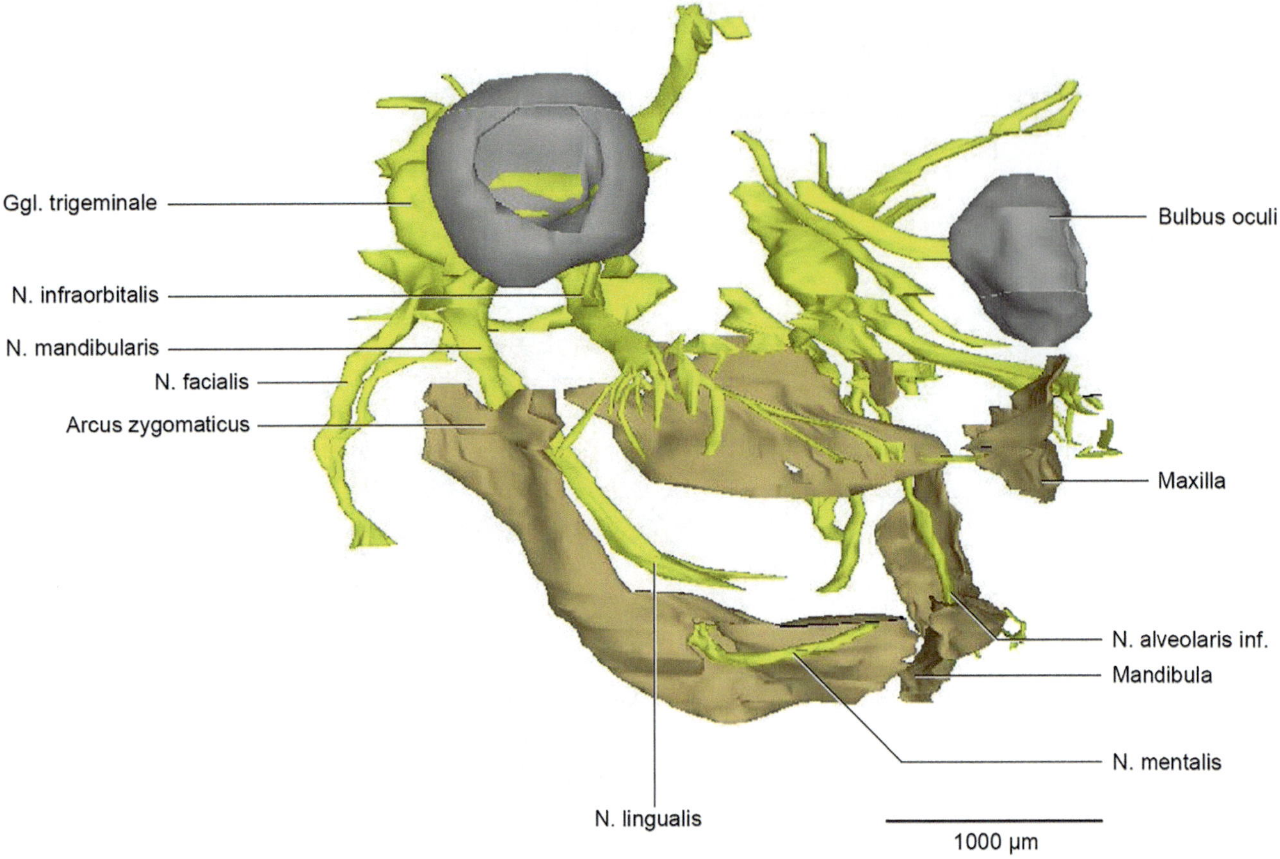

Fig. 13.10 Human embryo, 25 mm CRL, head, 45° lateral view, bone, nerves, and eyeballs

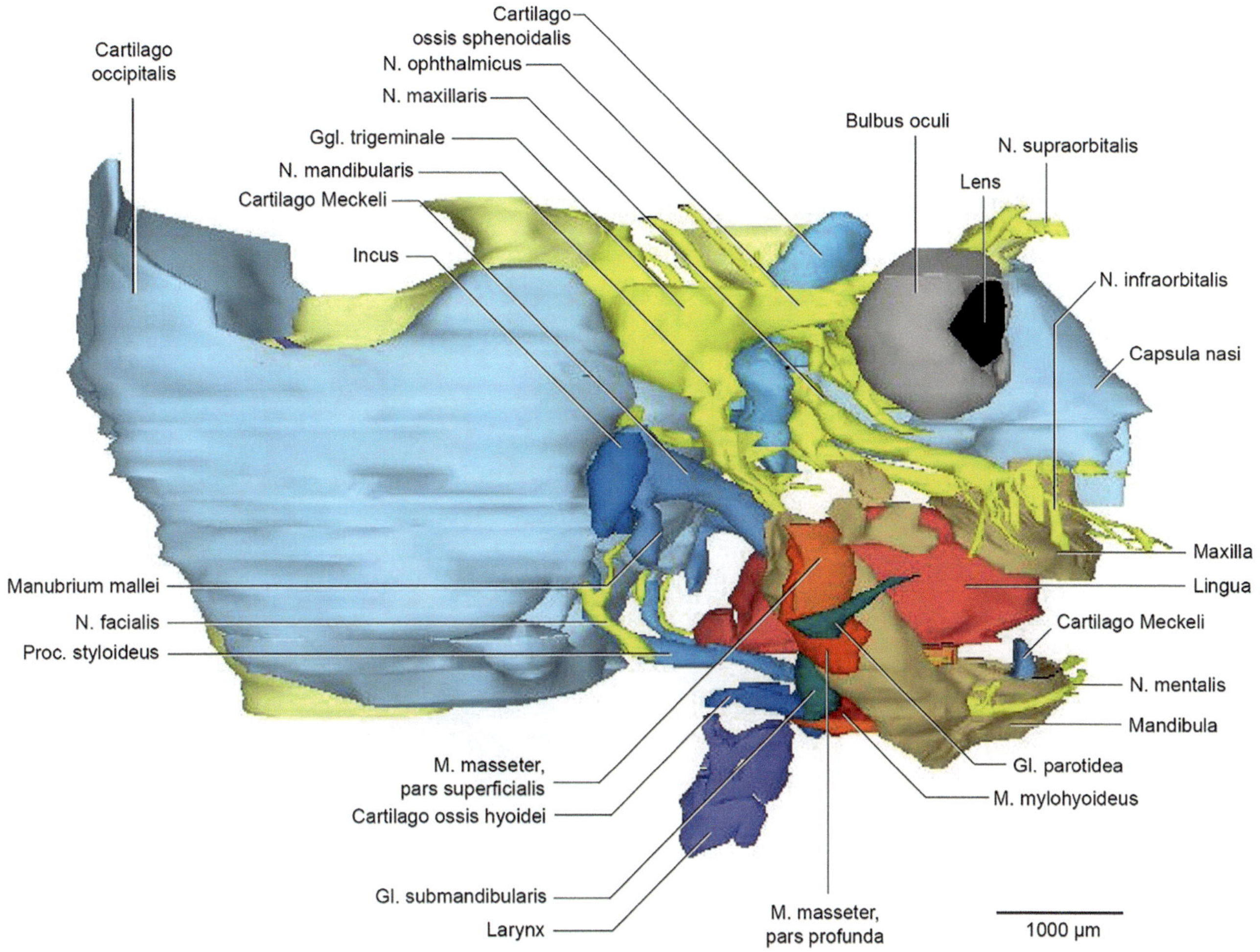

Fig. 13.11 Human embryo, 25 mm CRL, head, lateral view, skin removed, layer 1 of preparation

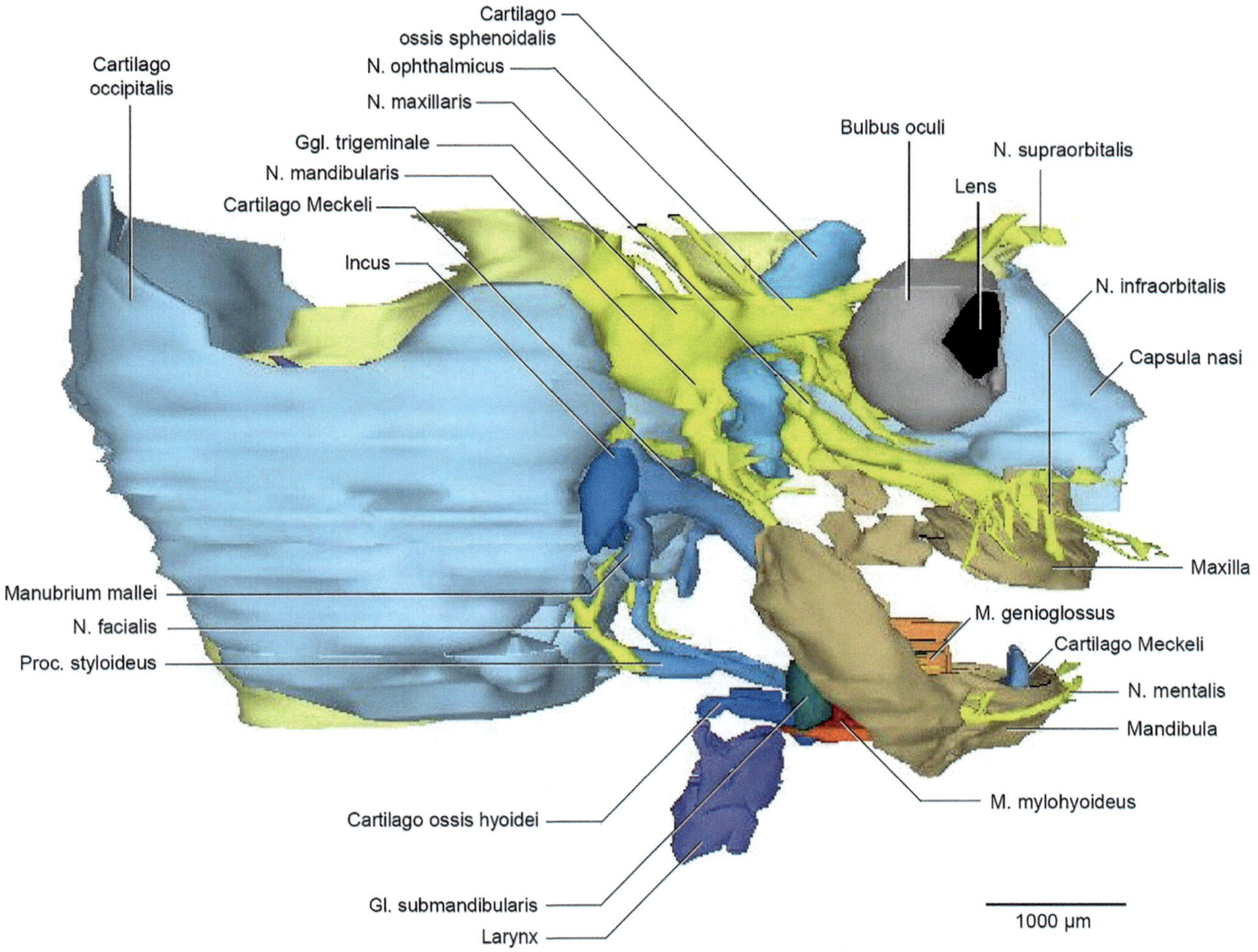

Fig. 13.12 Human embryo, 25 mm CRL, head, lateral view, layer 2 of preparation

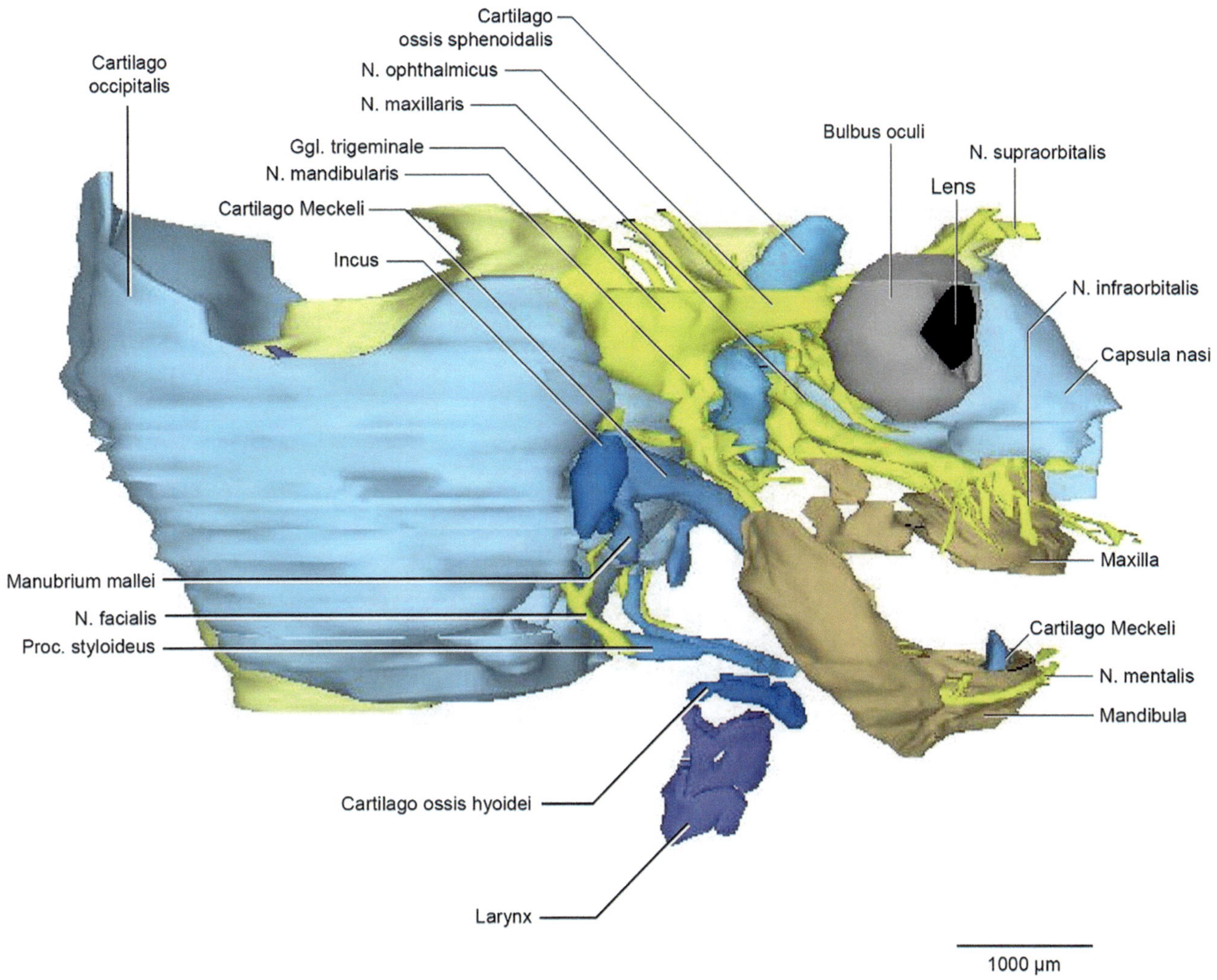

Fig. 13.13 Human embryo, 25 mm CRL, head, lateral view, layer 3 of preparation

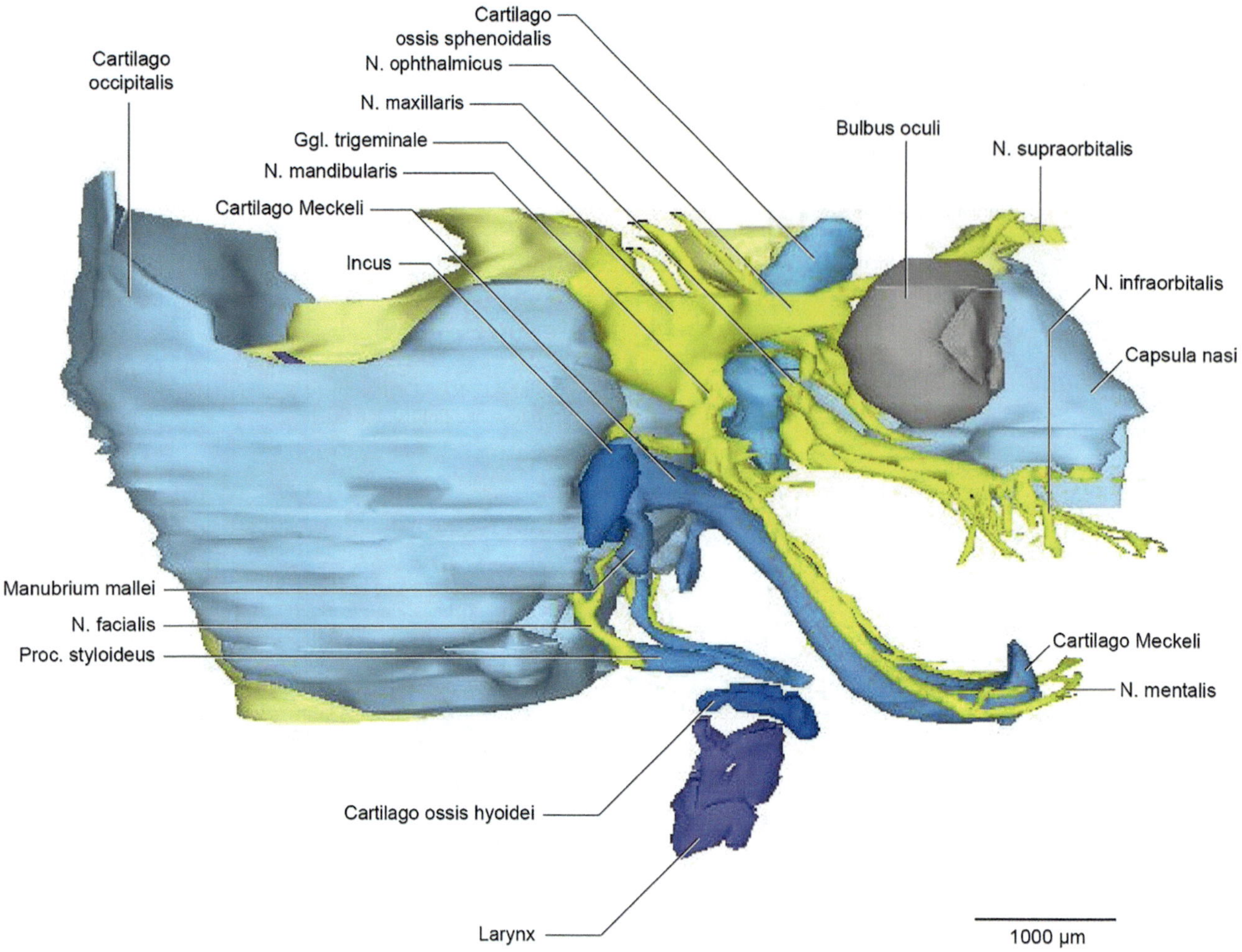

Fig. 13.14 Human embryo, 25 mm CRL, head, lateral view, cartilage, nerves, and eyeballs

Fig. 13.15 Human embryo, 25 mm CRL, head, lateral view, bone, nerves, and eyeballs

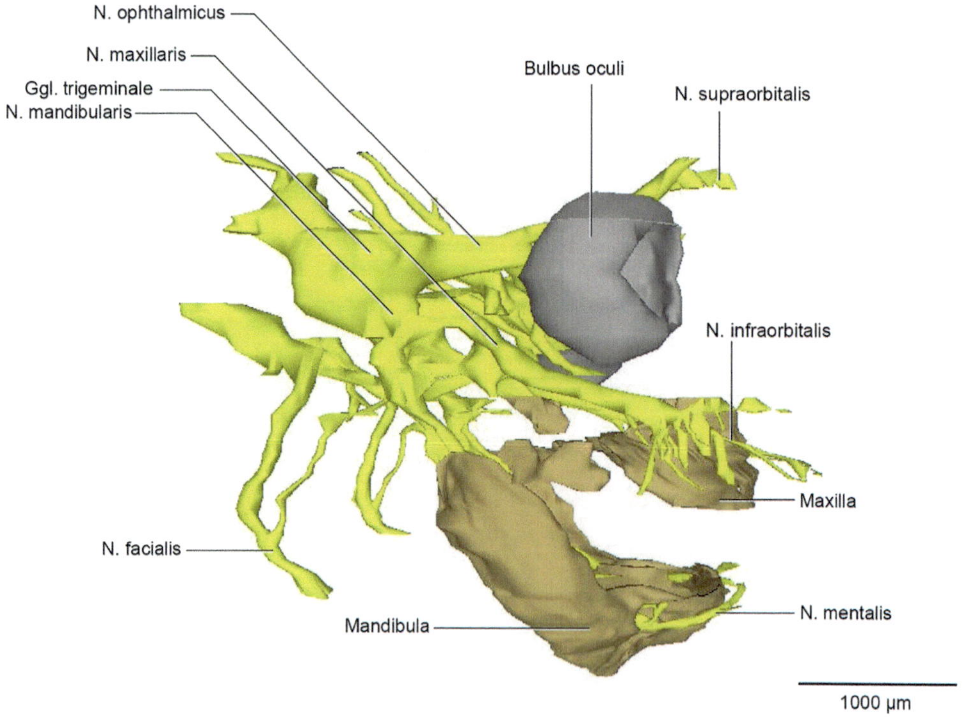

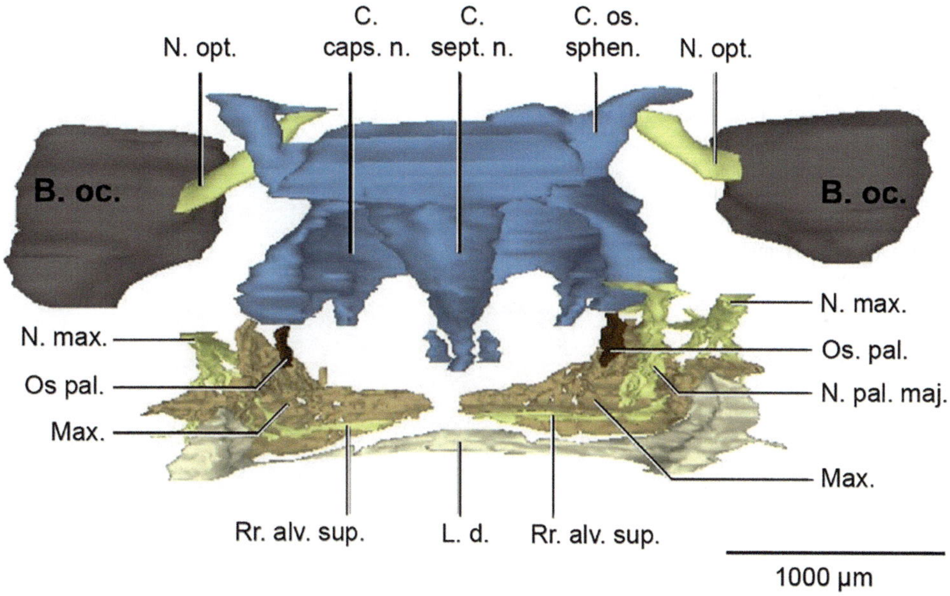

Fig. 13.16 Human embryo, 25 mm CRL, maxillary and nasal region, dorsal view. N. opt: N. opticus, optical n. B. oc.: Bulbus oculi, eyeball. N. max: N. maxillaris, maxillary n. Os pal.: Os palatinum, palatal bone. Max: Maxilla. Rr. alv. sup.: Rami alveolares sup., superior alveolar rami. L. d.: Lamina dentalis, dental lamina. C. caps. n.: Cartilago capsula nasi, cartilaginous nasal capsule. C. sept. n.: Cartilago septum nasi, cartilaginous nasal septum. C. os sphen.: Cartilago ossis sphenoidalis, cartilaginous primordium of sphenoid bone

Incisive Canal

The primary palate has begun to ossify before the secondary palate is even closed by connective tissue. This stage of development is still a long way from the bony formation of an incisive canal.

Nasopalatine Nerve and Nasopalatine Artery

The reconstruction of this stage of development enables a detailed study of the course and positional relationship of the nasopalatine nerve and the nasopalatine artery because no overlying bone structures obstruct the direct view. The nasopalatine nerve originates from the pterygopalatine ganglion and runs from there rostro-caudally and medially in the mucosa of the nasal septum and enters the primary palatal bulge caudally. It then bulges dorsally into the still common oronasal cavity. Lateral to the nasopalatine nerve, there runs the nasopalatine artery. Its position is very close to the nasopalatine nerve in the most caudal section and in the direction of the pterygopalatine ganglion with a distance of less than 10 µm. In between, it has a more laterally offset course (Figs. 13.17, 13.18, 13.19, 13.20, 13.21, 13.22, 13.23, 13.24, 13.25, 13.26, 13.27, 13.28, 13.29, 13.30, 13.31, 13.32 and 13.33).

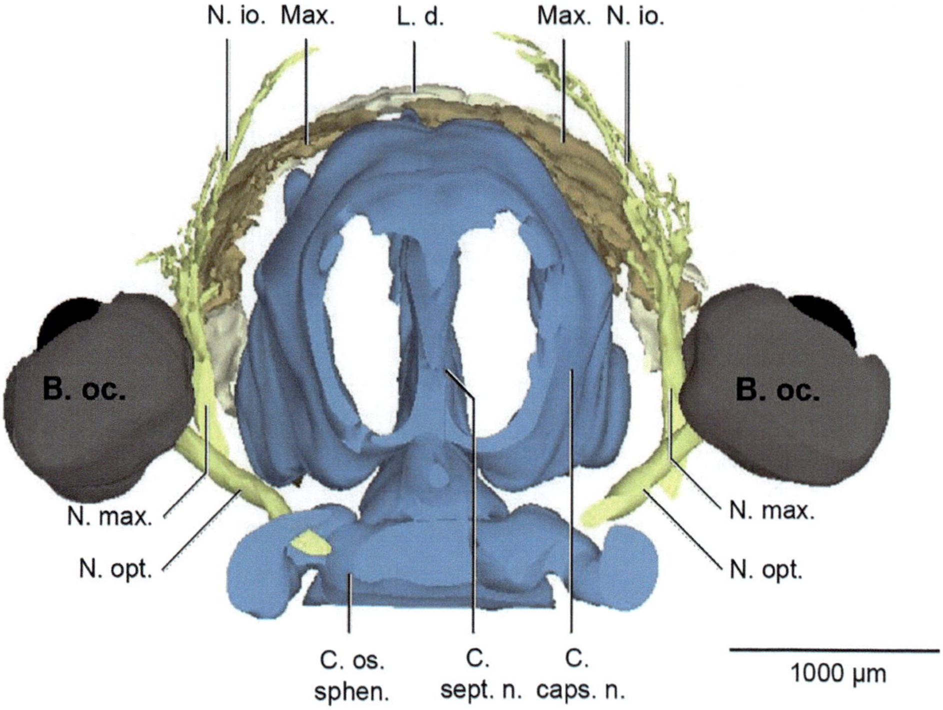

Fig. 13.17 Human embryo, 25 mm CRL, maxillary region, cranial view. N. opt: N. opticus, optical n. B. oc.: Bulbus oculi, eyeball. N. max: N. maxillaris, maxillary n. Os pal.: Os palatinum, palatal bone. Max: Maxilla. Rr. alv. sup.: Rami alveolares sup., superior alveolar rami. L. d.: Lamina dentalis, dental lamina. C. caps. n.: Cartilago capsula nasi, cartilaginous nasal capsule. C. sept. n.: Cartilago septum nasi, cartilaginous nasal septum. C. os sphen.: Cartilago ossis sphenoidalis, cartilaginous primordium of sphenoid bone. N. i.o.: Nervus infraorbitalis, infraorbital n.

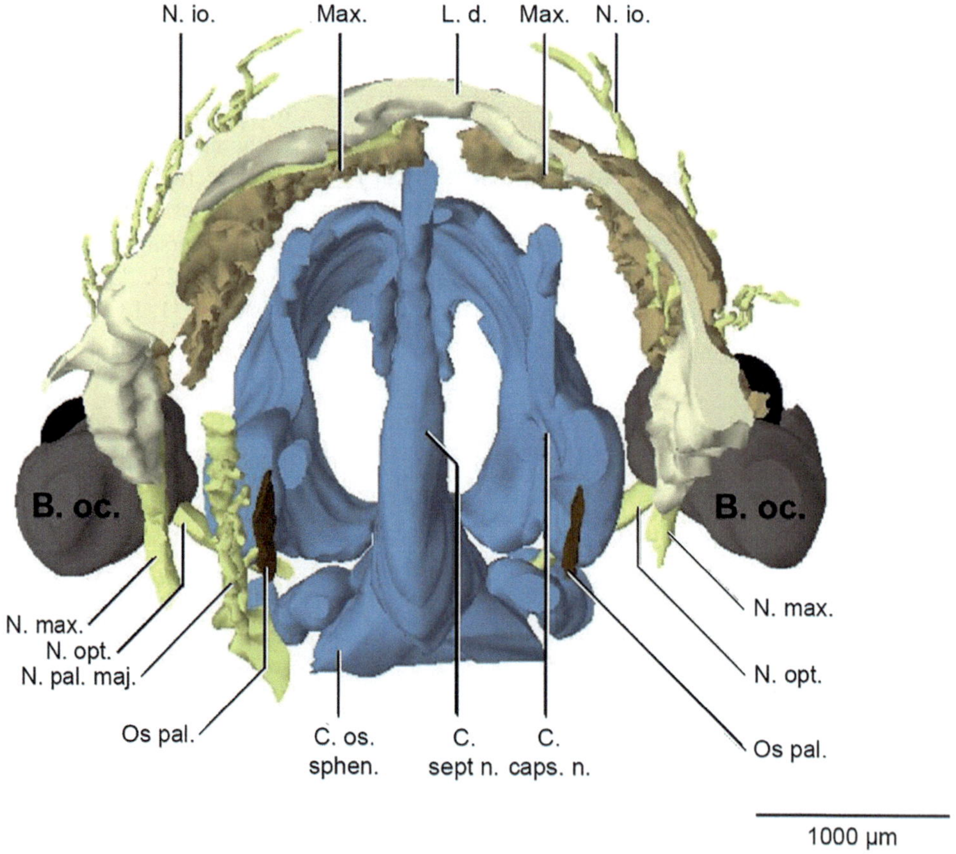

Fig. 13.18 Human embryo, 25 mm CRL, maxillary region, caudal view. N. opt: N. opticus, optical n. B. oc.: Bulbus oculi, eyeball. N. max: N. maxillaris, maxillary n. Os pal.: Os palatinum, palatal bone. Max: Maxilla. Rr. alv. sup.: Rami alveolares sup., superior alveolar rami. L. d.: Lamina dentalis, dental lamina. N. pal. maj.: N. palatinus major, major palatine n C. caps. n.: Cartilago capsula nasi, cartilaginous nasal capsule. C. sept. n.: Cartilago septum nasi, cartilaginous nasal septum. C. os sphen.: Cartilago ossis sphenoidalis, cartilaginous primordium of sphenoid bone. N. i.o.: Nervus infraorbitalis, infraorbital n.

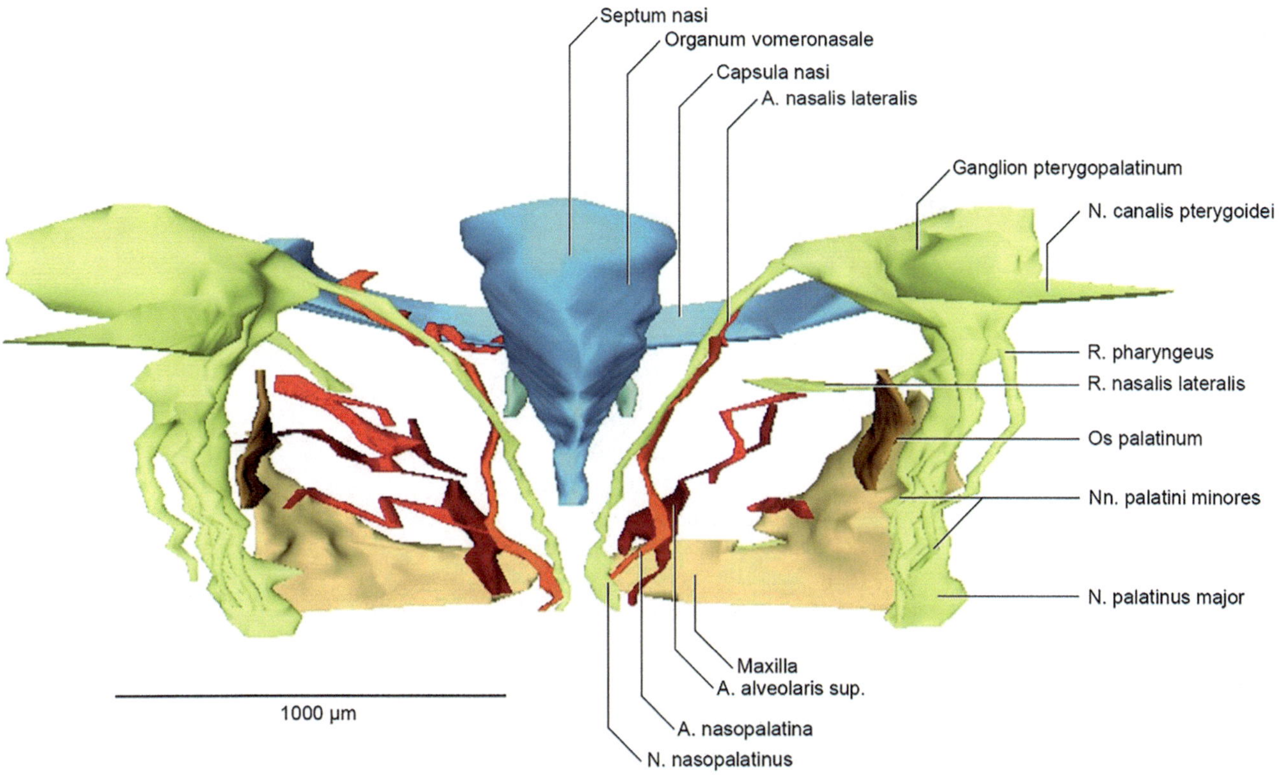

Fig. 13.19 Human embryo, 25 mm CRL, maxillary region. Nerves and vessels in the forming incisive canal. dorsal view

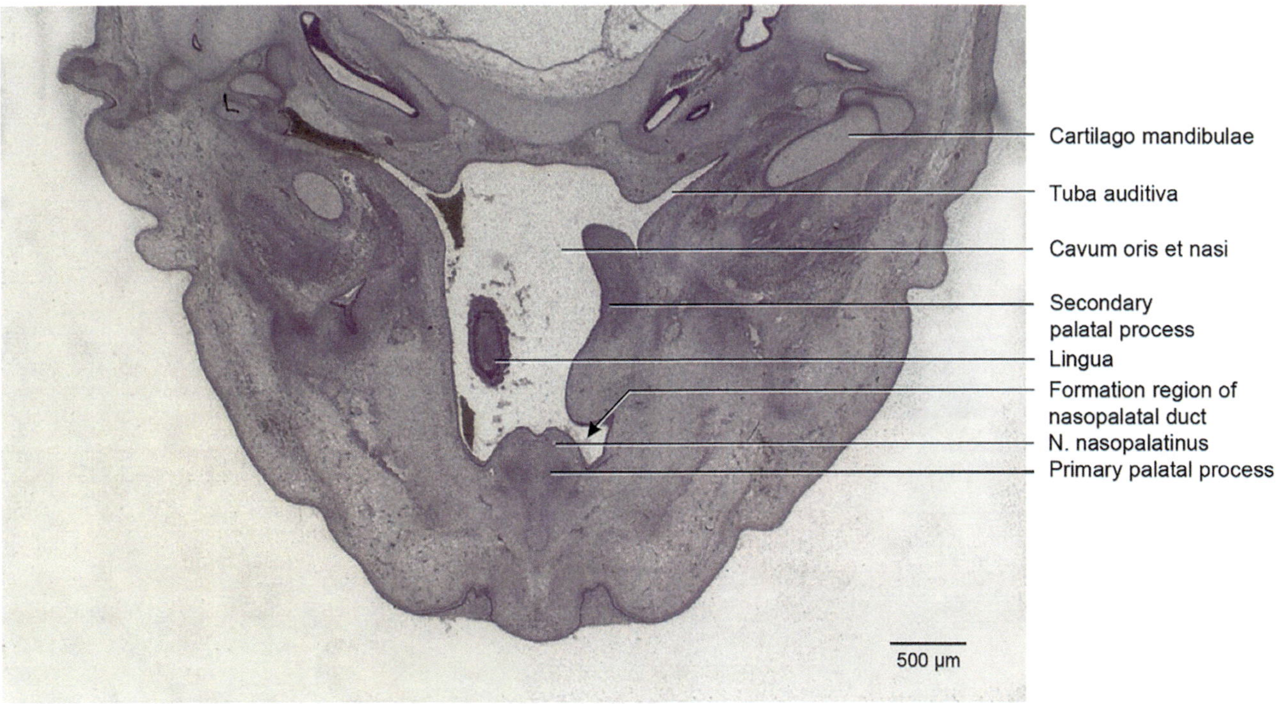

Fig. 13.20 Human embryo, 25 mm CRL, horizontal section at the height of the later palatal roof; estimated beginning of growth of nasopalatal duct. Staining: Hematoxilyn-eosin

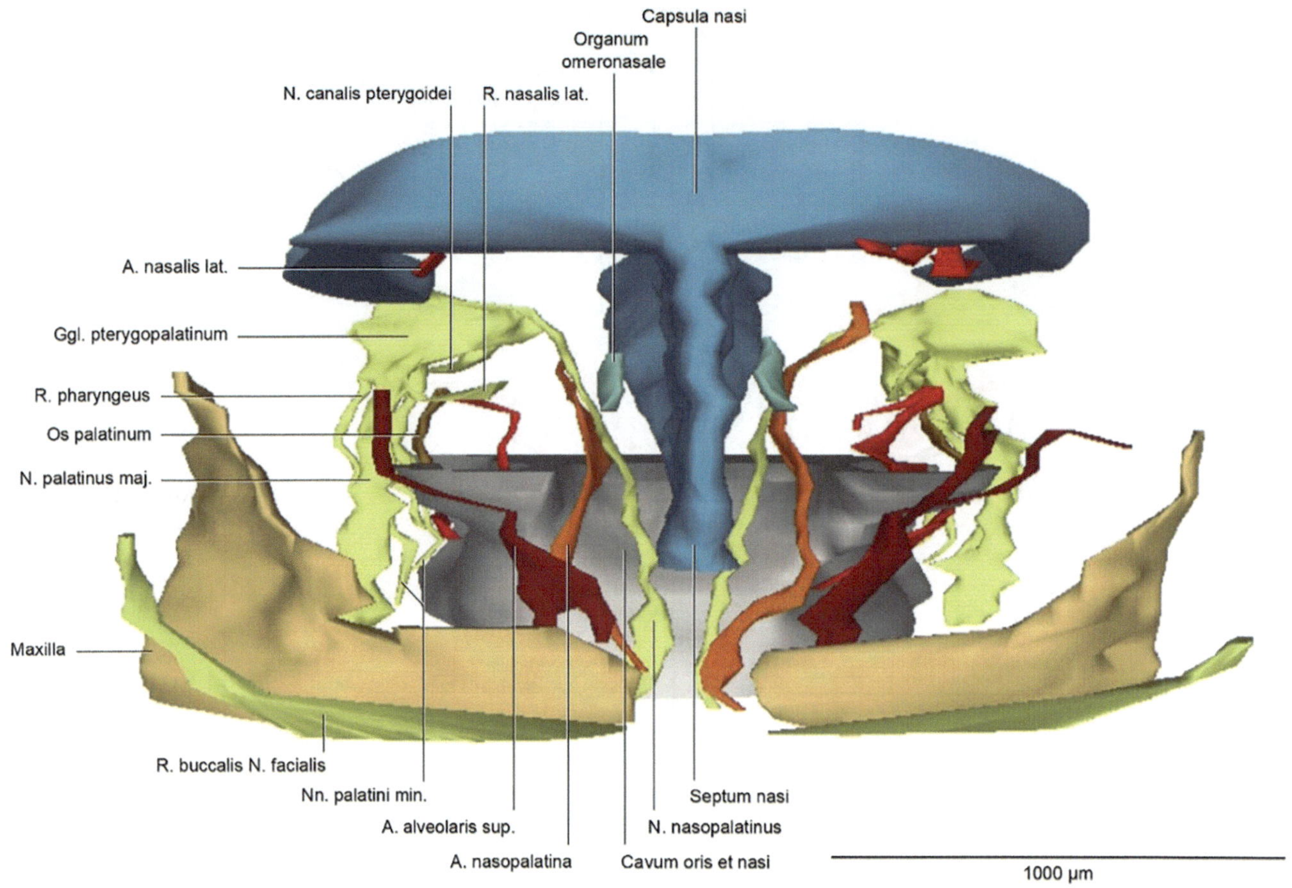

Fig. 13.21 Human embryo, 25 mm CRL, frontal view, visualization of nasopalatine duct and incisive canal (*From* Radlanski et al. [3]; *with permission.*)

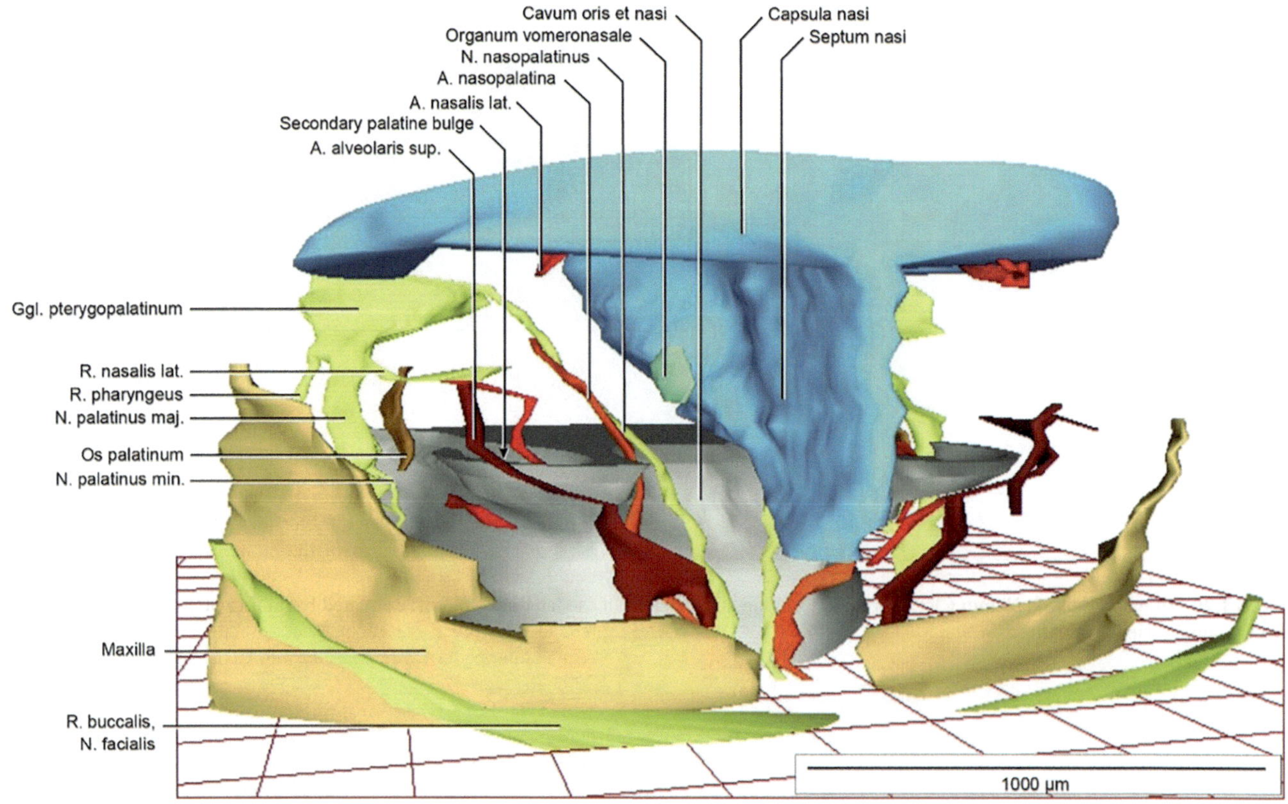

Fig. 13.22 Human embryo, 25 mm CRL, frontal and 45° lateral view, visualization of nasopalatine duct and incisive canal

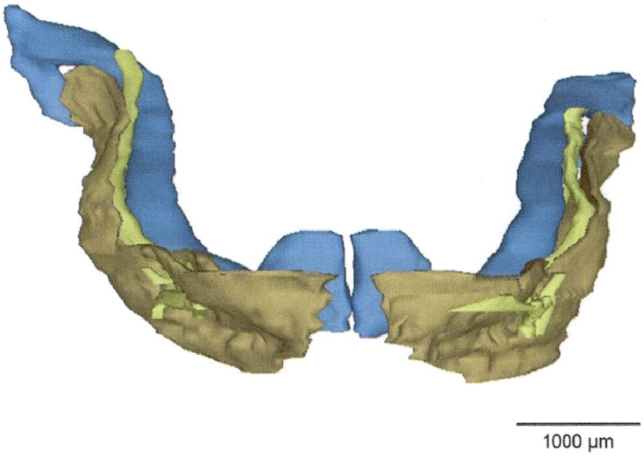

Fig. 13.23 Human embryo, 25 mm CRL, mandible, anterior view. Ochre: bone; blue: Meckel's cartilage; yellow: alveolar inf. n.

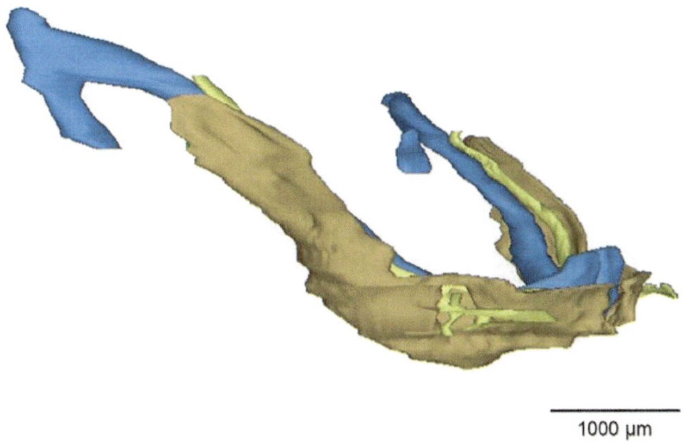

Fig. 13.24 Human embryo, 25 mm CRL, mandible, lateral view. Ochre: bone; blue: Meckel's cartilage; yellow: alveolar inf. n.

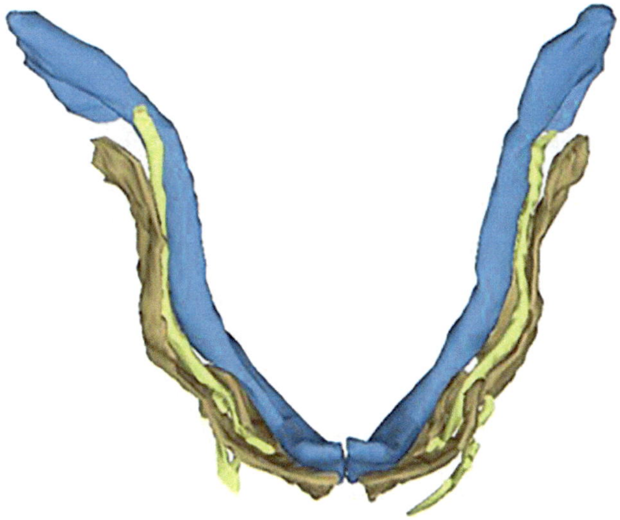

Fig. 13.25 Human embryo, 25 mm CRL, mandible, cranial view. Ochre: bone; blue: Meckel's cartilage; yellow: alveolar inf. n.

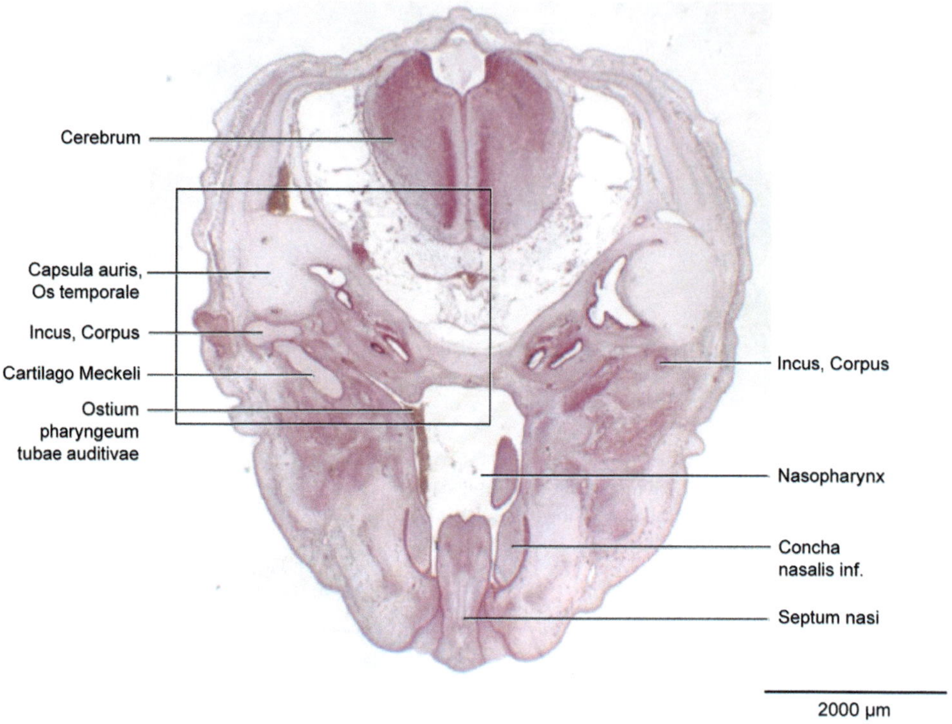

Fig. 13.26 Human embryo, 25 mm CRL, horizontal section, right ear region, overview of viscerocranium. Staining: Hematoxylin-eosin. Box: Fig. 13.24

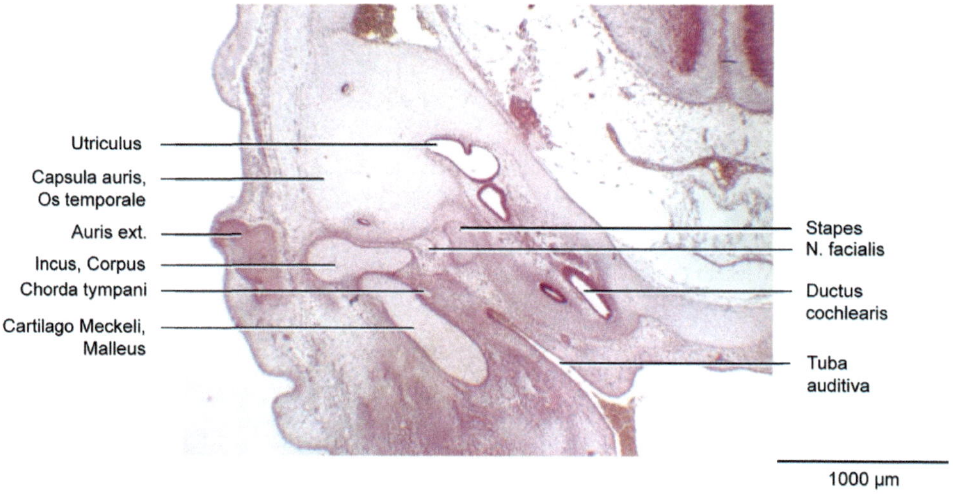

Fig. 13.27 Human embryo, 25 mm CRL, horizontal section, right ear region. Detail from Fig. 13.24 (box)

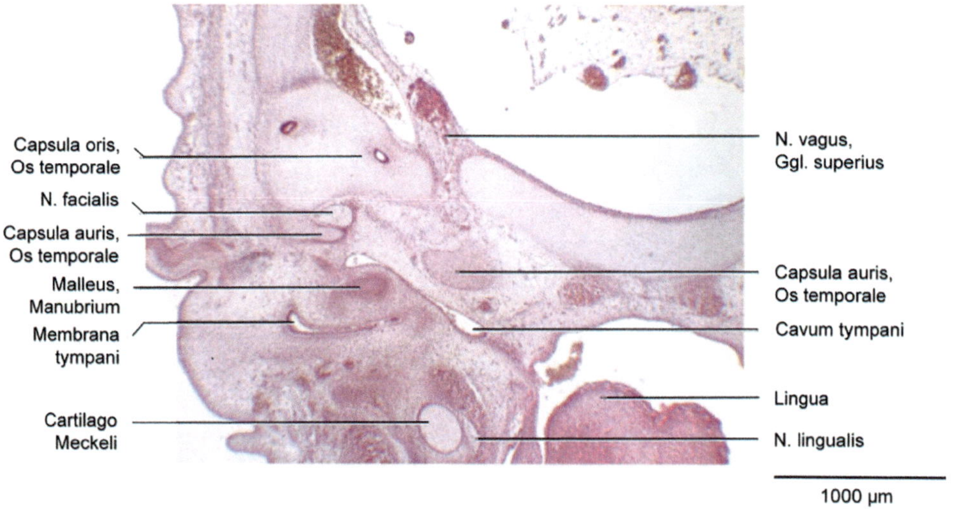

Fig. 13.28 Human embryo, 25 mm CRL, horizontal section, detailed presentation of ear region, this section is located caudally to section in Fig. 13.24

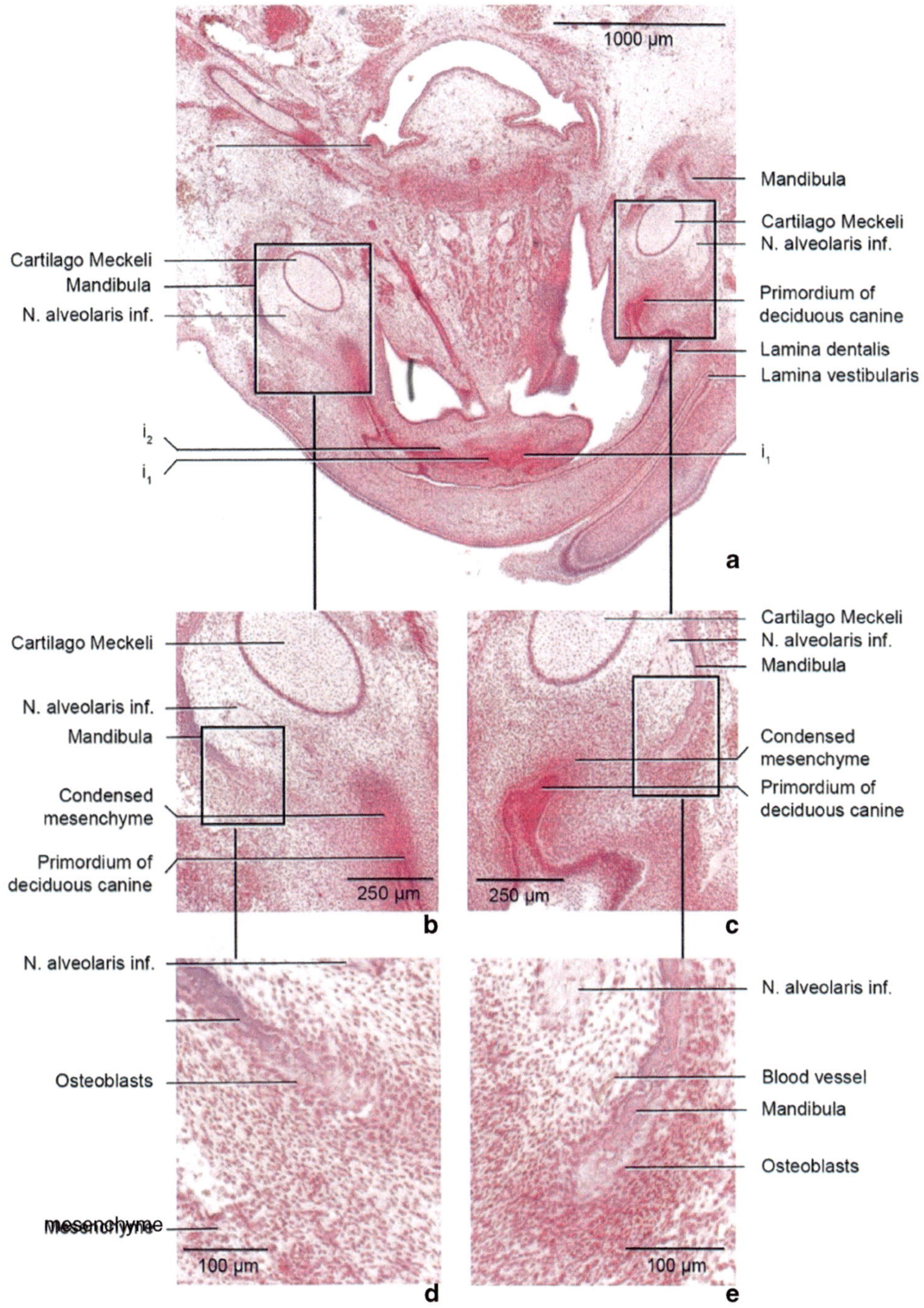

Fig. 13.29 Human embryo, 25 mm CRL, horizontal section. Staining: Hematoxylin-eosin. The lower part of the image points toward anterior, the right part of the image points to left-lateral. (**a**) horizontal section through the region of the deciduous dental primordia of the mandible; (**b**) right part at the height of the canine primordium; (**c**) left part at the height of the canine primordium; (**d**) vestibular lamina of the right mandible; (**e**) vestibular lamina of the left part of the mandible

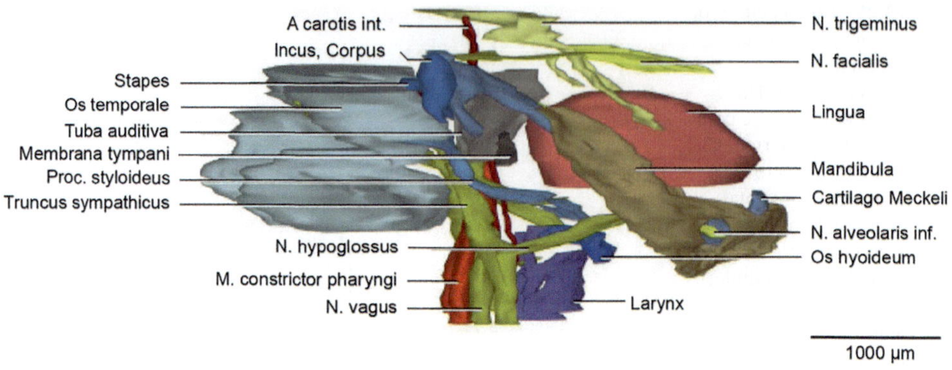

Fig. 13.30 Human embryo, 25 mm CRL, jaw-joint region, lateral view

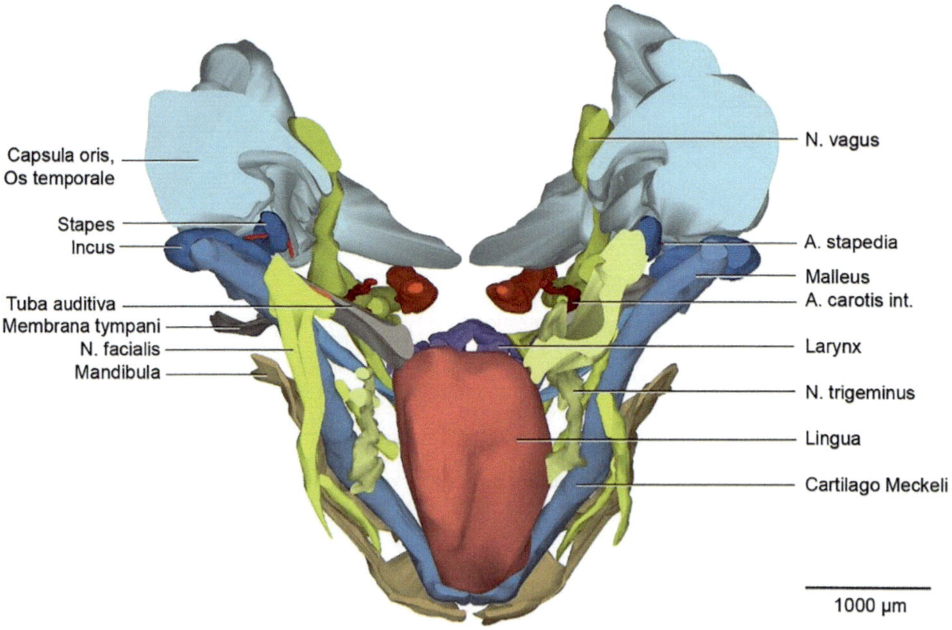

Fig. 13.31 Human embryo, 25 mm CRL, jaw-joint region, cranial view

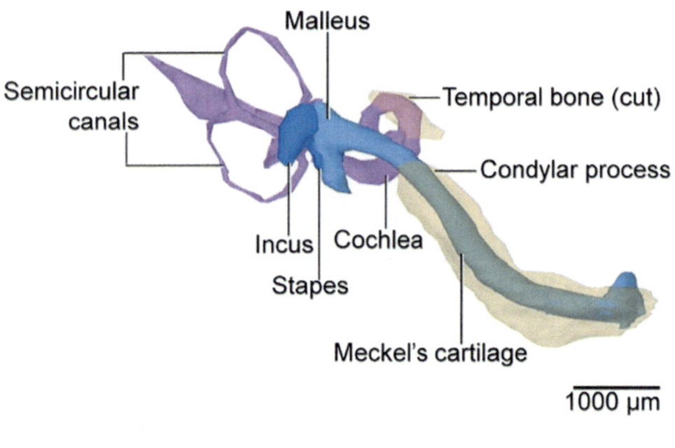

Fig. 13.32 Human embryo, 25 mm CRL, 3D reconstruction of selected parts of the temporomandibular joint and middle ear. Mandible and temporal bone made transparent. (*From* Fernández Rubio and Radlanski [2]; *with permission.*)

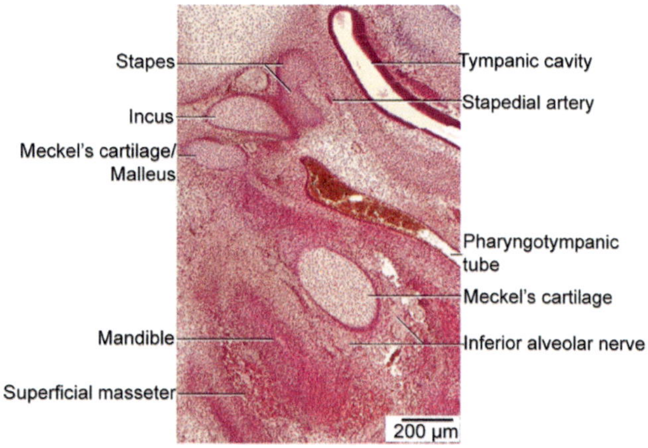

Fig. 13.33 Histological section of the region of the right middle ear and selected parts of the temporomandibular joint. Horizontal plane. Staining: Hematoxylin-eosin

Fig. 13.34 Human embryo, 25 mm CRL, right auditory organ, cranial view, Meckel's cartilage cut and visualized transparent

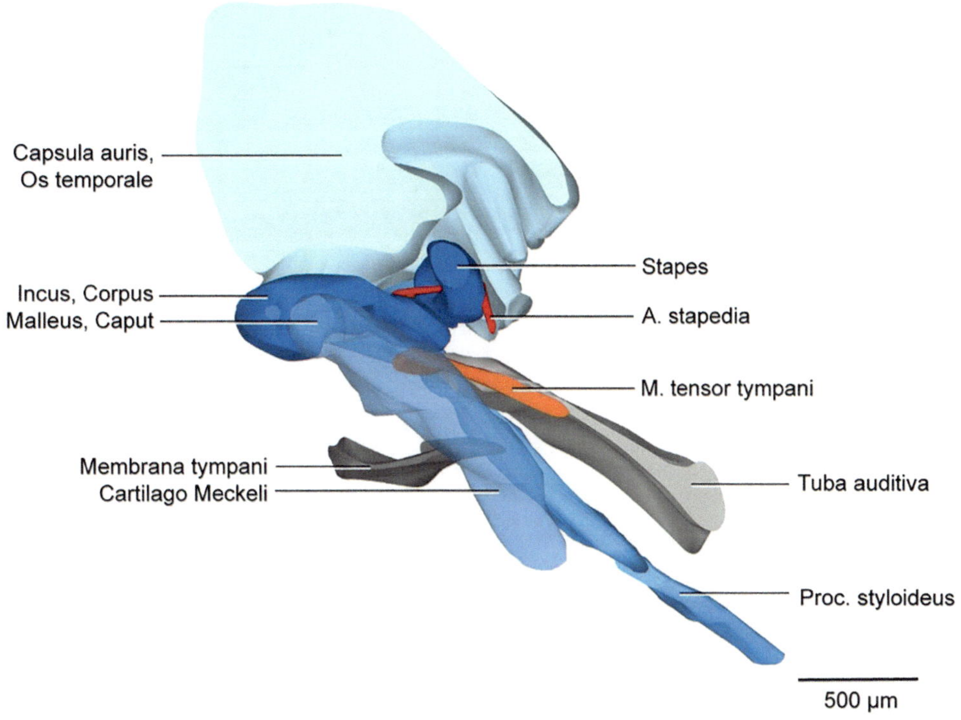

Temporomandibular Joint

Since the masticatory musculature is already well developed at this stage of development, we can reasonably expect movements of the mandible by now [1]. Still, the incudomalleolar joint serves as the articular connection of the jaw bones because the mandible does not yet show any condylar process, which will only make articular contact with the mandibular fossa later, by the 24th week of prenatal development [2]. At least the beginnings of a zygomatic arch are already recognizable (Fig. 13.34).

References

1. Giambartolomei LA, Brunotto M, Gomez de Ferrais ME. Human temporomandibular joint disc: anatomy and measurements in prenatal development. Acta Odontol Latino-Americana. 2011;24:98–103.
2. Fernández Rubio EM, Radlanski RJ. Development of the human primary and secondary jaw joint. Ann Anat. 2024;251:1–8.
3. Radlanski RJ, Emmerich S, Renz H. Prenatal morphogenesis of the human incisive canal. Anat Embryol. 2004;208:265–71.

Further Reading

Beckert K. Beiträge zur pränatalen Morphogenese der Region des Mittelohres des Menschen—Computergestützte 3D-Rekonstruktionen und histologische Untersuchungen menschlicher Embryonen/Feten von 19 bis 150 mm CRL (7. bis 18. Woche). Med. Diss. Charité-Universitätsmedizin Berlin; 2010.

Germelmann AR (née Than) Beitrag zur pränatalen Morphogenese des Os hyoideum des Menschen – Computergestützte 3D-Rekonstruktionen und histologische Untersuchungen menschlicher Embryonen/Feten von 19 – 150 mm CRL (7. bis 18. Woche) Med. Diss. Charité-Universitätsmedizin Berlin; 2008

Radlanski RJ, Renz H. An Atlas of the human prenatal craniofacial development. Eur J Oral Sci. 2010;118:321–4.

Embryo 29 mm CRL (ca. 8th wk): OLL080785

14

Mandible

The mandible here is a rather wide V and in this reconstructed individual the two halves of the mandible are quite straight. This may reasonably be attributed to individual differences in the individual specimens (Figs. 14.1, 14.2, 14.3, 14.4 and 14.5).

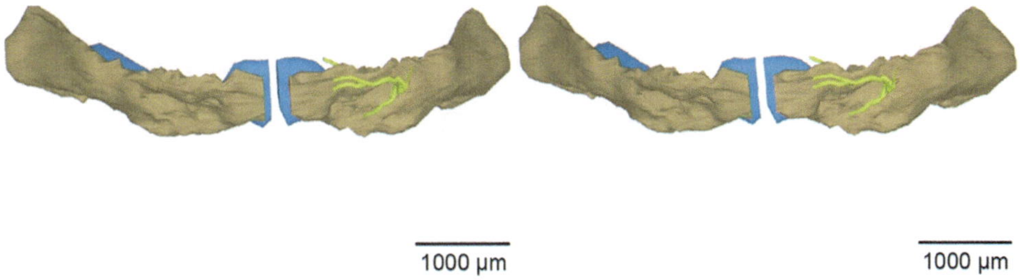

Fig. 14.1 Stereoscopical pair of images. Human embryo 29 mm CRL (ca. 8th week), mandible (ochre), Meckel's cartilage (blue), and left mental nerve (yellow), frontal view

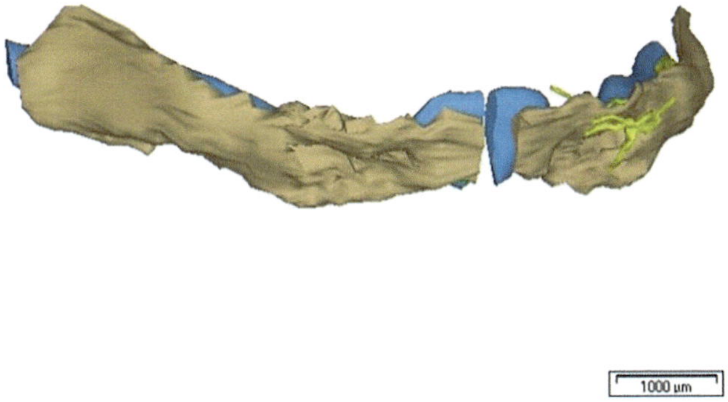

Fig. 14.2 Human embryo, 29 mm CRL, mandible, lateral view

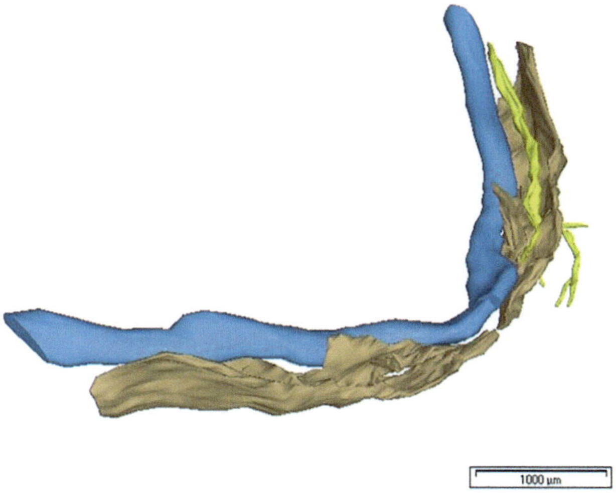

Fig. 14.3 Human embryo, 29 mm CRL, mandible, lateral-cranial view

R. J. Radlanski, *Atlas of Human Prenatal Craniofacial Development*, https://doi.org/10.1007/978-3-031-88812-0_14

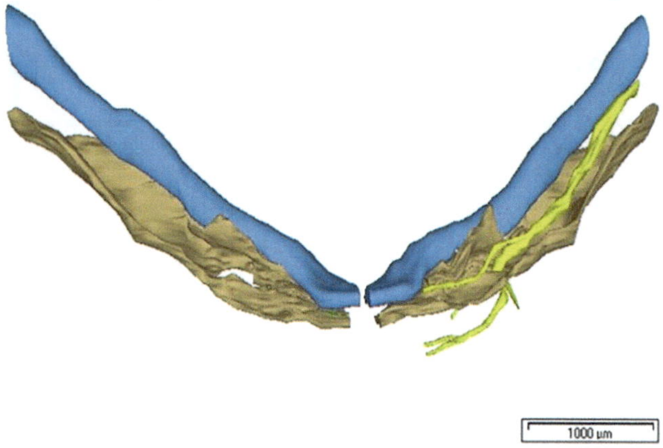

Fig. 14.4 Human embryo, 29 mm CRL, mandible, cranial view

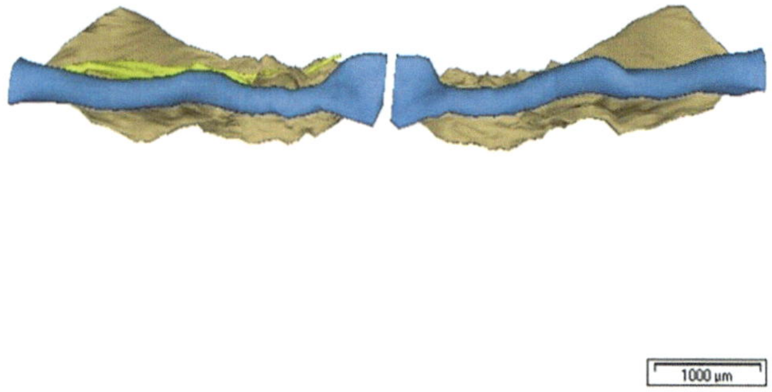

Fig. 14.5 Human embryo, 29 mm CRL, mandible, dorsal view

Further Reading

Radlanski RJ, Renz H, Müller U, Schneider RS, Marcucio RS, Helms JA. Prenatal morphogenesis of the human mental foramen. Eur J Oral Sci. 2002;110:452–9.

Radlanski RJ, Renz H, Klarkowski MC. Prenatal development of the human mandible. 3D reconstructions, morphometry and bone remodelling pattern, sizes 12-117 mm CRL. Anat Embryol (Berl). 2003;207:221–32.

Radlanski RJ, Renz H, Lajvardi S, Schneider R. A Bone remodeling during prenatal morphogenesis of the human mental foramen. Eur J Oral Sci. 2004;112:301–10.

Radlanski RJ. Oral structure and biology. Batavia: Quintessence Publishing; 2018.

Sperber GH. Craniofacial embryology. Hamilton: B.C. Decker; 2001.

Embryo 30 mm CRL (ca. 8th wk): T61_060765

15

Overall Remarks

The histological section (Fig. 15.1) shows the developmental states and positional relationships of the ear region and the area anterior to it. The temporal bone and its otic capsule are still cartilaginous. The cartilaginous precursors of the auditory ossicles malleus, incus, and stapes (its base) are in the section plane. The semicircular canals of the labyrinth are also sectioned. Anteriorly, the sphenoid bone is visible, and laterally, a small portion of the mandible is sectioned with the masseter and medial pterygoid muscles encircling this bony area. For orientation, the eye is shown at the bottom of the image (Fig. 15.2).

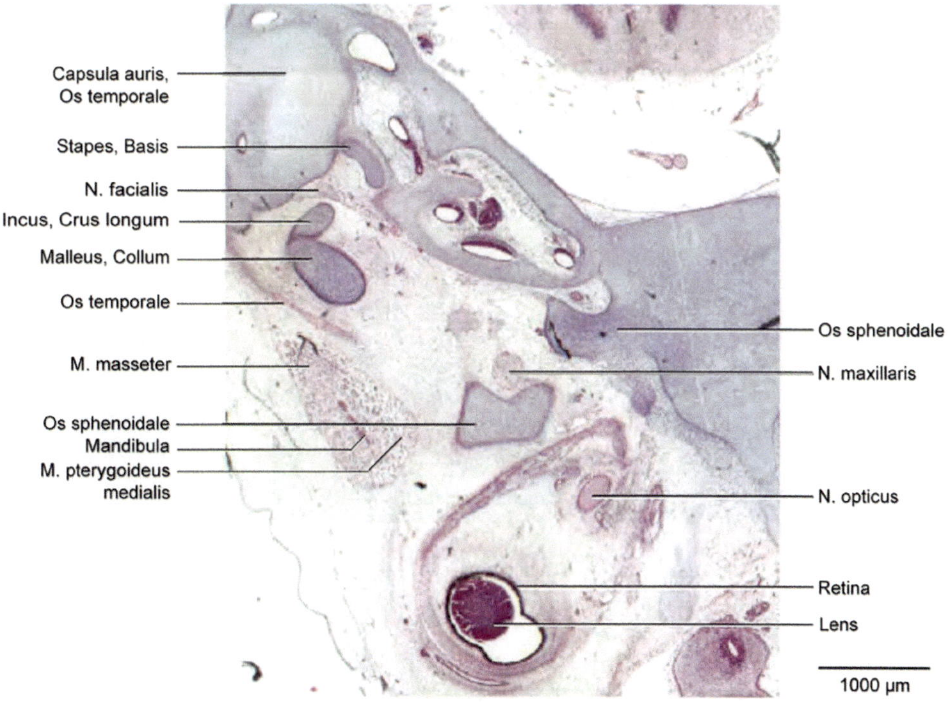

Fig. 15.1 Human embryo, 30 mm CRL, histological section (staining: Hematoxylin-eosin) at the level of the stapes, right side

R. J. Radlanski, *Atlas of Human Prenatal Craniofacial Development*, https://doi.org/10.1007/978-3-031-88812-0_15

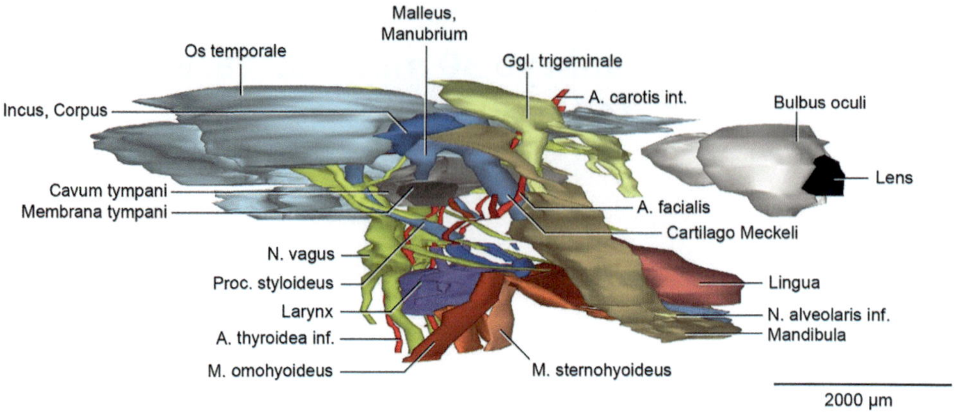

Fig. 15.2 Human embryo, 30 mm CRL, survey of the viscerocranium, lateral view

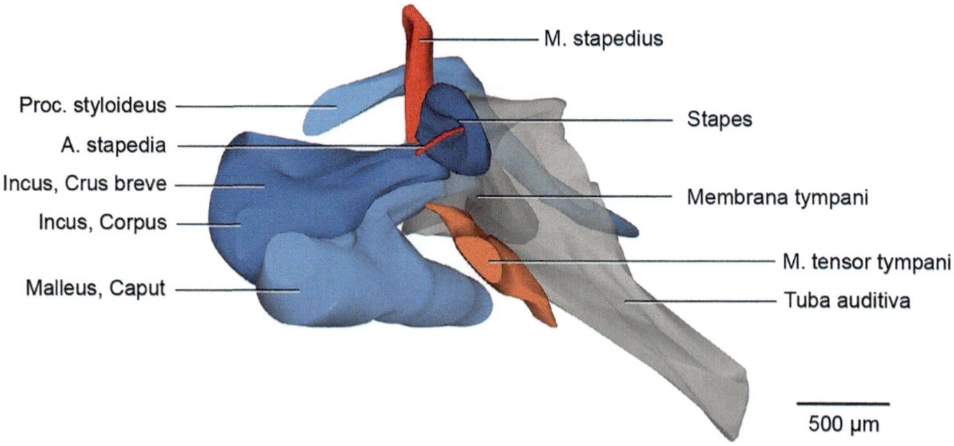

Fig. 15.3 Human embryo, 30 mm CRL, visualization of the right auditory system, cranial view

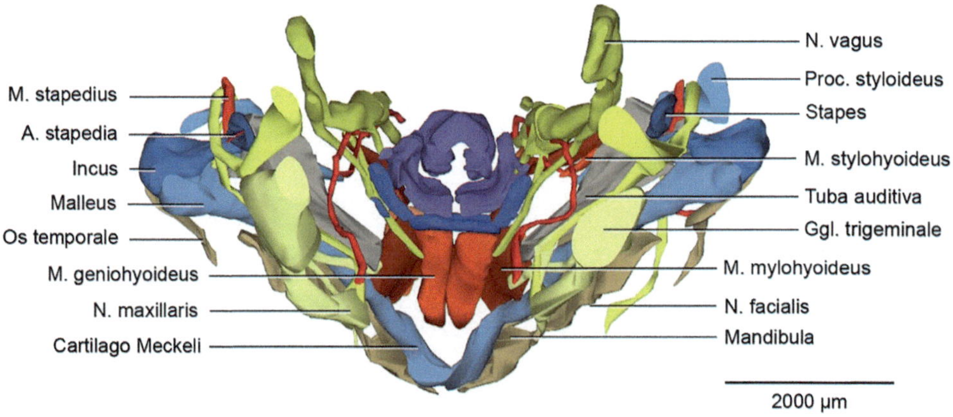

Fig. 15.4 Human embryo, 30 mm CRL, survey of the viscerocranium, cranial view

Middle Ear Region

In fact, the 3D reconstructions show the spatial relationships of the anatomical structures in a more comprehensible way. The reconstructions in Figs. 15.2, 15.3 and 15.4 focus on the region of the developing acoustic organ. At this stage, the tympanic membrane is already visible and in the tympanic cavity the cartilaginous anlagen of the auditory ossicles are in close proximity. The stapedial artery still passes through the stirrup-like opening, and the location and course of the stapedius and tensor tympani muscles can be seen in Fig. 15.3. It is not known whether sound conduction already occurs at this time. Instead, the incudomalleolar joint functions as a hinge for the movement of the mandible. After all, there is no definite secondary mandibular joint yet.

Floor of the Mouth

In the region of the floor of the mouth, the paired, sagittal course of the geniohyoid muscle from the symphysis region to the hyoid bone is well developed—although the hyoid bone is also still only present as a cartilaginous precursor. The mylohyoid muscle, one level superior to the geniohyoid muscle, runs horizontally between the two spans of Meckel's cartilage. At this stage of development, it still inserts on the cartilage. Only later, at a size of 68 mm CRL, about the 10th week, when Meckel's cartilage begins to be embraced also lingually by bone, does the insertion zone render into a bony one [1]. The V- or U-shaped course of Meckel's cartilage described several times so far is also repeated when the continuity of styloid process and hyoid bone is followed in their course. This is obvious because of the origin from the 1st, resp. 2nd visceral arch. However, it must be pointed out that only the lesser horn of the hyoid originates from the 2nd visceral arch. The greater horn of that hyoid bone develops from tissue of the 3rd visceral arch (Figs. 15.5, 15.6, 15.7, and 15.8) [2].

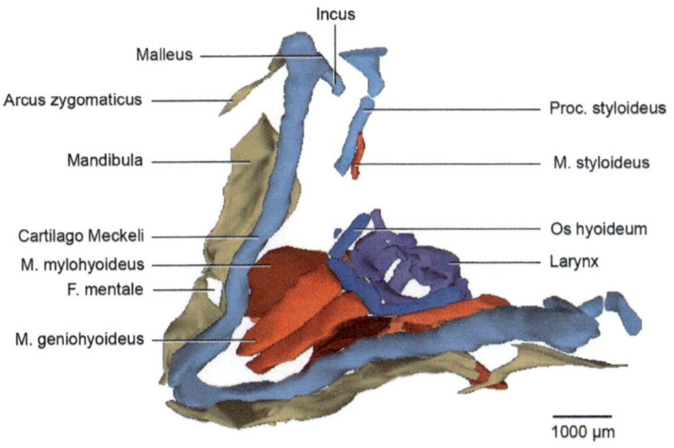

Fig. 15.5 Human embryo, 30 mm CRL, visualization of the floor of the mouth, lateral, 45°posterior and 45° caudal view

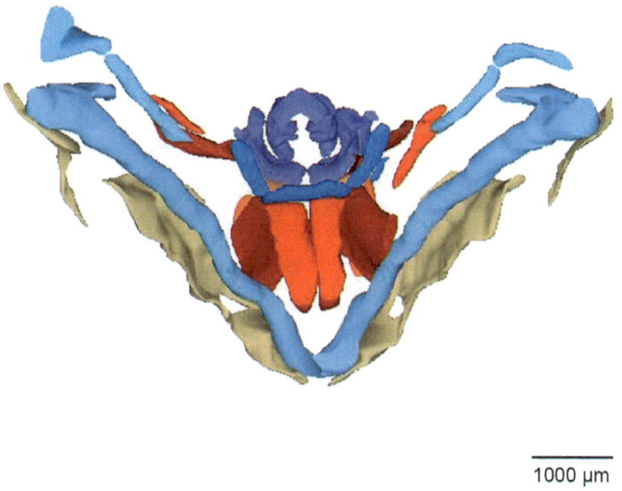

Fig. 15.6 Human embryo, 30 mm CRL, visualization of the floor of the mouth, cranial view

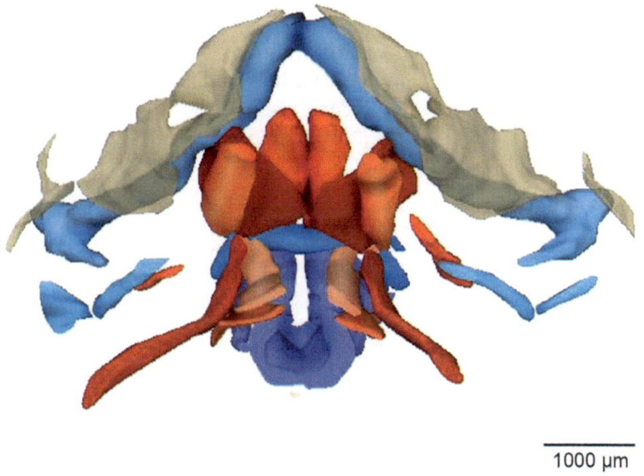

Fig. 15.7 Human embryo, 30 mm CRL, visualization of the floor of the mouth (mandible transparent), caudal view

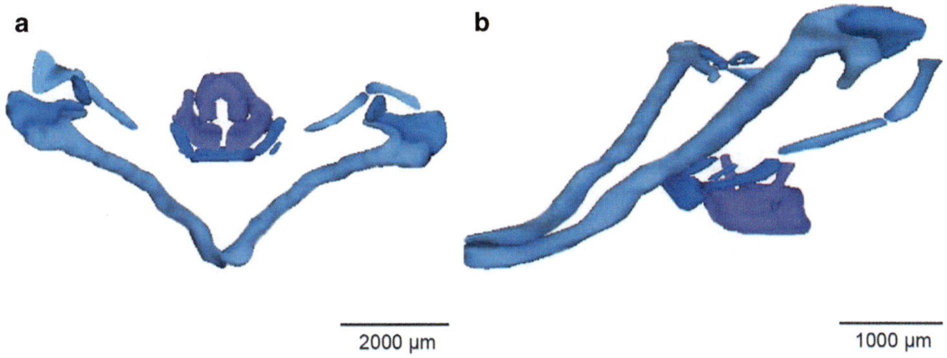

Fig. 15.8 (**a, b**) Human embryo, 30 mm CRL, visualization of cartilaginous part of the mandible, lateral view. (**c**) Human embryo, 30 mm CRL, primordium of hyoid bone, cranial view

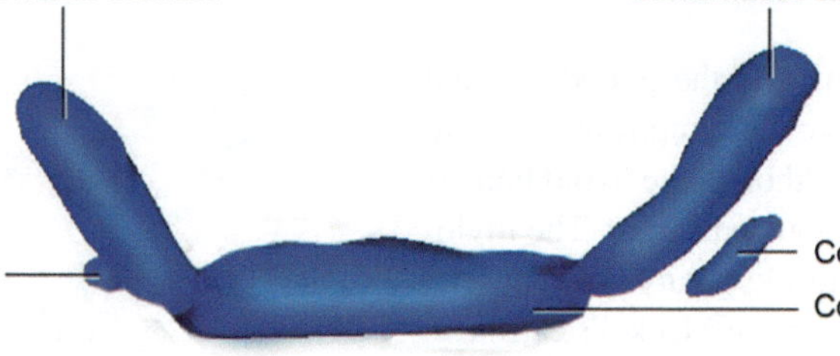

Fig. 15.8 (continued)

References

1. Radlanski RJ, Renz H, Tabatabai A. Prenatal development of the muscles in the floor of the mouth in human embryos and fetuses from 6.9 to 76 mm CRL. Ann Anat. 2001;183:511–8.
2. Germelmann AR (née Than). Beitrag zur pränatalen Morphogenese des Os hyoideum des Menschen—Computergestützte 3D-Rekonstruktionen und histologische Untersuchungen menschlicher Embryonen/Feten von 19–150 mm CRL (7. bis 18. Woche) Med. Diss. Charité-Universitätsmedizin Berlin 2008

Further Reading

Kjaer I, Keeling JW. Fischer-Hansen B The prenatal human cranium—normal and pathologic development. Copenhagen: Munksgaard; 1999.

Radlanski RJ, Kjaer I, Vastardis H. Renz H Morphometrische Untersuchungen zur Fetalentwicklung der menschlichen Mandibula. Fortschr Kieferorthop. 1994;55:77–83.

16 Embryo 31 mm CRL (ca. 8th wk): PIP161189

Most of the structures described so far do not show any significant changes at this stage of development. The dental primordia, however, have now passed the bud stage and are now—at least in this specimen—in their cap stages (Figs. 16.1, 16.2, 16.3, 16.4, 16.5 and 16.6).

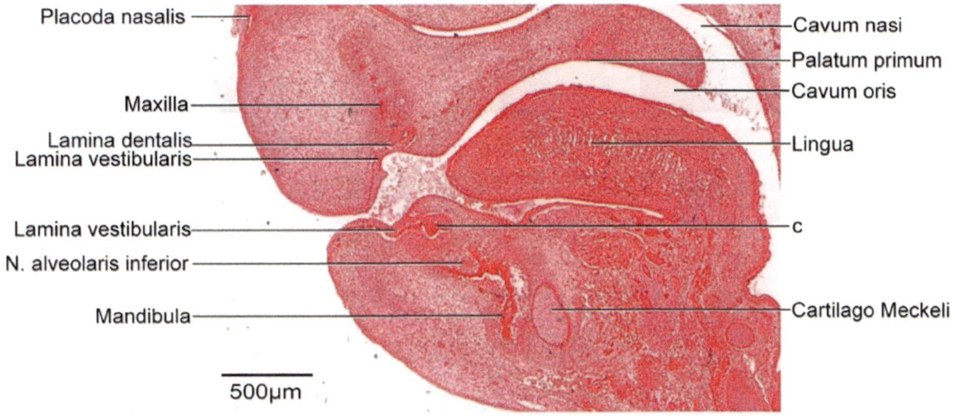

Fig. 16.1 Human embryo, 31 mm CRL, sagittal histological section (staining: Hematoxylin-eosin) of the oral region

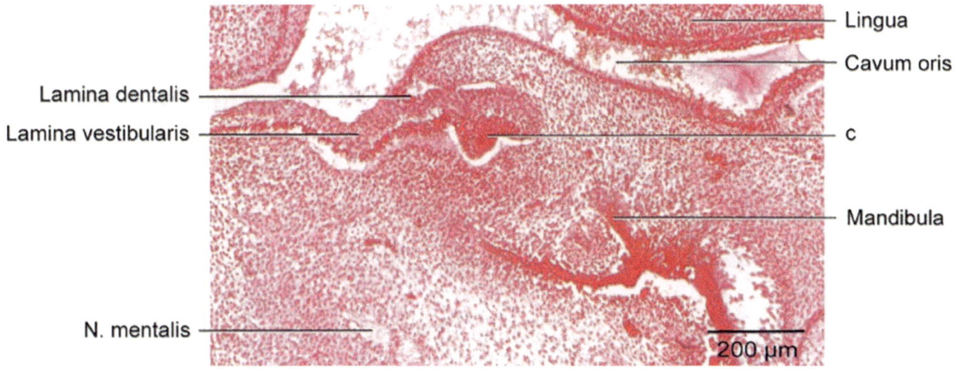

Fig. 16.2 Human embryo, 31 mm CRL, sagittal histological section (staining: Hematoxylin-eosin), at level of the lower deciduous canine primordium (c)

R. J. Radlanski, *Atlas of Human Prenatal Craniofacial Development*, https://doi.org/10.1007/978-3-031-88812-0_16

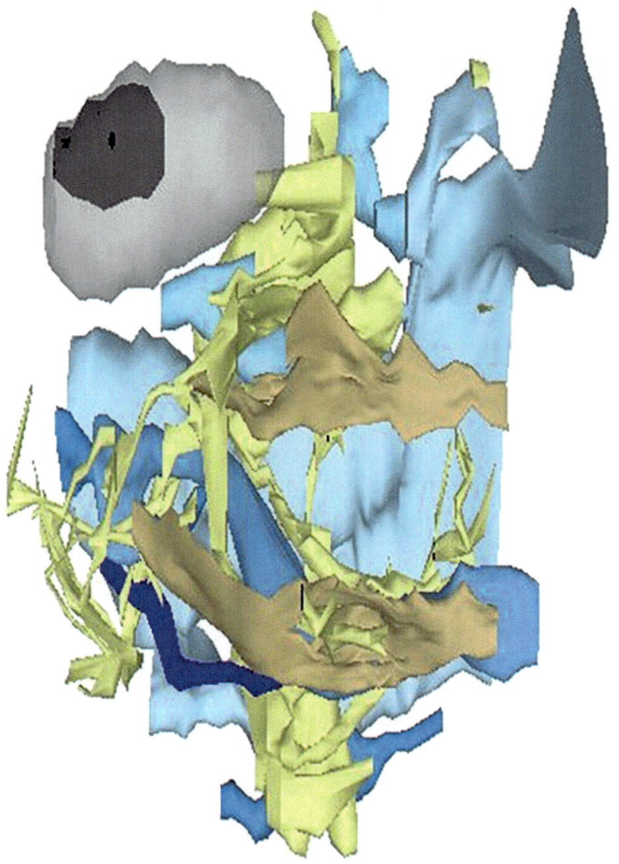

Fig. 16.3 Human embryo, 31 mm CRL, right half of head, frontal view

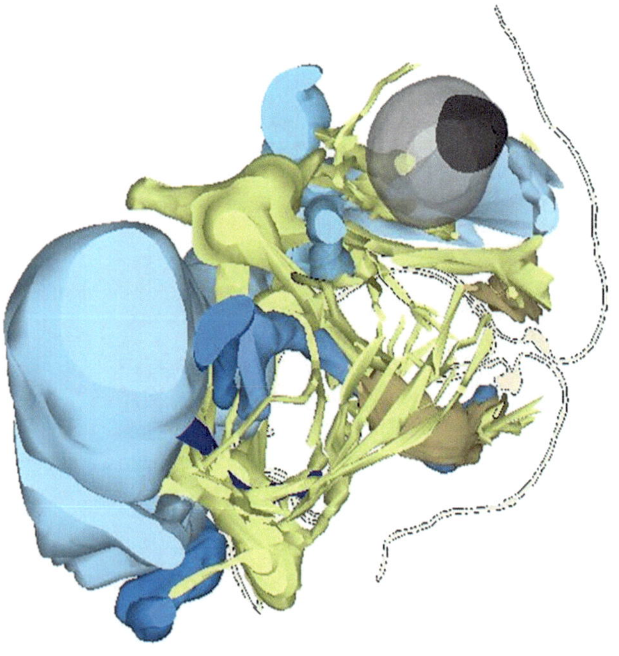

Fig. 16.4 Human embryo, 31 mm CRL, right half of head, lateral view

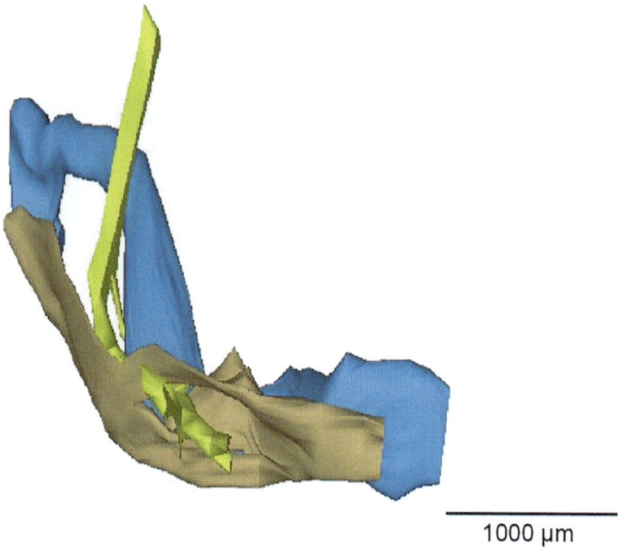

Fig. 16.5 Human embryo, 31 mm CRL, mandible, right half, frontal view

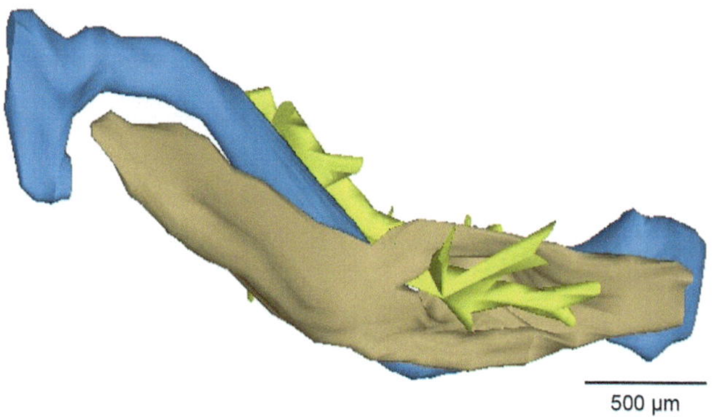

Fig. 16.6 Human embryo, 31 mm CRL, mandible, right half, lateral view

Further Reading

Beckert K. Beiträge zur pränatalen Morphogenese der Region des Mittelohres des Menschen – Computergestützte 3D-Rekonstruktionen und histologische Untersuchungen menschlicher Embryonen/Feten von 19 bis 150 mm CRL (7. bis 18. Woche). Med Diss Charité-Universitätsmedizin Berlin 2010.

Radlanski RJ, Renz H, Müller U, Schneider RS, Marcucio RS, Helms JA. Prenatal morphogenesis of the human mental foramen. Eur J Oral Sci. 2002;110:452–9.

Radlanski RJ, Renz H, Klarkowski MC. Prenatal development of the human mandible. 3D reconstructions, morphometry and bone remodelling pattern, sizes 12-117 mm CRL. Anat Embryol (Berl). 2003;207:221–32.

Radlanski RJ, Renz H, Lajvardi S, Schneider RA. Bone remodeling during prenatal morphogenesis of the human mental foramen. Eur J Oral Sci. 2004;112:301–10.

Embryo 37 mm CRL (ca. 9th wk): DOR100785

17

Overall Remarks

In this stage of development, the extensions of the bony structures have gained more size. The expansion of the maxilla has increased. Thus, its frontal process has become higher, laterally it has also extended further, so that it has neared the zygomatic arch up to about 250 µm. The mandible now also shows the first signs of a coronoid process although a condylar process is not yet visible (Figs. 17.1, 17.2, 17.3, 17.4, and 17.5).

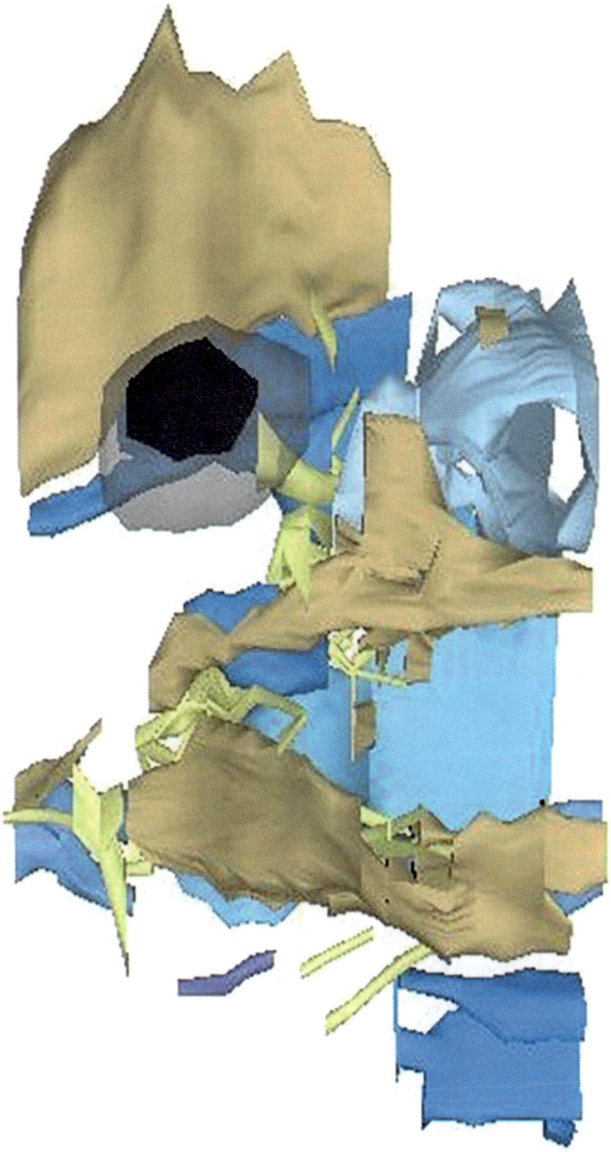

Fig. 17.1 Human embryo, 37 mm CRL, right half of head, frontal view

R. J. Radlanski, *Atlas of Human Prenatal Craniofacial Development*, https://doi.org/10.1007/978-3-031-88812-0_17

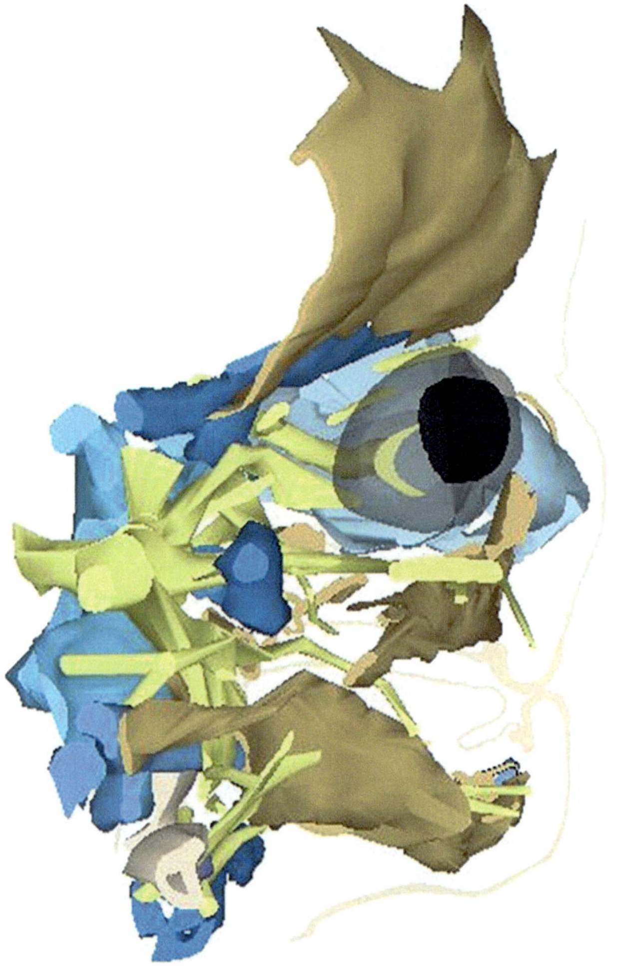

Fig. 17.2 Human embryo, 37 mm CRL, right half of head, lateral view

Fig. 17.3 Human embryo, 37 mm CRL, mandible, lateral view

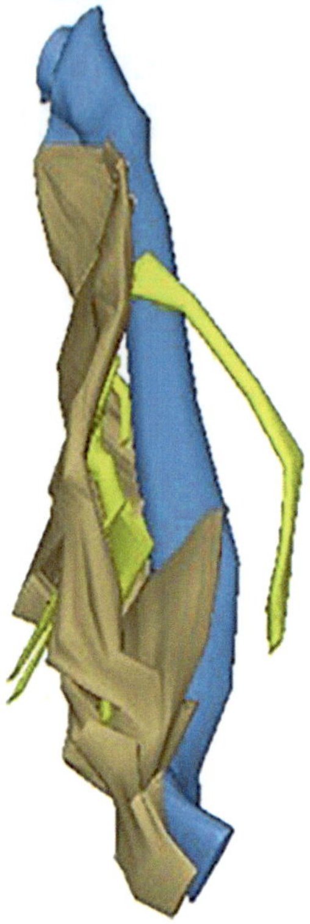

Fig. 17.4 Human embryo, 37 mm CRL, mandible, cranial view

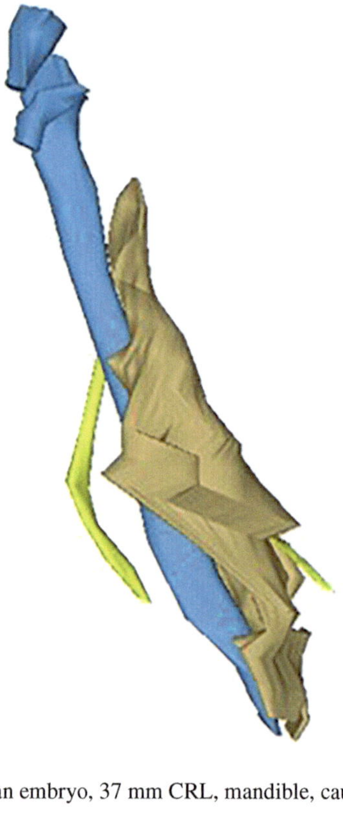

Fig. 17.5 Human embryo, 37 mm CRL, mandible, caudal view

Dental Primordia

The primordia of the incisors (the first dentition) in the maxilla have reached the mature bud stage, the primordia for the canines and the first molars are already somewhat further along and show the cap stage. The primordium of the second molar naturally lags somewhat behind and shows a thickening of its epithelium. The mesenchyme around the epithelial primordia is markedly condensed (Figs. 17.6, 17.7 and 17.8).

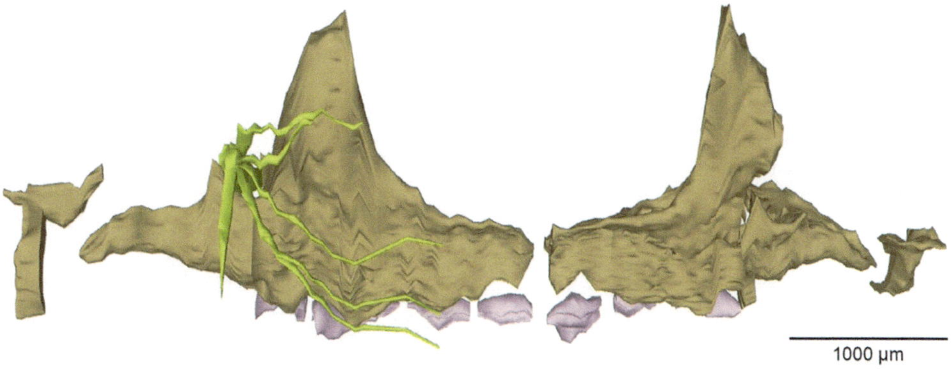

Fig. 17.6 Human embryo, 37 mm CRL, maxilla, infraorbital n., and dental primordia, frontal view

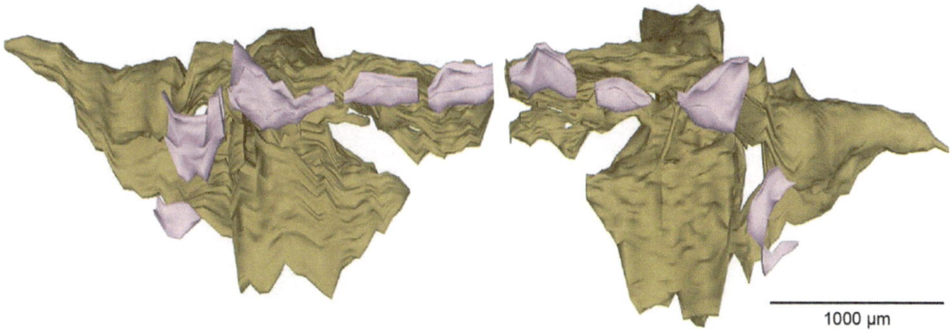

Fig. 17.7 Human embryo, 37 mm CRL, maxilla and dental primordia, caudal view

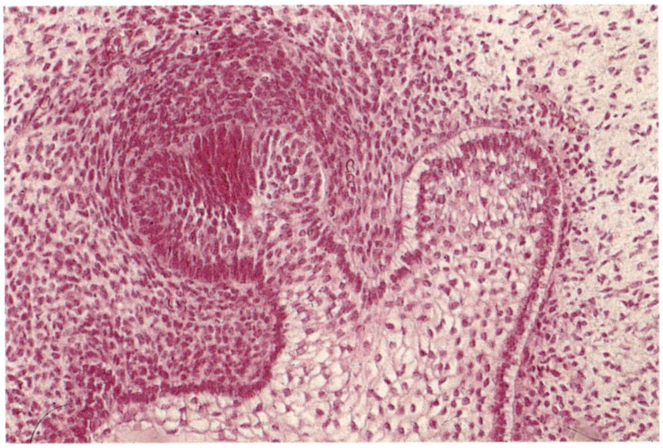

Fig. 17.8 Human embryo, 37 mm CRL, vertical section at level of dental primordium i^1Staining: Hematoxylin-eosin (*From* Radlanski [1]; *with permission.*)

Maxilla

The maxillary bony structures in the region of the incisor primordia extend as close as 100 μm toward the epithelium of the vestibular and dental laminae. The distance between the canine primordium and the bone is somewhat greater, 200 μm, and this is despite the fact that the canine structure has already developed into the cap stage.

In contrast to the mandible, where bone formation tends to be more planar and contiguous, maxillary bone appears as a multitude of individual islands of bone that only later, during further development, come into a continuity with each other (Figs. 17.9, 17.10, and 17.11).

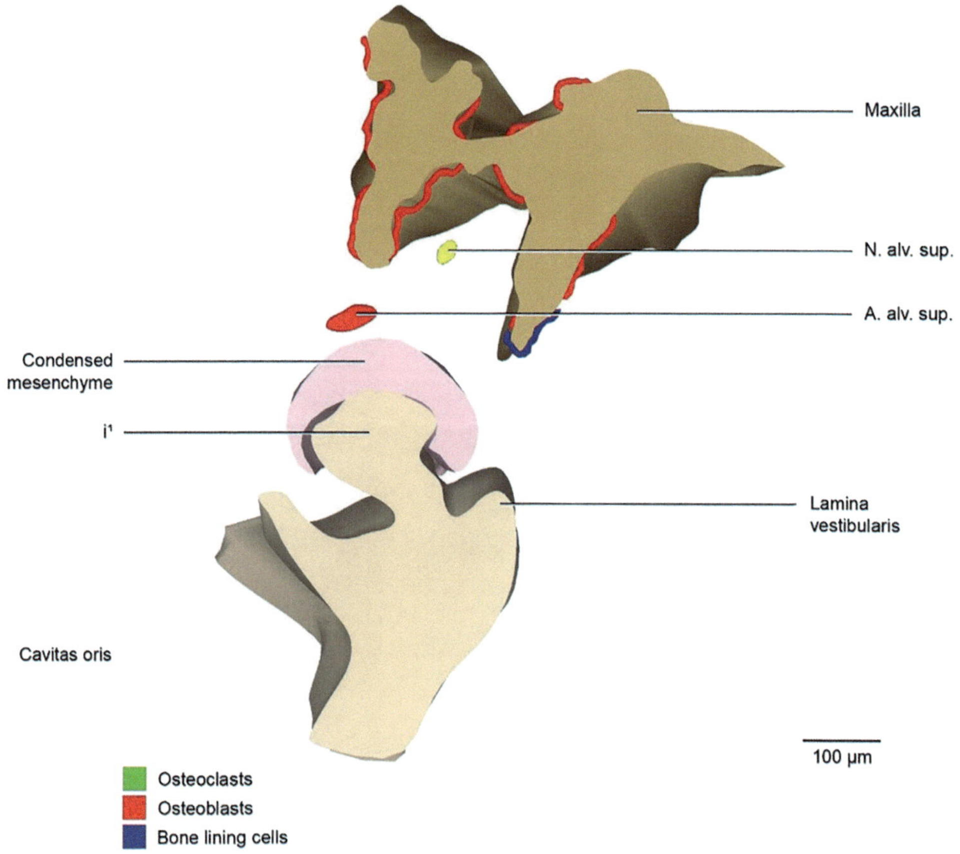

Fig. 17.9 Human embryo, 37 mm CRL, maxilla and dental primordium i^1, sectioned medially, medial view, with bone remodeling pattern superimposed

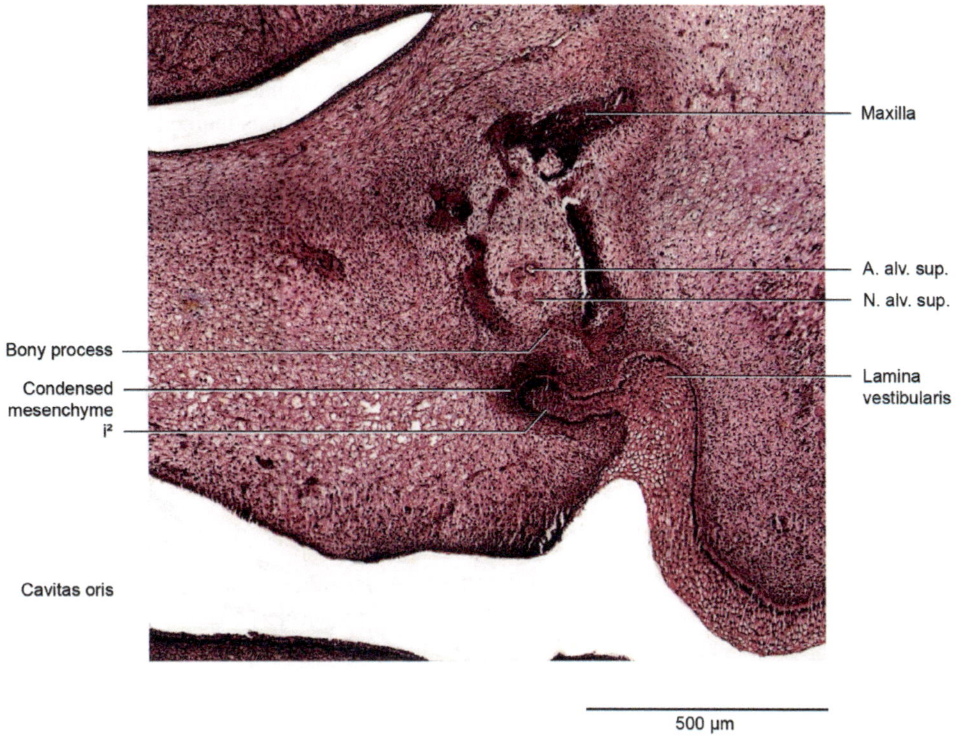

Fig. 17.10 Human embryo, 37 mm CRL, vertical section at level of dental primordium i^2, staining: Hematoxylin-eosin

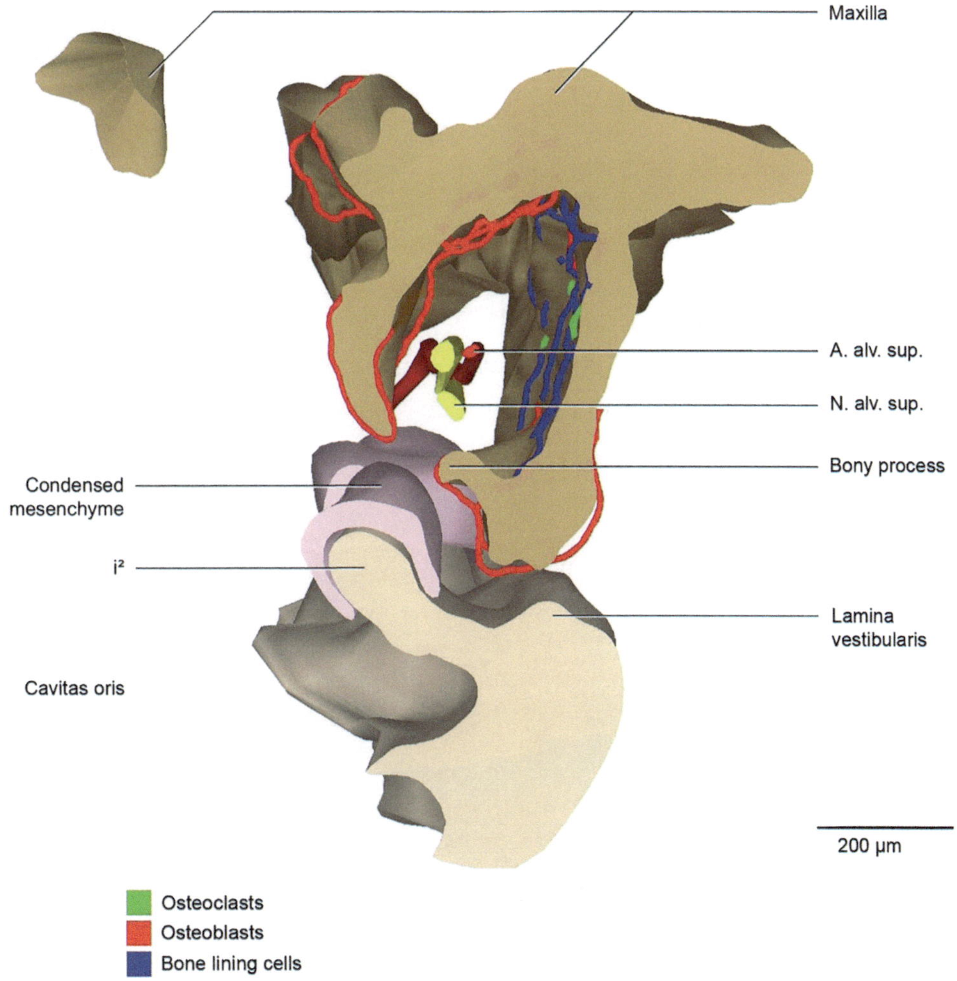

Fig. 17.11 Human embryo, 37 mm CRL, maxilla and dental primordium i^2, sectioned medially, medial view, with bone remodeling pattern superimposed

Maxilla, Bone Remodeling

In histological sections, it is possible to conclude the activity of bone formation by counting the number of osteoblasts, osteoclasts, and lining cells on the bone surface. In the 3D reconstructions, the regions with preferential bone resorption are marked in green, bone apposition is marked in red, and the inactive areas, where lining cells predominate, are marked in blue. Thus, it can be seen, that most areas of the maxillary bone extensions in the direction of the dental primordia are characterized by the apposition of bone—this is shown by the areas marked in red in the 3D images. The groove above the dental primordium is therefore not (yet) due to resorption of bone at this stage. The vascular and nerve trunks, which will later supply the dental and peridental structures, already run within this bone cavity (Figs. 17.12 and 17.13).

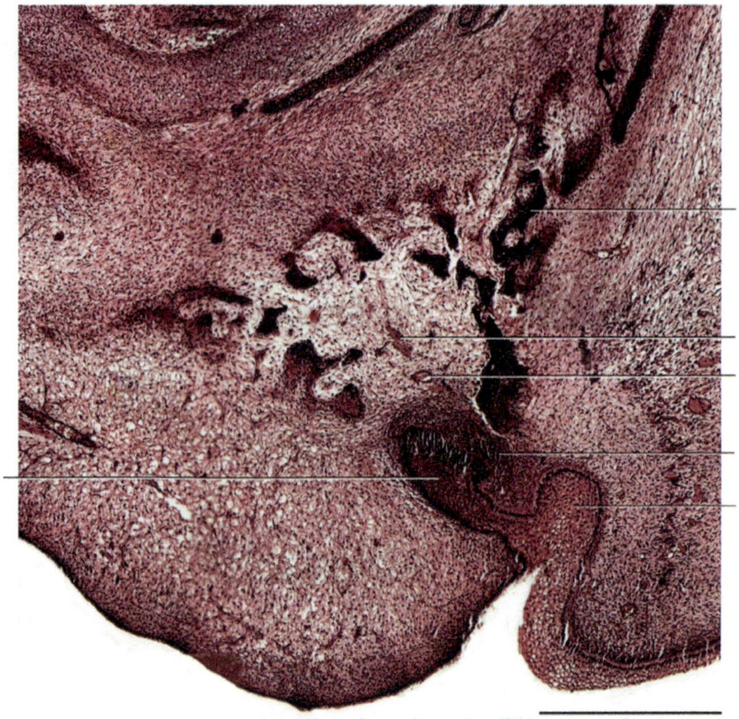

Fig. 17.12 Human embryo, 37 mm CRL, vertical section at level of the deciduous canine primordium, staining: Hematoxylin-eosin

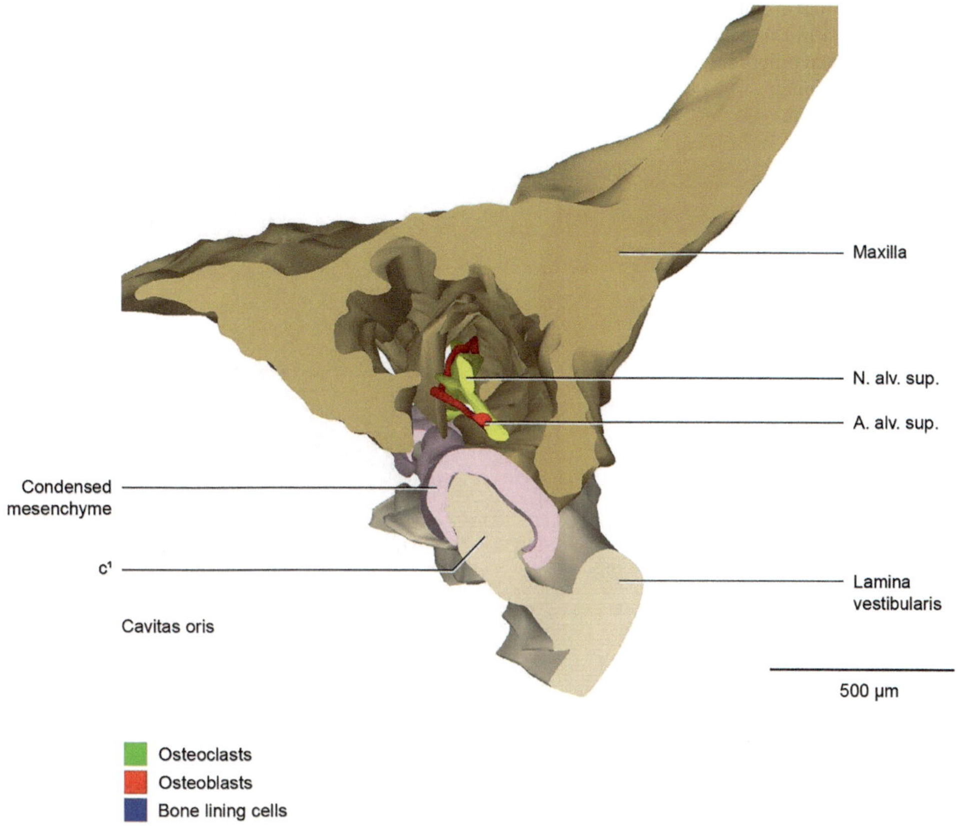

Fig. 17.13 Human embryo, 37 mm CRL, maxilla and dental primordium c^1, sectioned medially, medial view, with bone remodeling pattern superimposed

Vestibular Lamina

In the image (Fig. 17.14), apart from the molar primordia and their surrounding bone, it can be seen, that the previously compact vestibular lamina is deepening due to a disintegration of the epithelial continuity. The reasons for this are still unclear. It could be that the epithelial cells perish, simply because they are too far away from the basal lamina and their nutrition. Apoptosis processes or even specific signal cascades are also discussed, which could trigger the deepening of the vestibular lamina (Fig. 17.15).

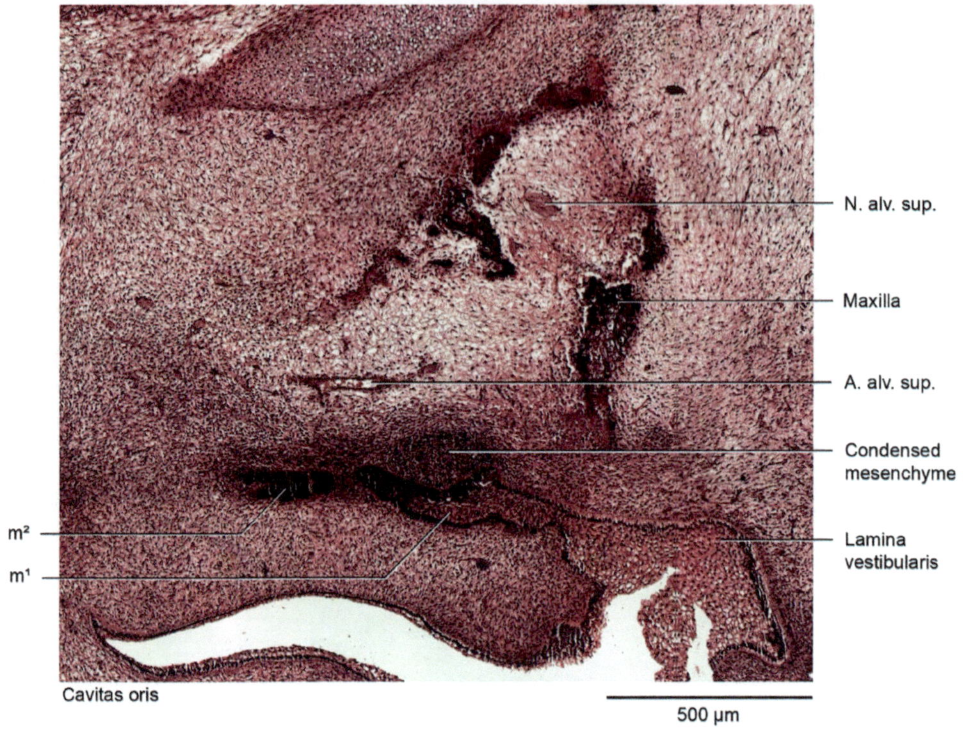

Fig. 17.14 Human embryo, 37 mm CRL, vertical section at level of dental primordium m^1 and m^2. Staining: Hematoxylin-eosin

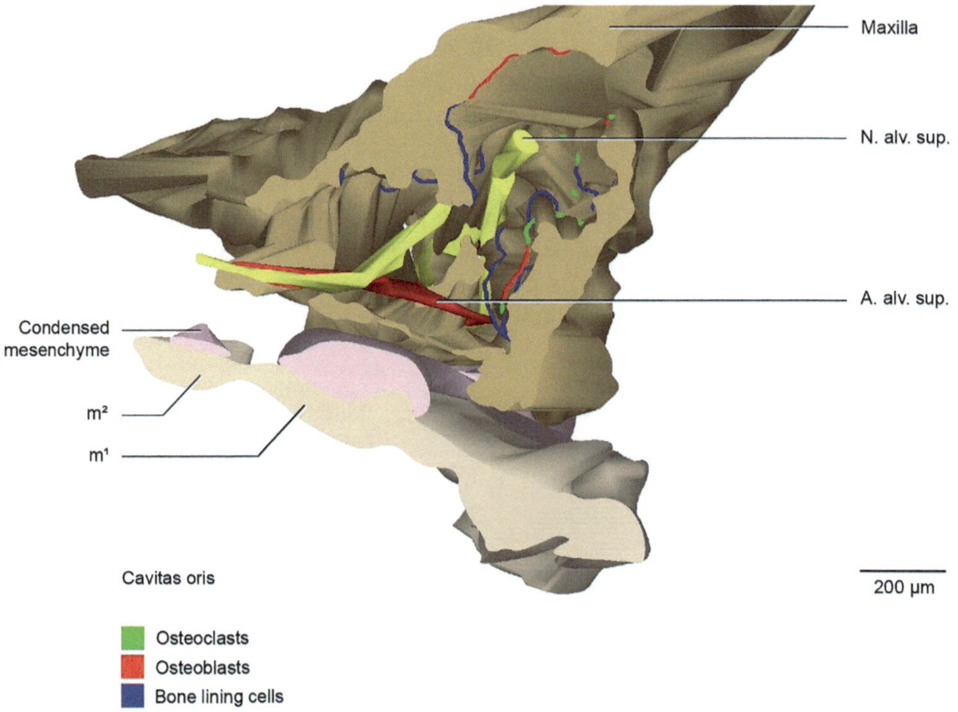

Fig. 17.15 Human embryo, 37 mm CRL, maxilla and dental primordia m^1 and m^2, sectioned medially, medial view, with bone remodeling pattern superimposed

Dental Primordia

This series of figures (Fig. 17.16) shows at the top (a) the histological section through the dental primordium c^1. Only in the 3D reconstruction does the expansion of the individual bone islands into a continuous maxilla become clear (b). In addition, the bone remodeling pattern is superimposed: active bone apposition prevails, as shown by the red markings.

In the bottom image (c), the dental primordia i^1 and i^2 have also been included in the reconstruction.

a

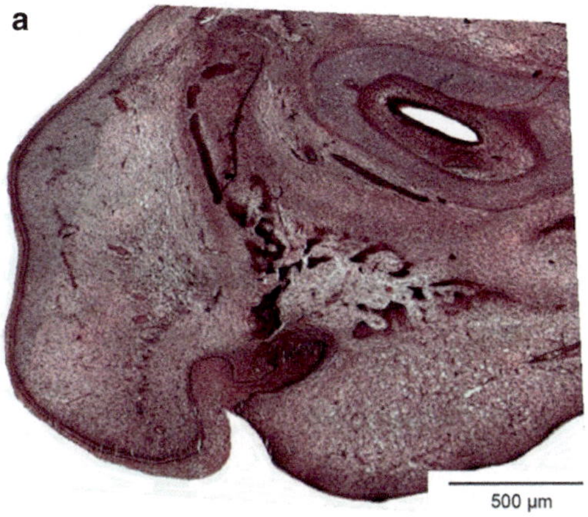

Fig. 17.16 (**a**) Human embryo, 37 mm CRL, sagittal section (staining: Hematoxylin-eosin) at level of dental primordium c^1.

b

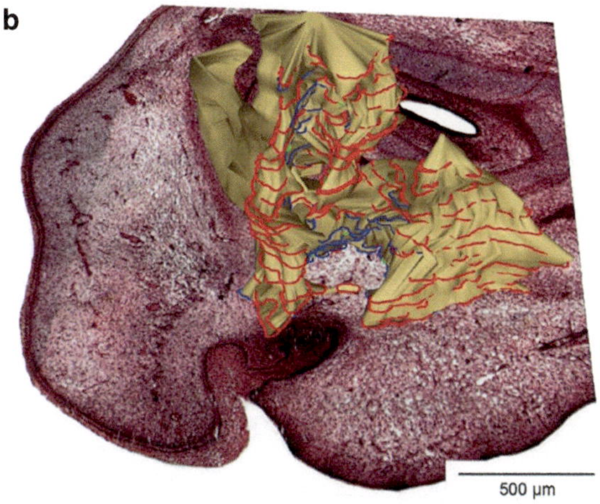

Fig. 17.16 (**b**) 3D-reconstruction of maxillary bone with bone remodeling pattern superimposed. red: apposition, green: resorption, blue: lining cells.

c

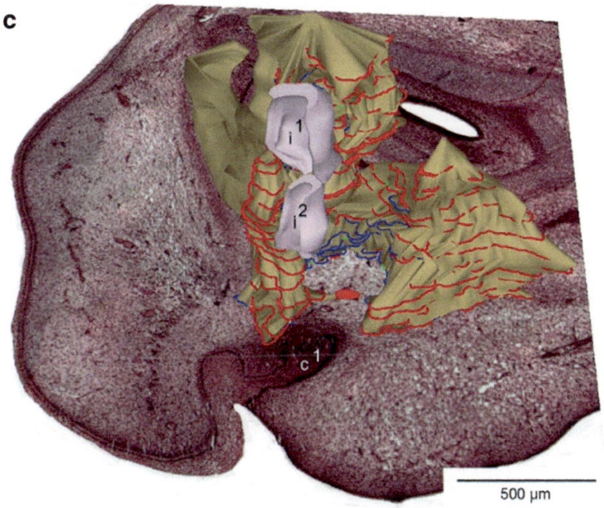

Fig. 17.16 (**c**) dental primordia i^1 and i^2 as 3D-reconstructions superimposed

Reference

1. Radlanski RJ. Oral structure and biology. Batavia, USA: Quintessence Publishing; 2018.

Further Reading

Kalinke U. Embryologische Untersuchungen zur Entwicklung des Alveolarknochens im Oberkiefer beim Menschen zur Zeit der vorgeburtlichen Entwicklungsstadien von 19–68 mm SSL. Med. Diss. Charité-Universitätsmedizin Berlin; 2014

Radlanski RJ. Contributions to the development of human deciduous tooth primordia. Chicago: Quintessence Publishing Co, Inc.; 1993.

Radlanski RJ, Renz H, Müller U, Schneider RS, Marcucio RS, Helms JA. Prenatal morphogenesis of the human mental foramen. Eur J Oral Sci. 2002;110:452–9.

Radlanski RJ, Renz H, Klarkowski MC. Prenatal development of the human mandible. 3D reconstructions, morphometry and bone remodelling pattern, sizes 12-117 mm CRL. Anat Embryol (Berl). 2003;207:221–32.

Radlanski RJ, Renz H, Lajvardi S, Schneider RA. Bone remodeling during prenatal morphogenesis of the human mental foramen. Eur J Oral Sci. 2004;112:301–10.

Embryo 41 mm CRL (ca. 9th wk): KUR030389

18

Overall Remarks

In this chapter, we can concentrate on the mandible in more detail. At this stage of development, it still consists of two individual bony plates that laterally follow the course of Meckel's cartilage. Anteriorly, they are still about 500 µm apart. The two cartilage braces are also separated from each other. From this arrangement, a mobility of the mental symphysis at this stage of development may be concluded. The bone formation runs almost still strictly laterally from the cartilage, only in the last third of its course the bone embraces the cartilage brace somewhat from caudal (Figs. 18.1, 18.2, 18.3, 18.4 and 18.5).

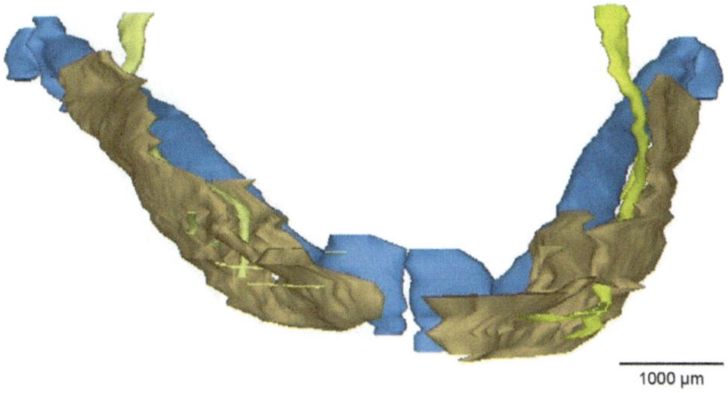

Fig. 18.1 Human embryo, 41 mm CRL, mandible (ochre), with Meckel's cartilage (blue), and inferior alveolar nerve (yellow), frontal view

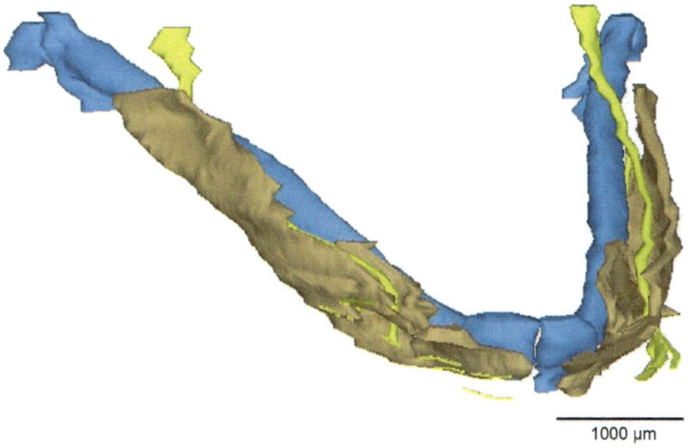

Fig. 18.2 Human embryo, 41 mm CRL, mandible, 45° lateral view

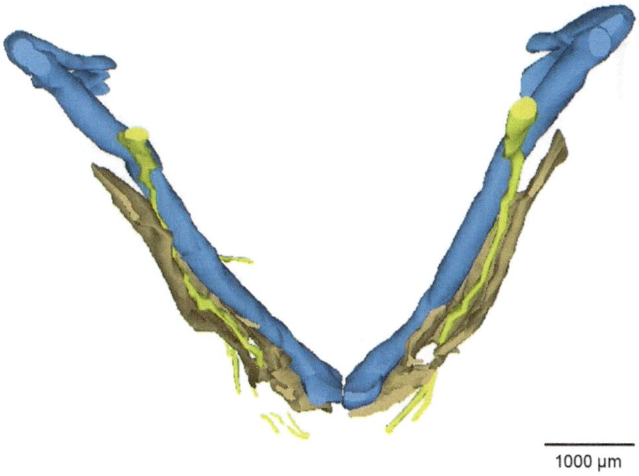

Fig. 18.3 Human embryo, 41 mm CRL, mandible, cranial view

R. J. Radlanski, *Atlas of Human Prenatal Craniofacial Development*, https://doi.org/10.1007/978-3-031-88812-0_18

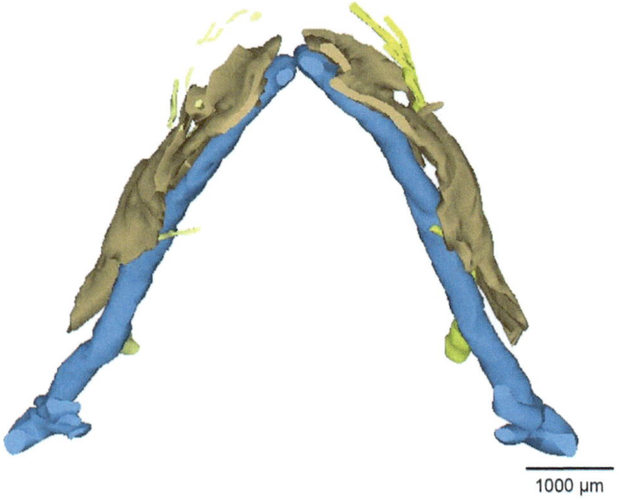

Fig. 18.4 Human embryo, 41 mm CRL, mandible, caudal view

Fig. 18.5 Human embryo, 41 mm CRL, mandible, dorsal view

Mandible

At the dorsal end of the mandible, neither the formation of a coronoid process nor a condylar process is recognizable at this stage of development. In this way, it becomes again clear, that the time sequence of the development can vary very strongly individually. In the previous stage (see Chap. 17), first indications of a coronoid process were already visible (Figs. 18.6, 18.7 and 18.8).

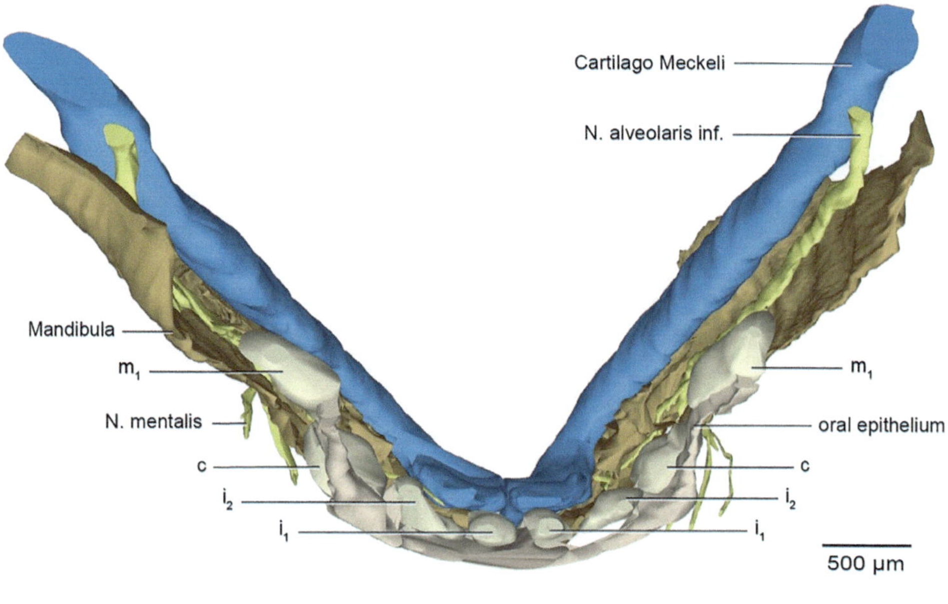

Fig. 18.6 Human embryo, 41 mm CRL, mandible with dental primordia, cranial view

Fig. 18.7 Human embryo, 41 mm CRL, mandible, dental primordia removed, bone remodeling pattern superimposed. Red: apposition, blue: lining cells

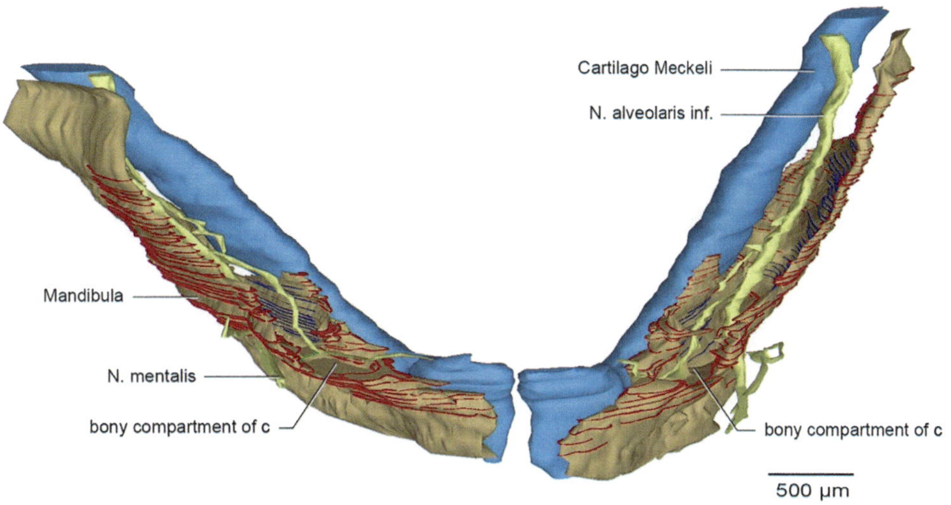

Fig. 18.8 Human embryo, 41 mm CRL, partial reconstruction of peridental bone around dental primordium of the lower left deciduous canine. Bone remodeling pattern superimposed, histological section (staining: Hematoxylin-eosin) underneath. Red: apposition of bone

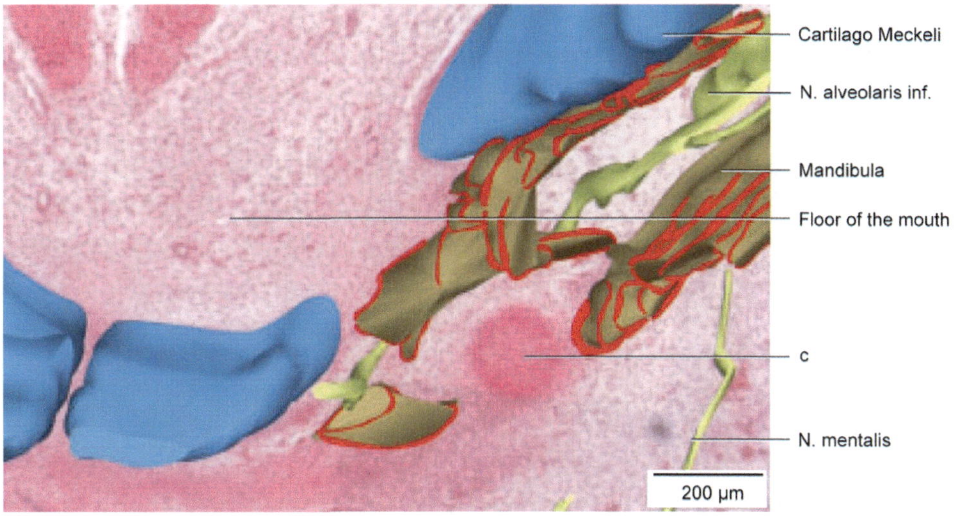

Dental Primordia

The mandible has created a groove laterally from the cartilaginous brace in which the dental primordia are located. The medial wall of the groove is particularly well developed medial to the primordia c and m_1. The cranial edges of the groove have been created due to bone apposition, as shown by the red marking. The inner surface of the bone is preferably inactive in terms of bone remodeling. On the bottom of the groove run the alveolar inferior nerve and its accompanying vessels (which are not shown here in the 3D reconstruction).

Further Reading

Radlanski RJ, Renz H, Müller U, Schneider RS, Marcucio RS, Helms JA. Prenatal morphogenesis of the human mental foramen. Eur J Oral Sci. 2002;110:452–9.

Radlanski RJ, Renz H, Klarkowski MC. Prenatal development of the human mandible. 3D reconstructions, morphometry and bone remodelling pattern, sizes 12–117 mm CRL. Anat Embryol (Berl). 2003;207:221–32.

Radlanski RJ, Renz H, Lajvardi S, Schneider R. A Bone remodeling during prenatal morphogenesis of the human mental foramen. Eur J Oral Sci. 2004;112:301–10.

Tsengelsaikhan N. Zur Entwicklung des Kieferknochens im Bereich der Zahnanlagen—Histologische und morphologische Untersuchung an der embryonalen Mandibula des Menschen während der frühen Entwicklungsstadien der Zahnanlagen (19–68 mm SSL) Med. Diss. Charité-Universitätsmedizin Berlin; 2014.

Embryo 53 mm CRL (ca. 9th wk): HAN040389

19

Overall Remarks

This embryo is so beautifully developed that we can show it in this chapter in many reconstructions and in different views in great detail. The orbitae are already preformed cartilaginously; the nasal capsule is impressive in the midface; the frontal bones and the maxilla have also already spread out considerably and the maxilla shows extensions of the frontal processes on either side of the nasal capsule, the zygomatic arches are recognizable laterally and the mandible has not yet fused in the symphysis, but already shows the beginnings of the condylar and coronoid processes at either of its dorsal ends. At this stage, however, there is still the primary temporomandibular joint, i.e., the malleus (dorsal end of Meckel's cartilage) articulates with the incus. The actual, the secondary temporomandibular joint does not yet exist. But the structures in the inner ear and the semicircular canals are also beautifully developed. All the muscles including of course the tongue are already well formed and also the salivary glands (Figs. 19.1, 19.2, 19.3, 19.4, 19.5, 19.6, 19.7, 19.8, 19.9, 19.10, 19.11, 19.12, 19.13, 19.14, and 19.15).

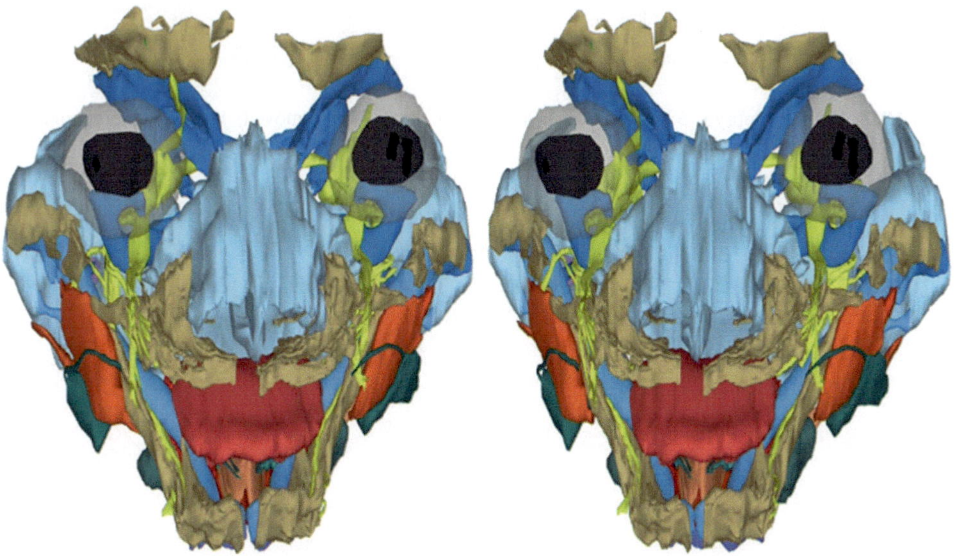

Fig. 19.1 Stereoscopical pair of images. Human embryo, 53 mm CRL, head, frontal view, skin removed

R. J. Radlanski, *Atlas of Human Prenatal Craniofacial Development*, https://doi.org/10.1007/978-3-031-88812-0_19

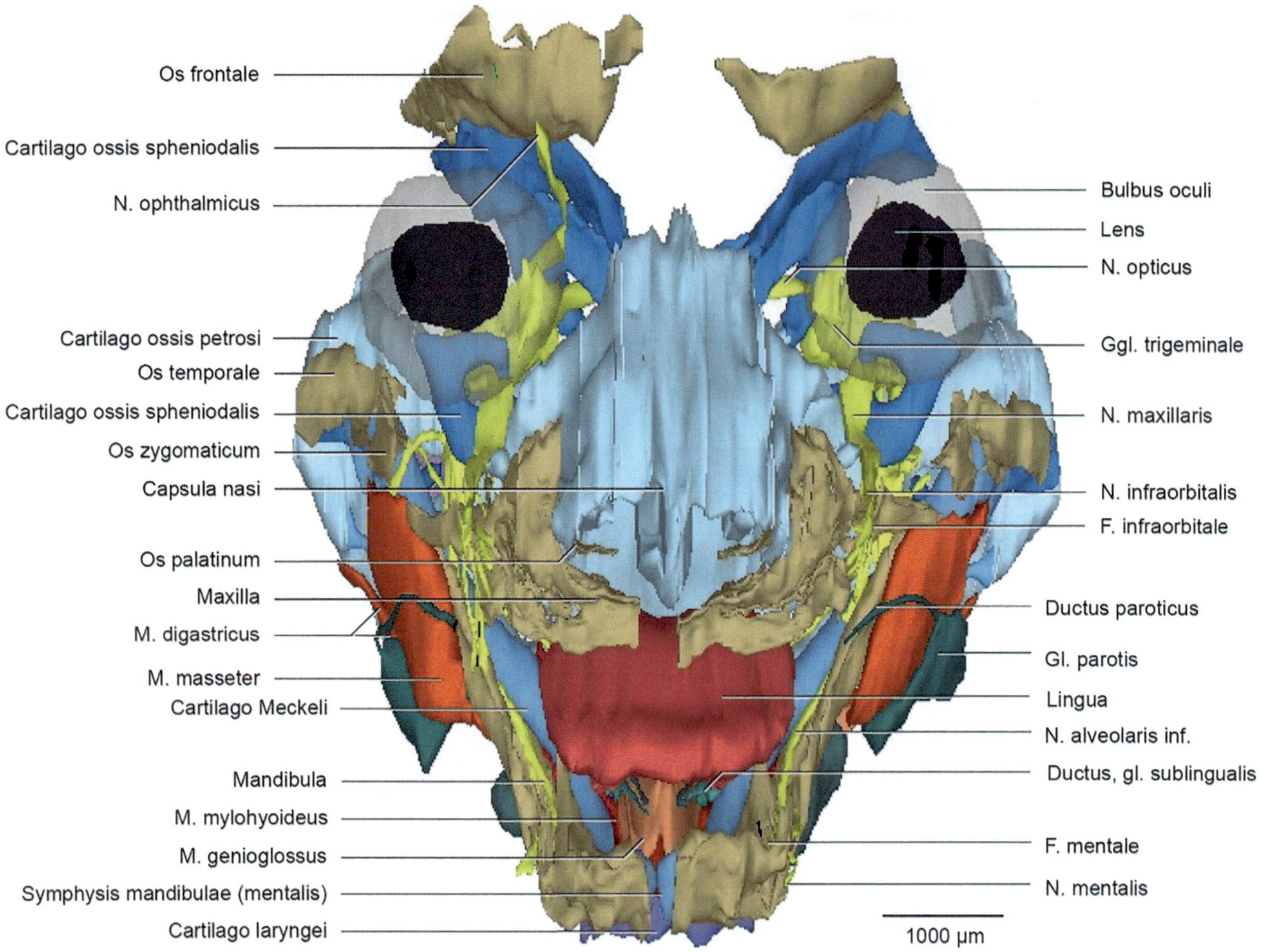

Fig. 19.2 Human embryo, 53 mm CRL, head, frontal view, skin removed, layer 1 of preparation (*From* Radlanski and Renz [5]; *with permission.*)

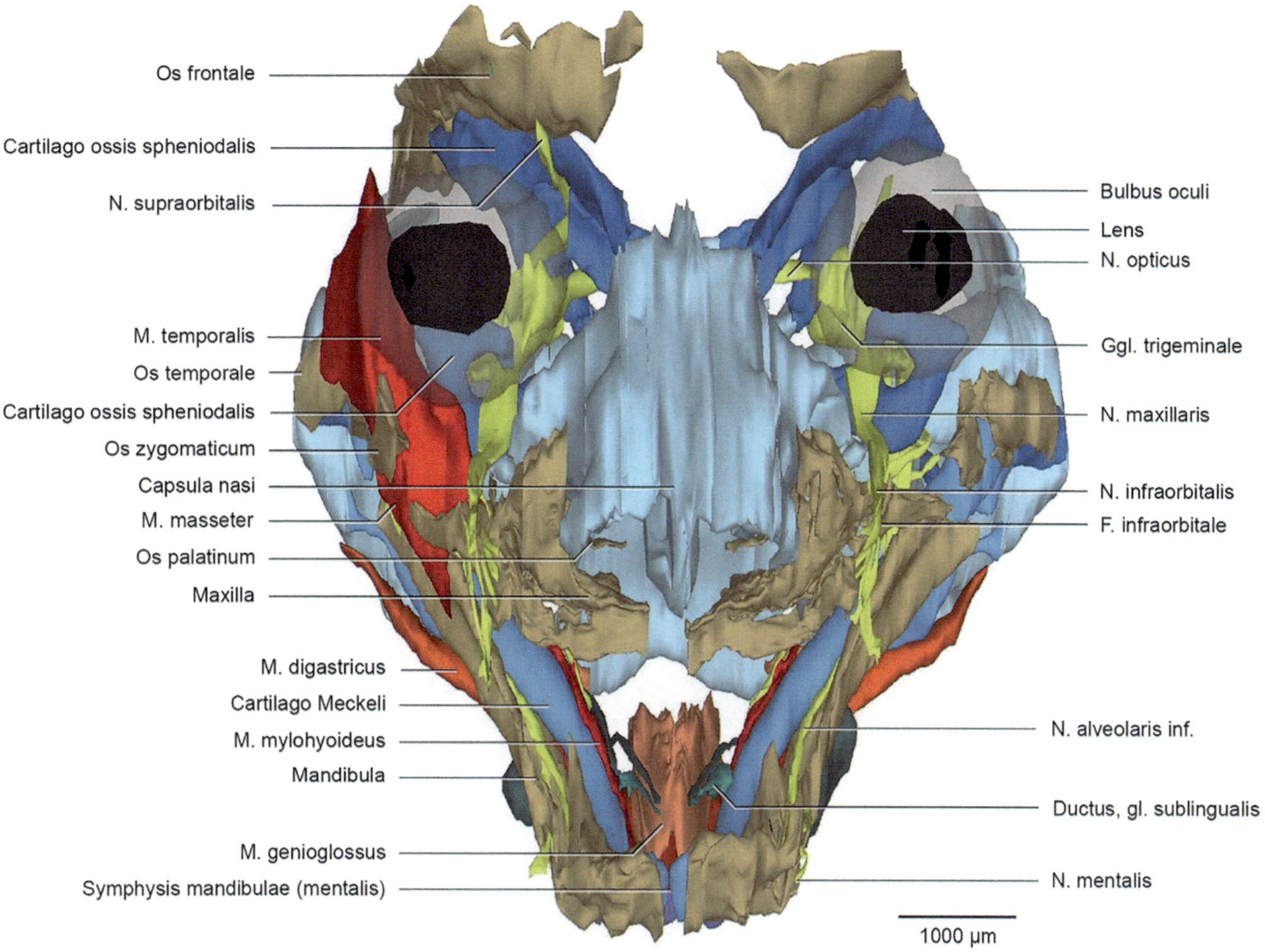

Fig. 19.3 Human embryo, 53 mm CRL, head, frontal view, layer 2 of preparation

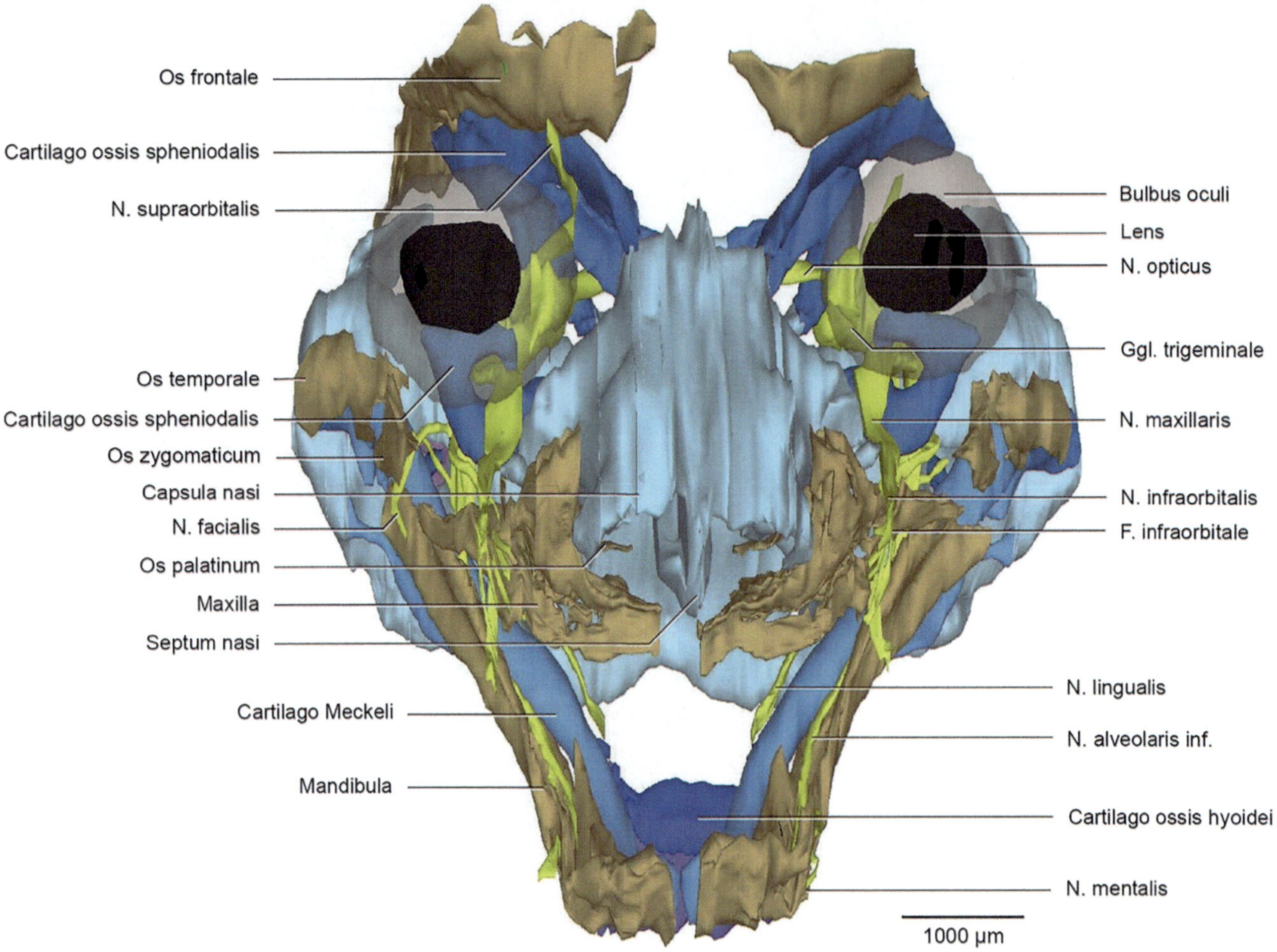

Fig. 19.4 Human embryo, 53 mm CRL, head, frontal view, layer 3 of preparation

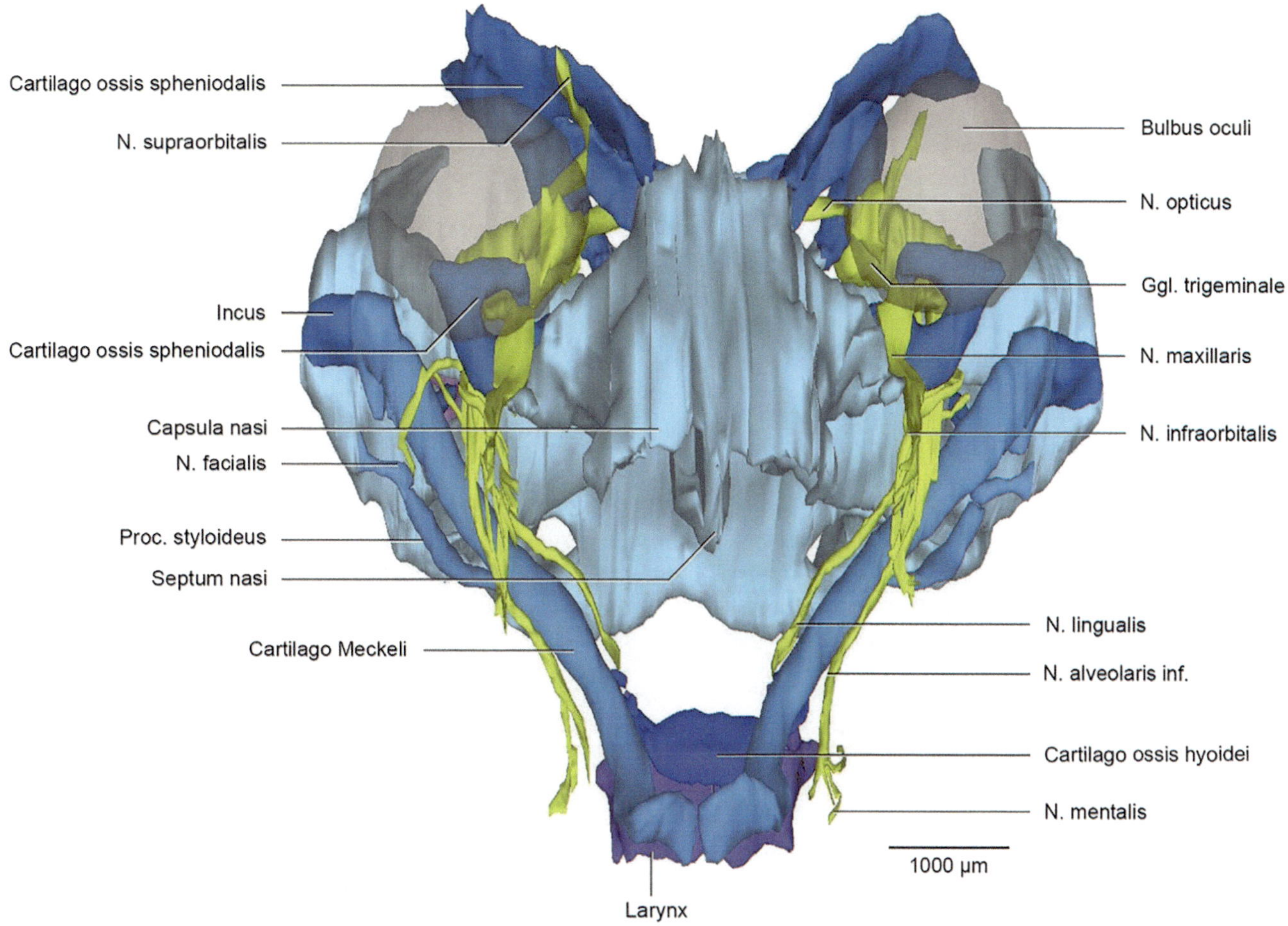

Fig. 19.5 Human embryo, 53 mm CRL, head, frontal view, cartilage, nerves, and eyeballs

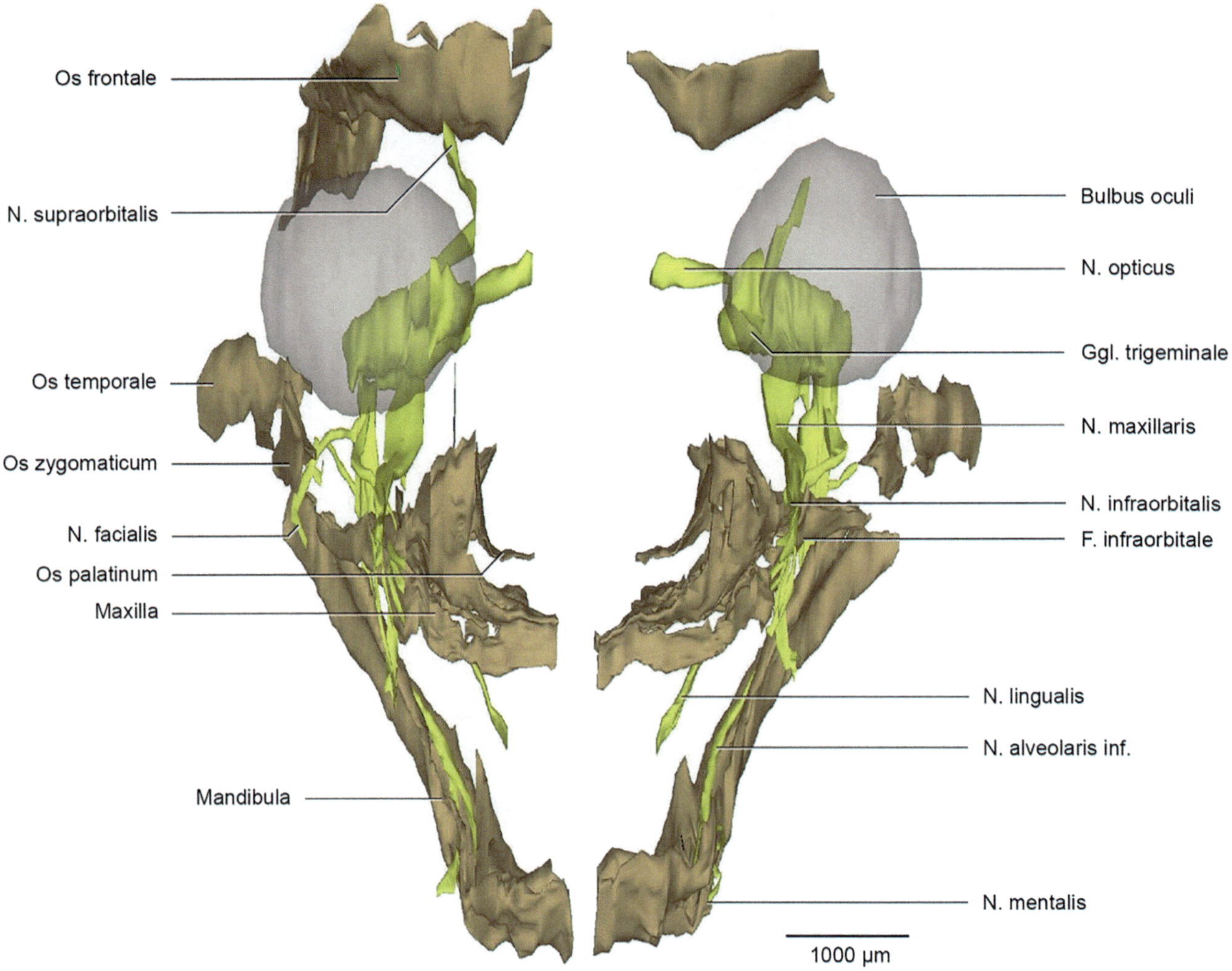

Fig. 19.6 Human embryo, 53 mm CRL, head, frontal view, bone, nerves, and eyeballs

Fig. 19.7 Stereoscopical pair of images. Human embryo, 53 mm CRL, head, 45° lateral view, skin removed

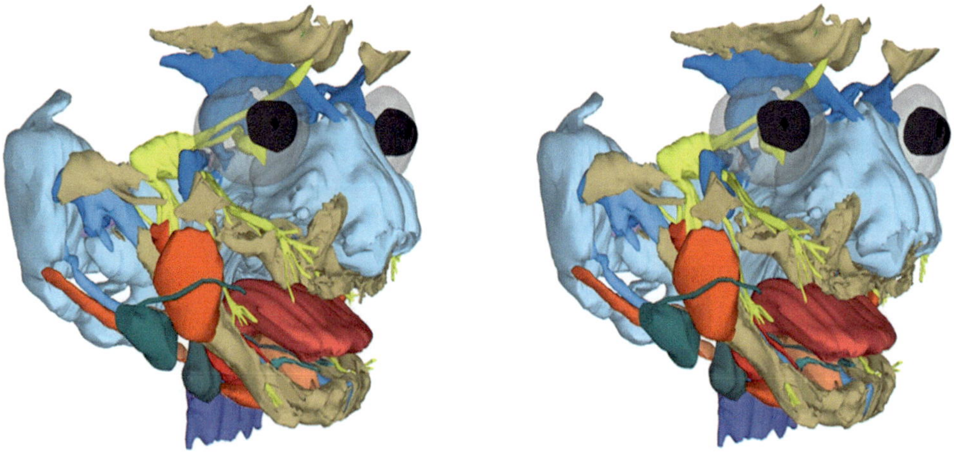

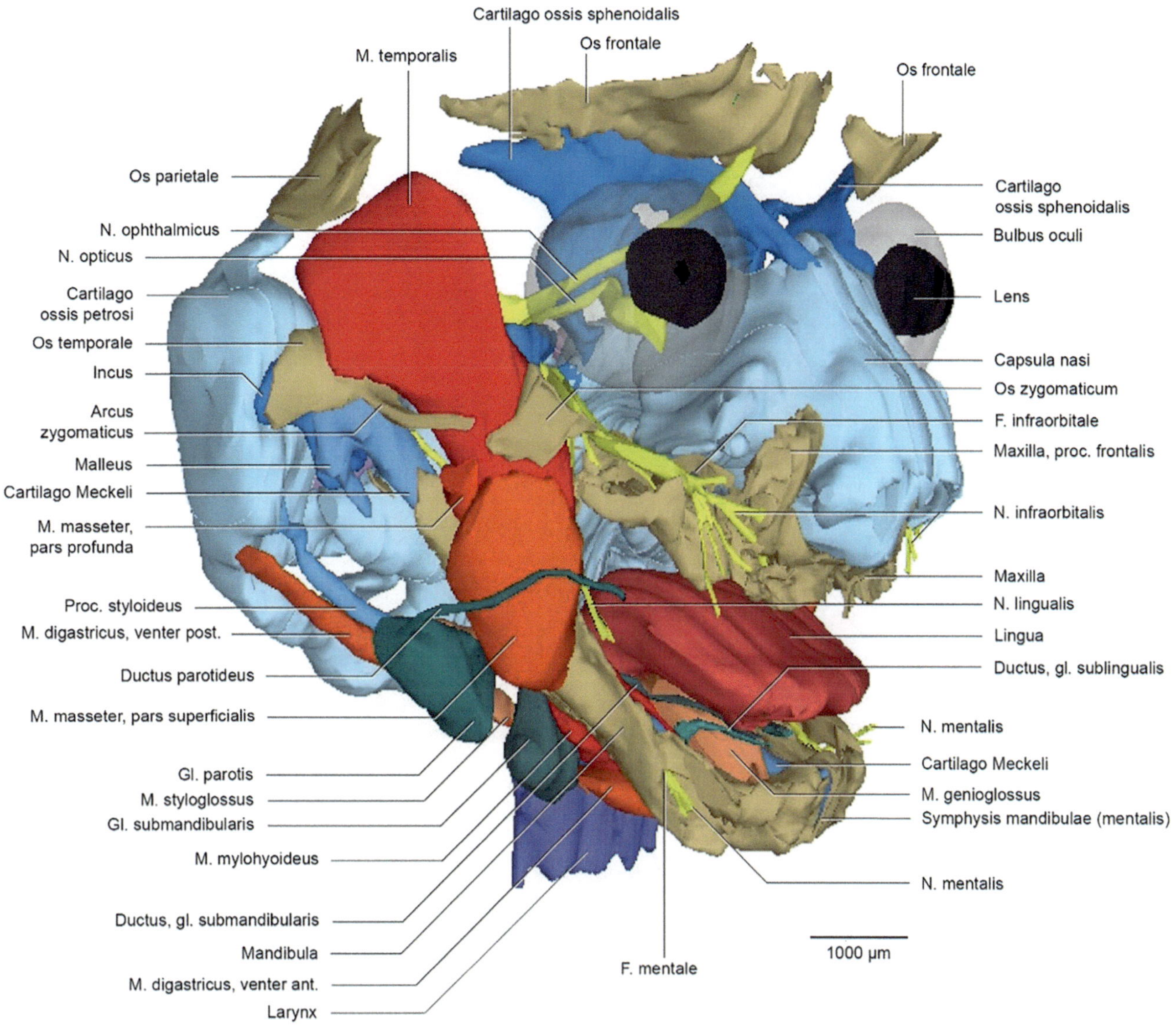

Fig. 19.8 Human embryo, 53 mm CRL, head, 45° lateral view, skin removed, layer 1 of preparation (From Radlanski [6]; *with permission.*)

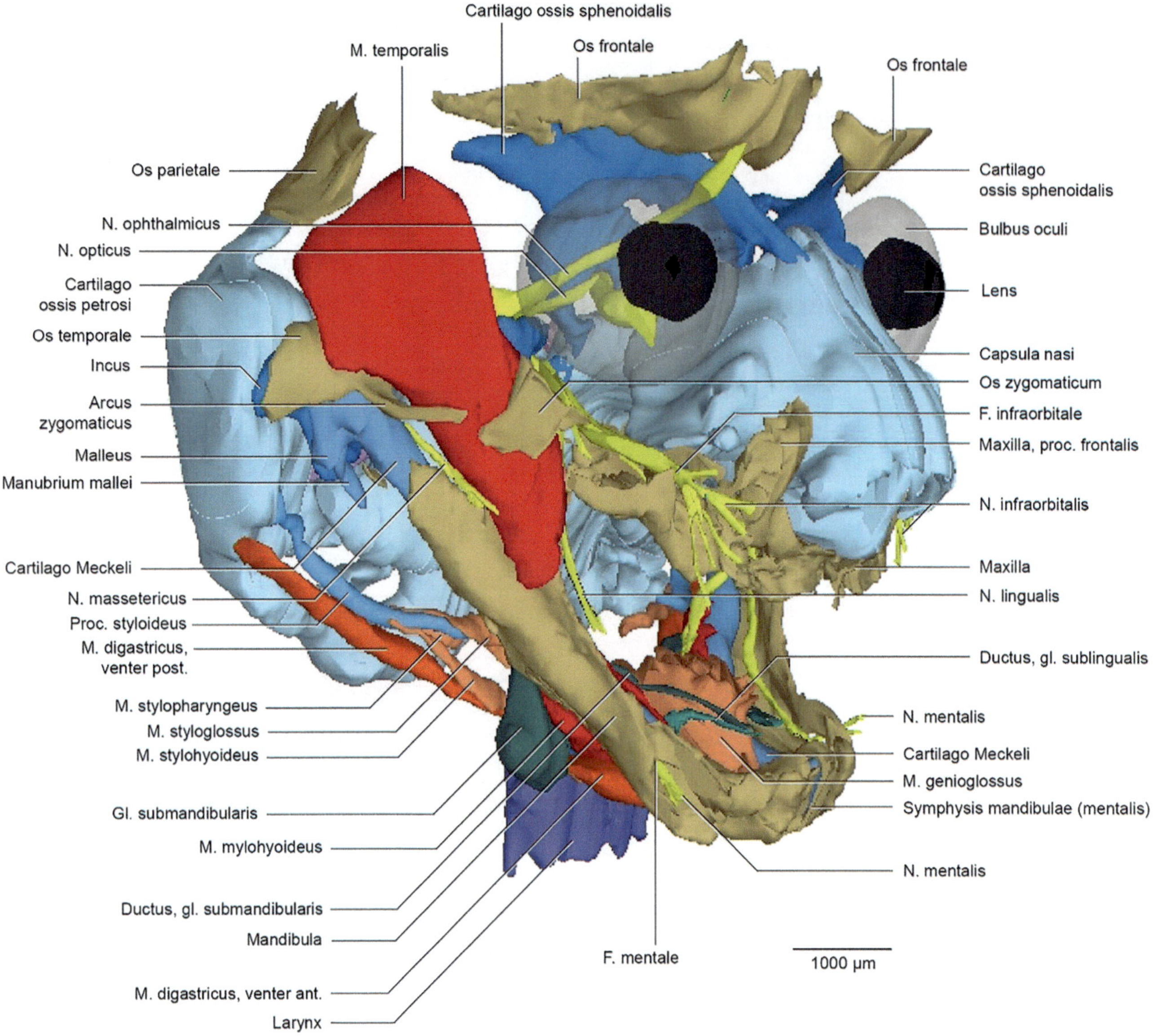

Fig. 19.9 Human embryo, 53 mm CRL, head, 45° lateral view, layer 2 of preparation

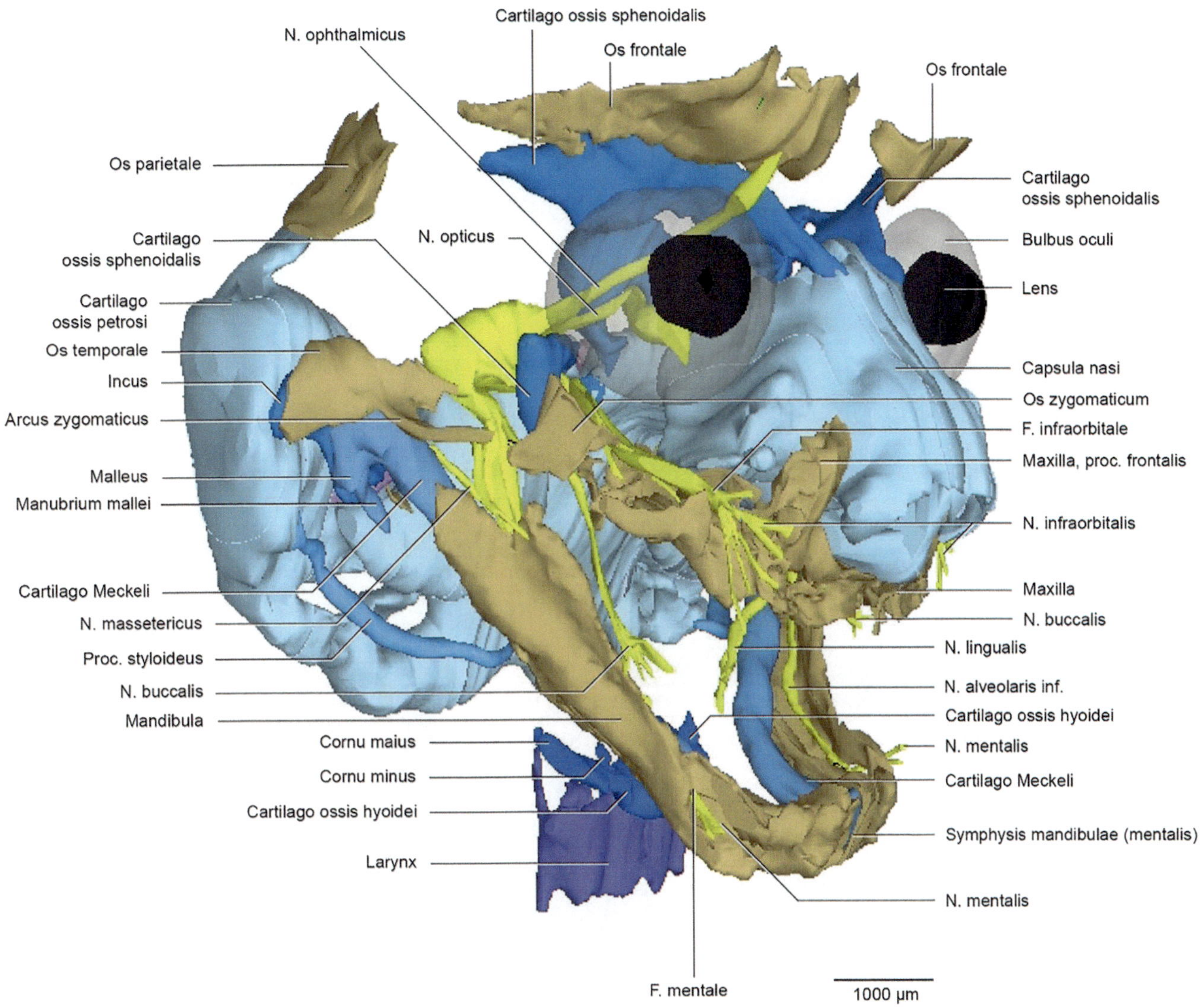

Fig. 19.10 Human embryo, 53 mm CRL, head, 45° lateral view, layer 3 of preparation

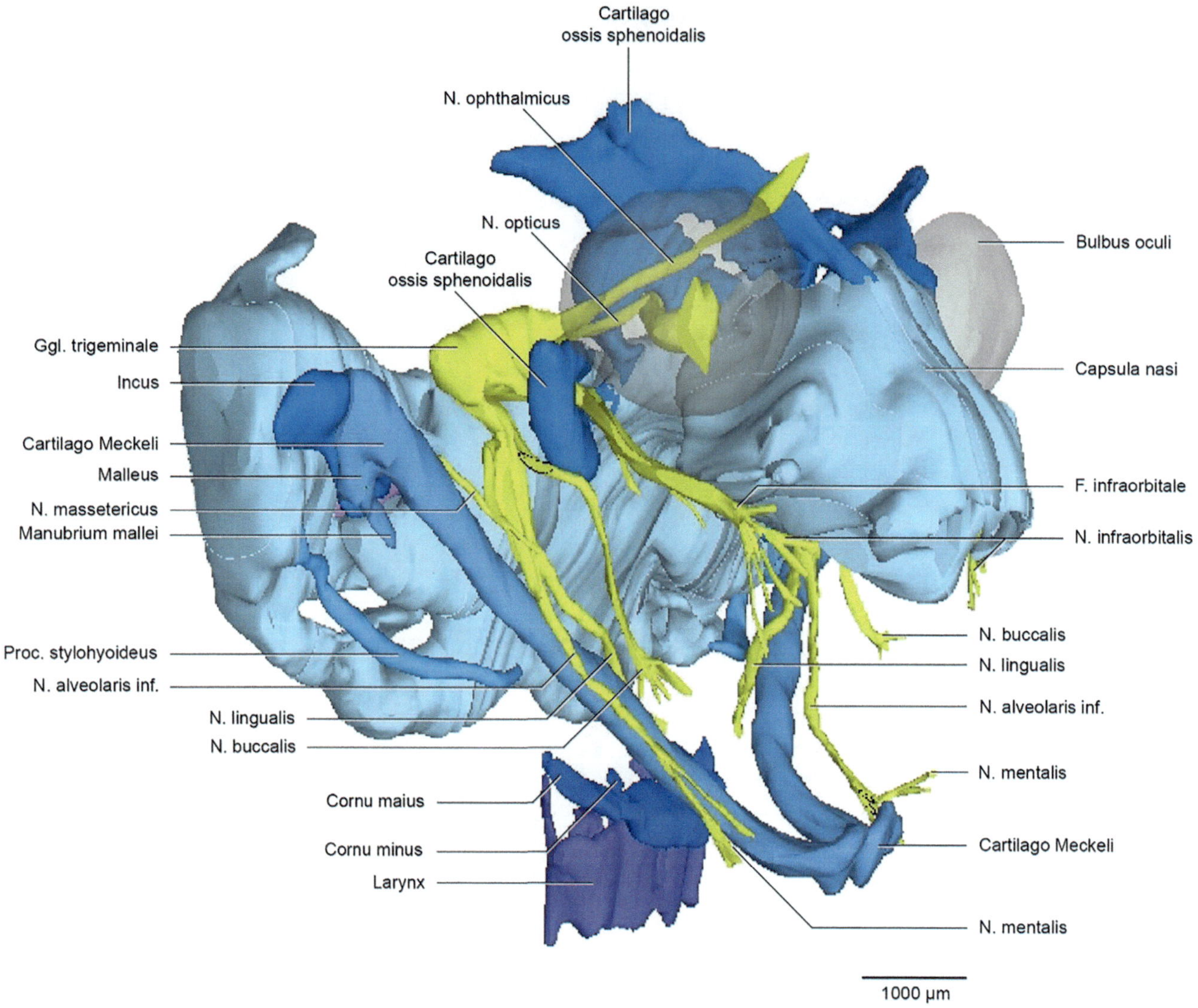

Fig. 19.11 Human embryo, 53 mm CRL, head, 45° lateral view, cartilage, nerves, and eyeballs

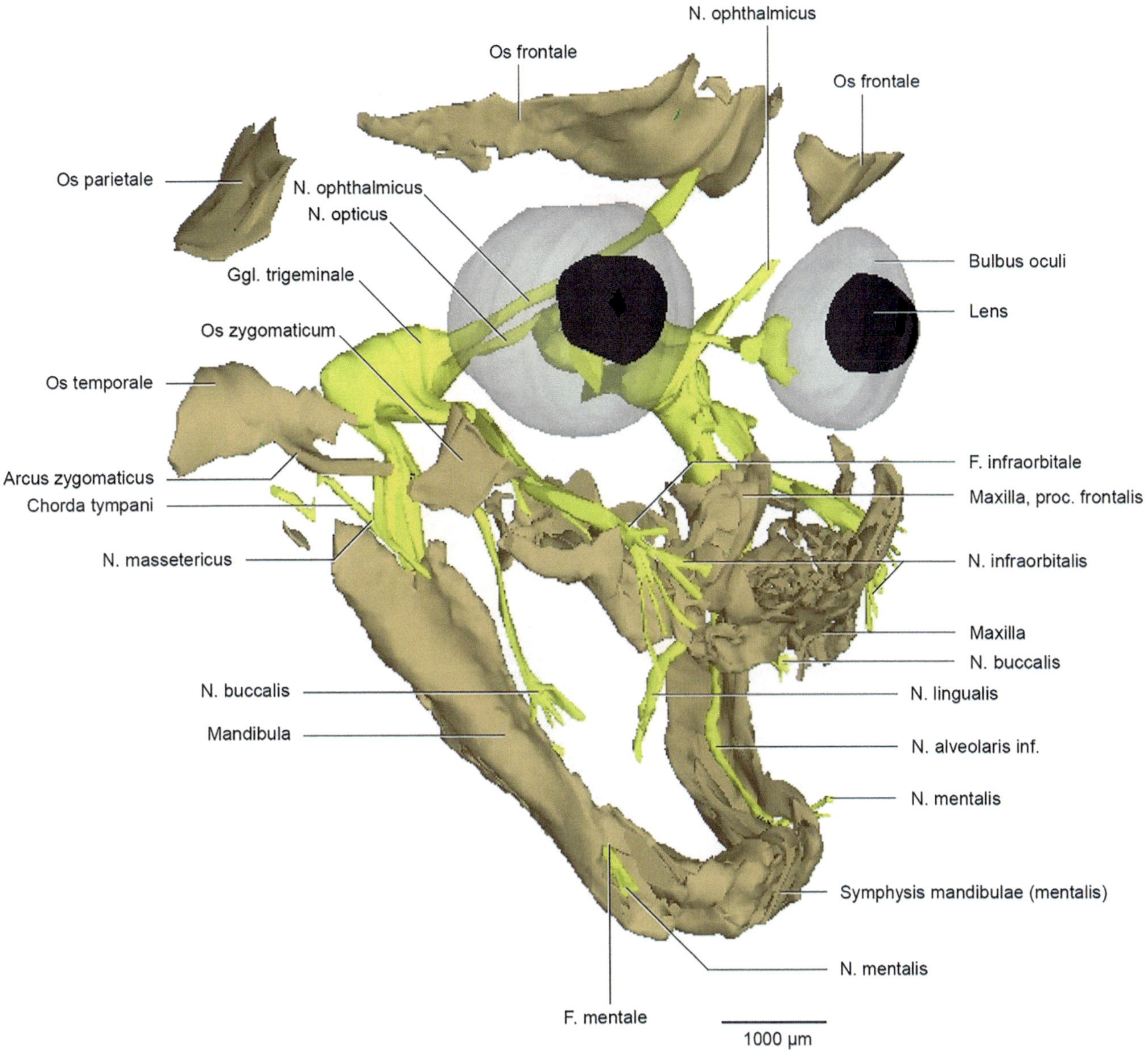

Fig. 19.12 Human embryo, 53 mm CRL, head, 45° lateral view, bone, nerves, and eyeballs

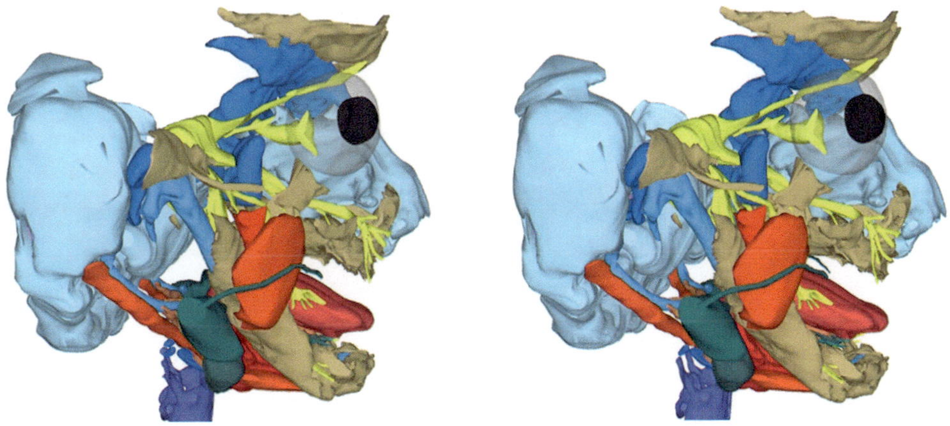

Fig. 19.13 Stereoscopical pair of images. Human embryo, 53 mm CRL, head, lateral view, skin removed

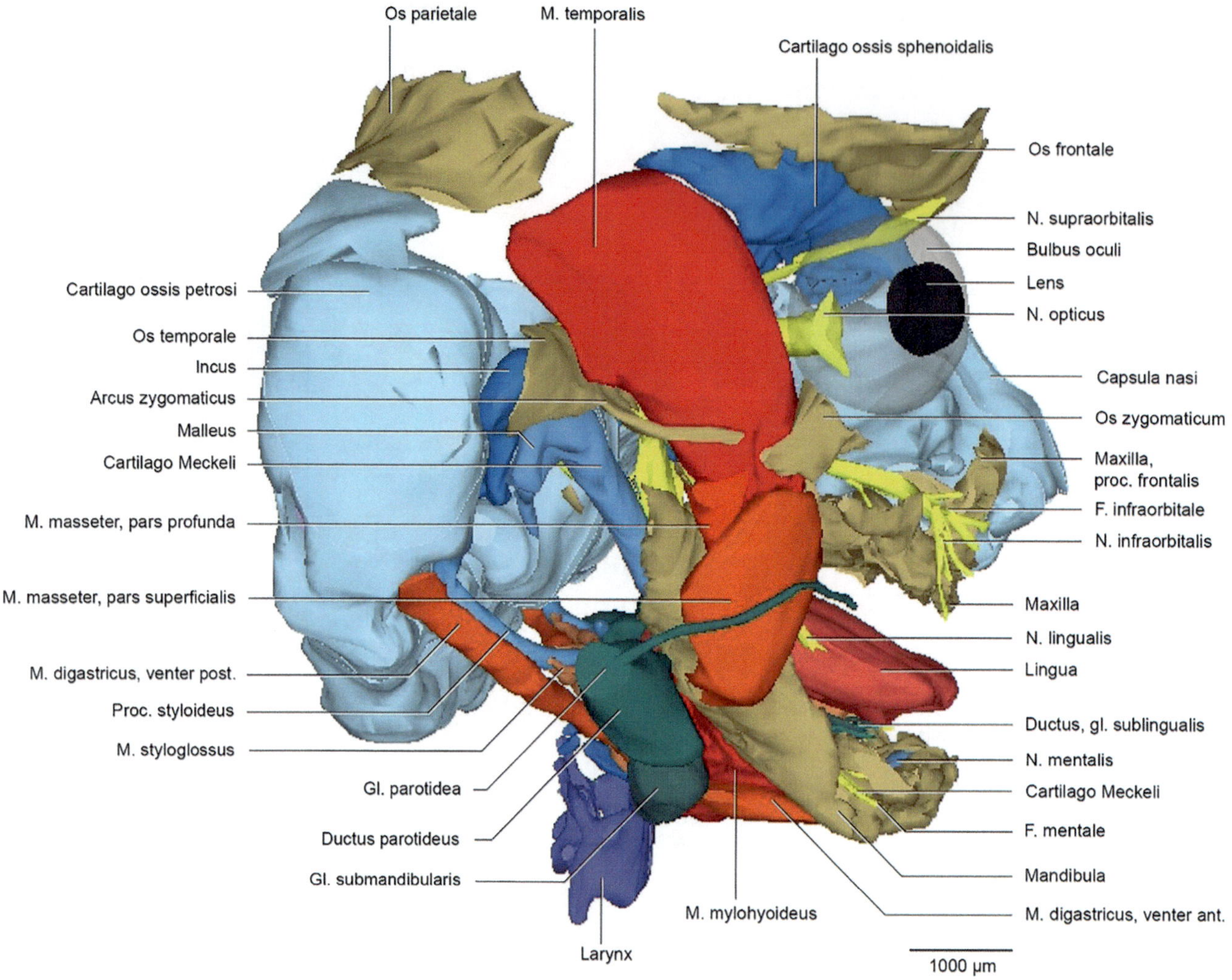

Fig. 19.14 Human embryo, 53 mm CRL, head, lateral view, skin removed, layer 1 of preparation

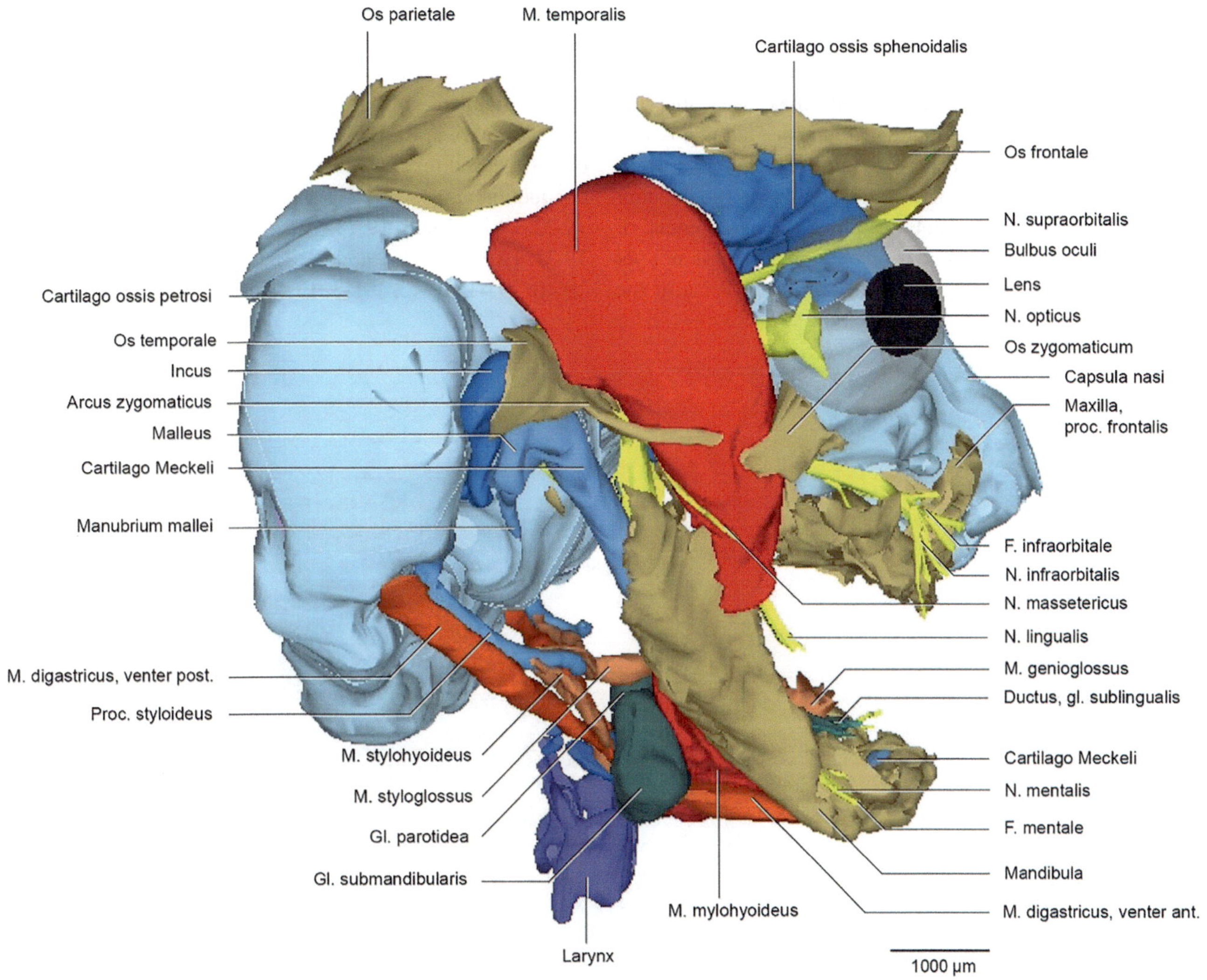

Fig. 19.15 Human embryo, 53 mm CRL, head, lateral view, layer 2 of preparation

Temporomandibular Joint

We know from ultrasound examinations that embryos at this stage of their development are already making swallowing movements and even sucking their thumbs. This is possible because all the facial and chewing muscles required for this are already developed [1]. However, and this is the peculiar and interesting aspect about the prenatal development of mammals: the actual, secondary mandibular joint is not yet available for the movements of the mandible because the condylar process is not yet sufficiently developed and there is still no mandibular fossa. Instead, Meckel's cartilage still exists in its full length and articulates with its posterior end, the later malleus, with the incus. This condition is addressed as the primary jaw joint [2]. It is only later that these primordia of the auditory ossicles come into the service of sound conduction (Figs. 19.16, 19.17, 19.18, 19.19, 19.20, 19.21, 19.22, 19.23, 19.24, 19.25, 19.26, 19.27, 19.28, 19.29, 19.30, 19.31, 19.32, 19.33, 19.34, 19.35, 19.36, 19.37, 19.38, 19.39, 19.40, 19.41, 19.42, 19.43, 19.44, 19.45 and 19.46) [3].

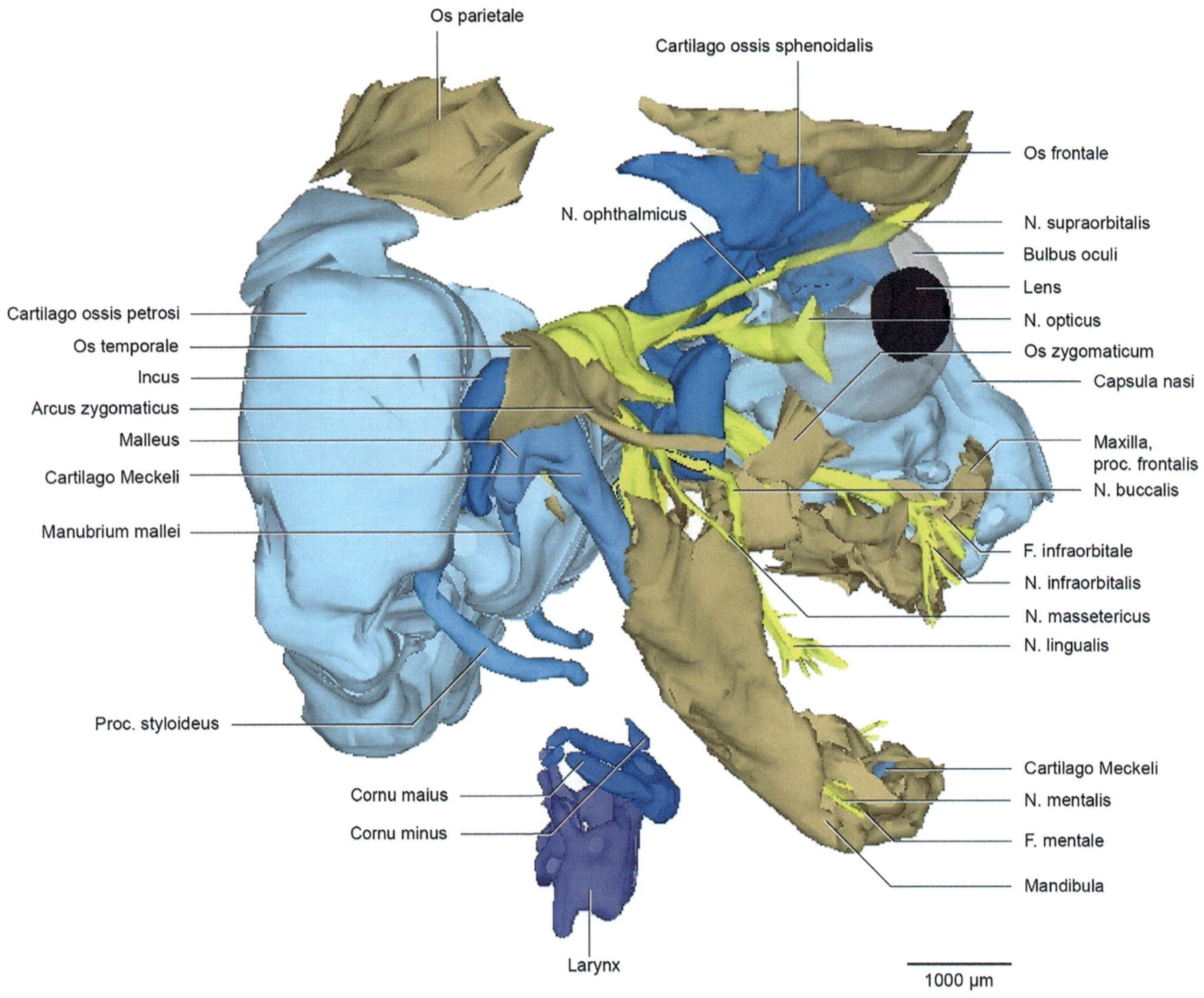

Fig. 19.16 Human embryo, 53 mm CRL, head, lateral view, layer 3 of preparation

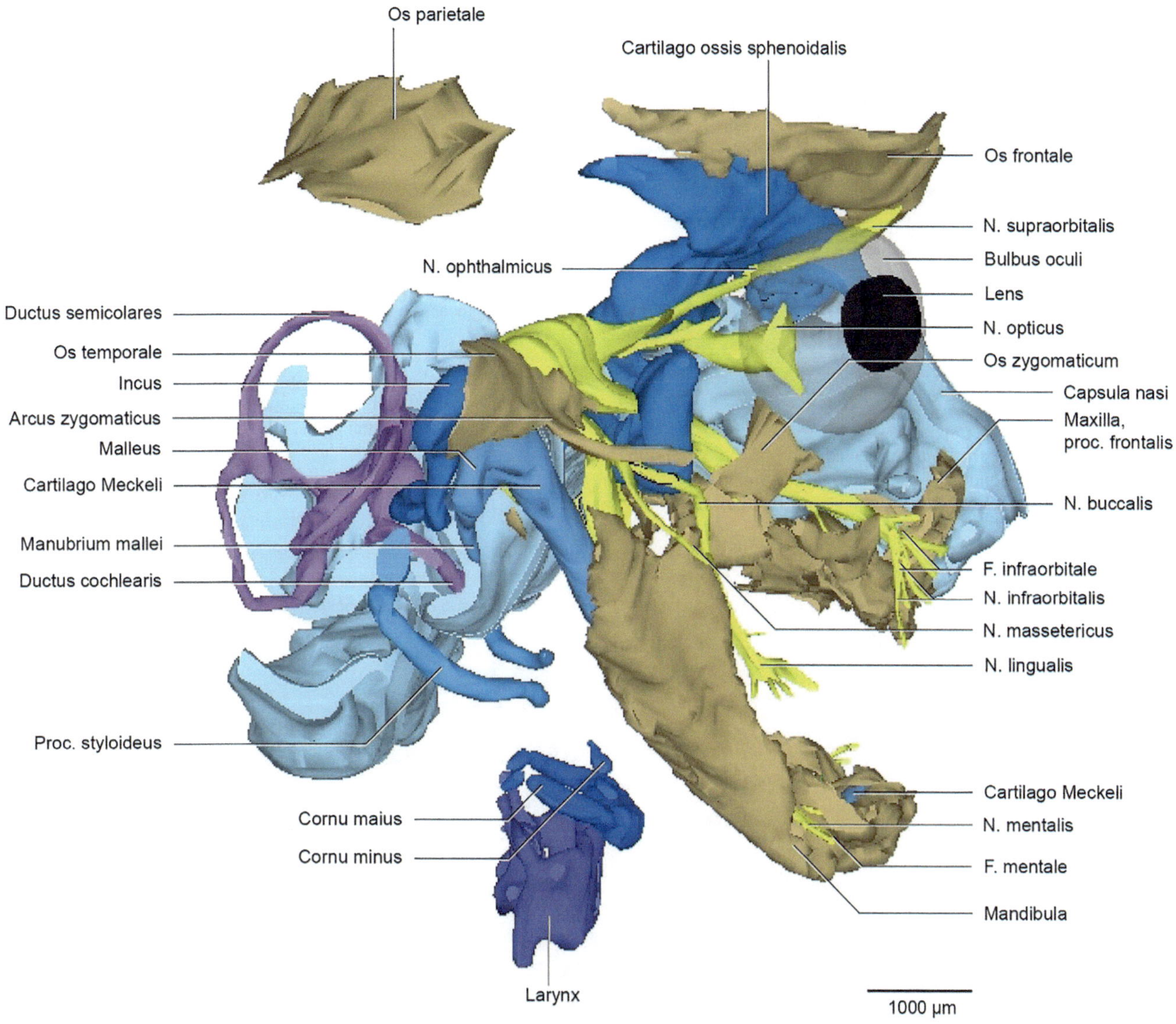

Fig. 19.17 Human embryo, 53 mm CRL, head, lateral view, layer 4 of preparation. From Radlanski (2018)

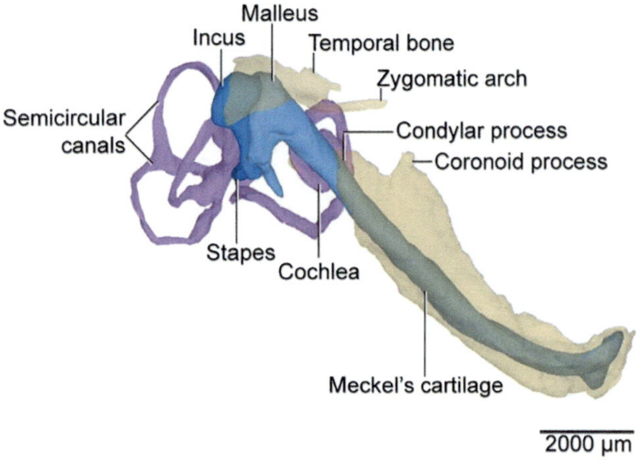

Fig. 19.18 3D reconstruction of selected parts of the temporomandibular joint, middle, and inner ear. Mandible and temporal bone made transparent (From Fernández Rubio and Radlanski [2]; *with permission.*)

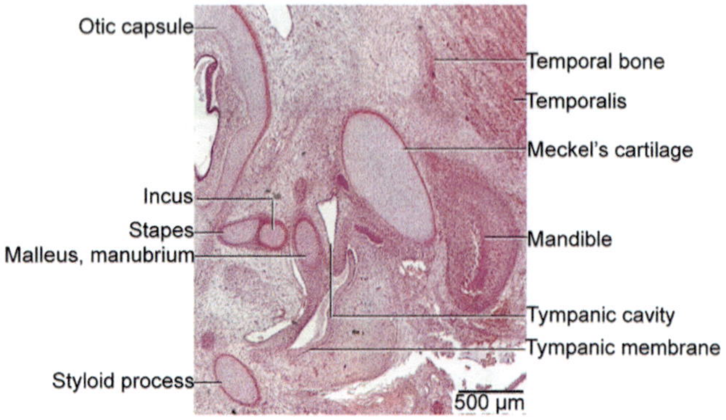

Fig. 19.19 Histological section of the region of the right middle ear and selected parts of the temporomandibular joint. Staining: Hematoxylin-eosin. Sagittal plane, HE-staining

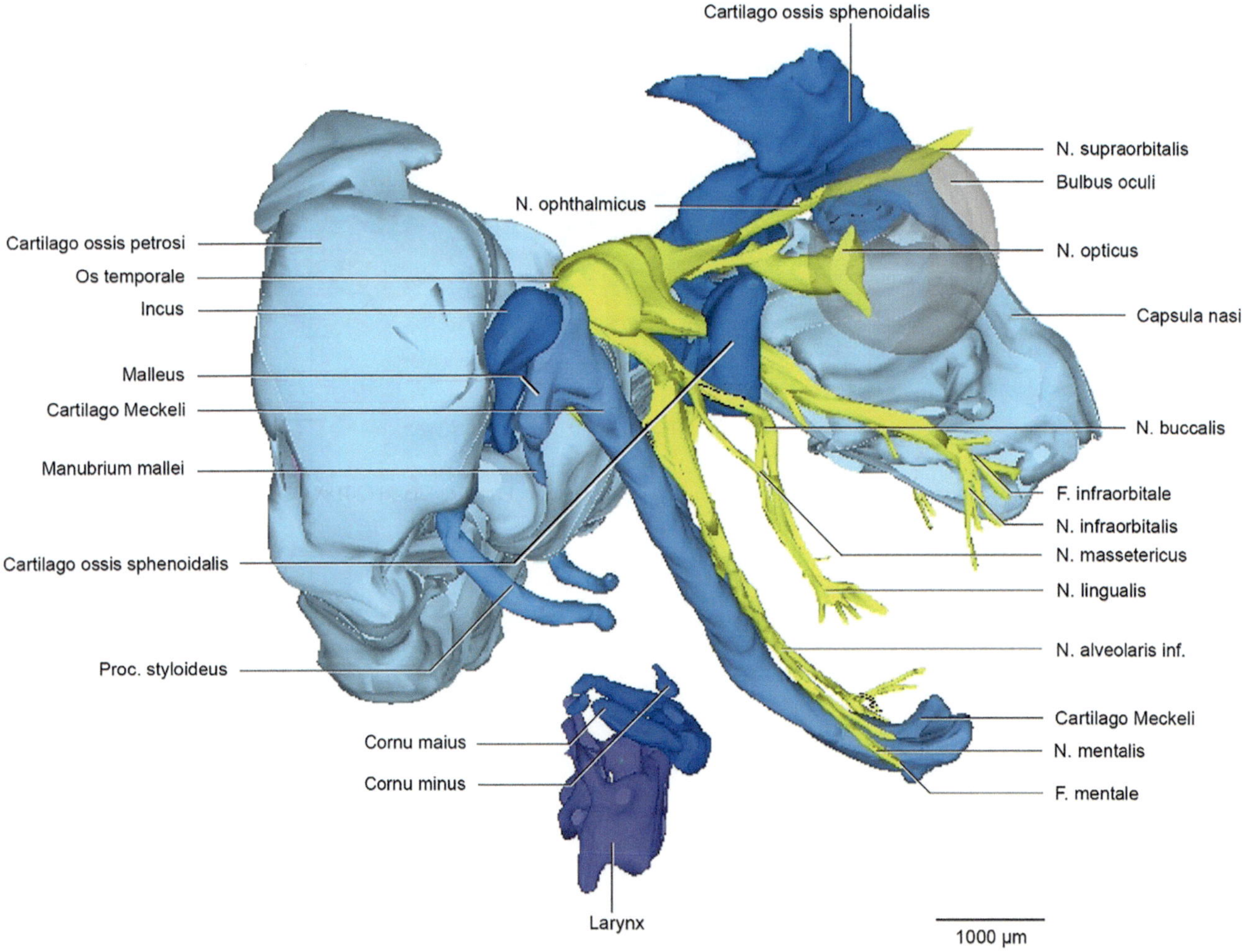

Fig. 19.20 Human embryo, 53 mm CRL, head, lateral view, cartilage, nerves, and eyeballs

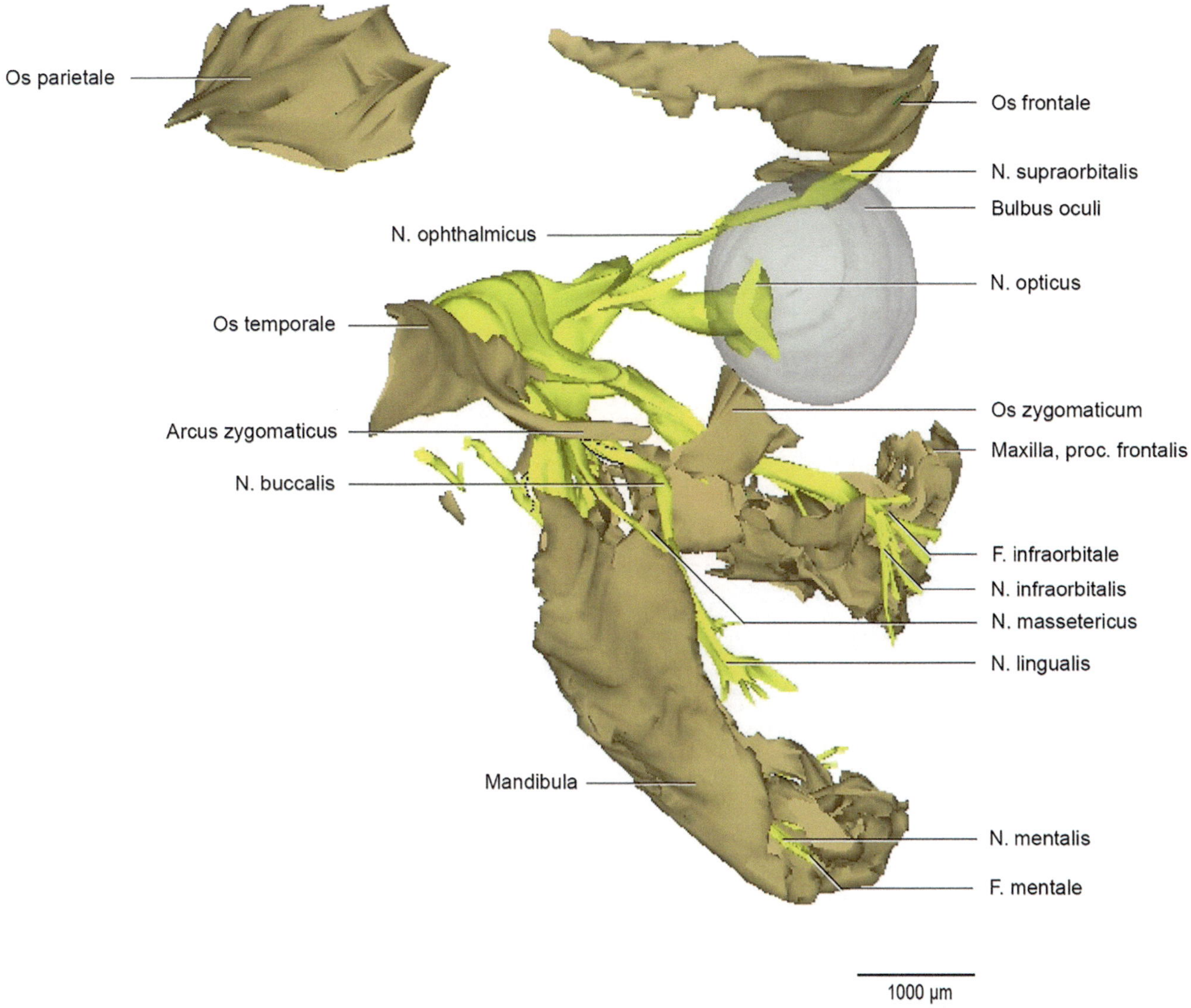

Fig. 19.21 Human embryo, 53 mm CRL, head, lateral view, bone, nerves, and eyeballs

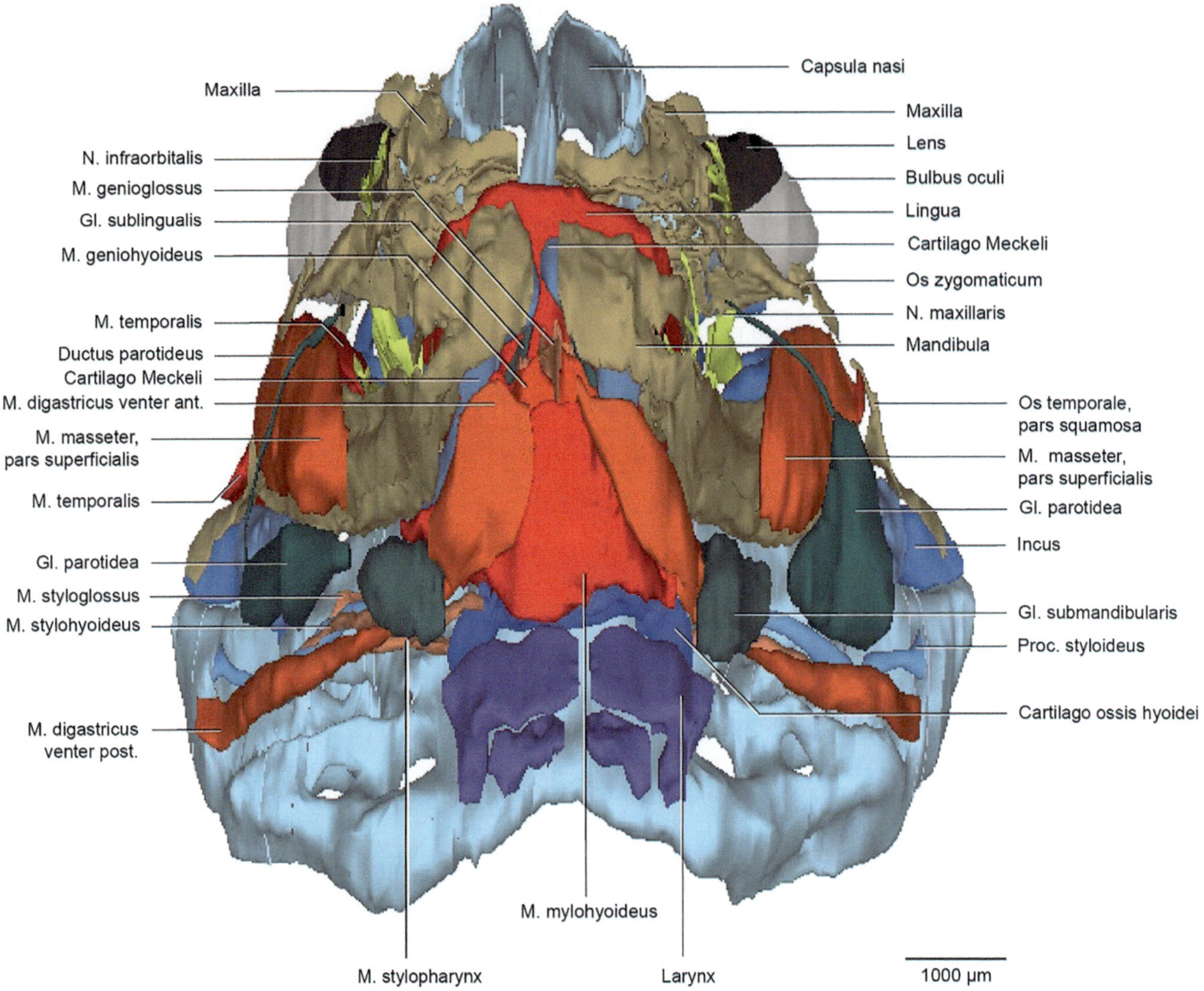

Fig. 19.22 Human embryo, 53 mm CRL, head, caudal view, skin removed

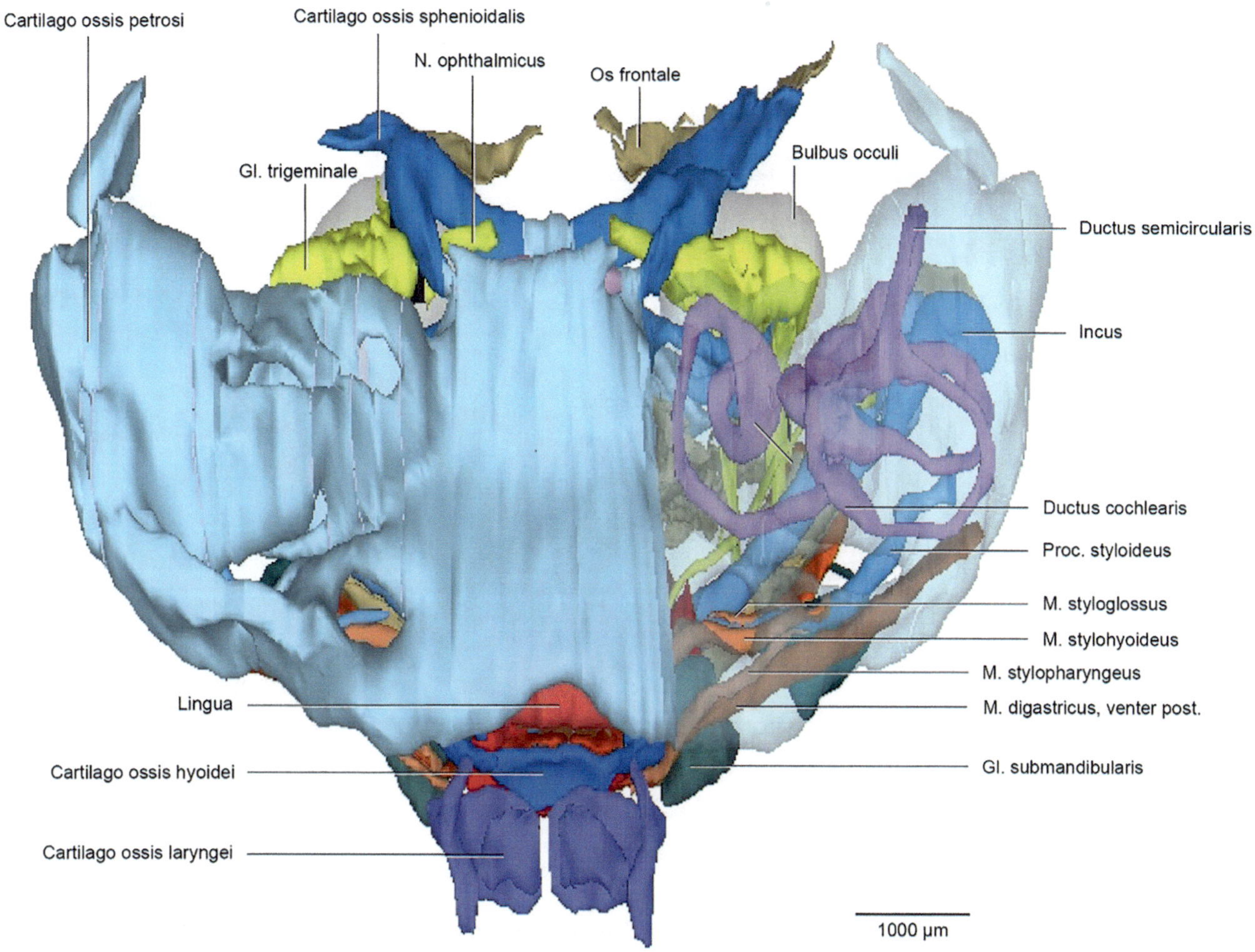

Fig. 19.23 Human embryo, 53 mm CRL, head, caudal view, cartilaginous skull base

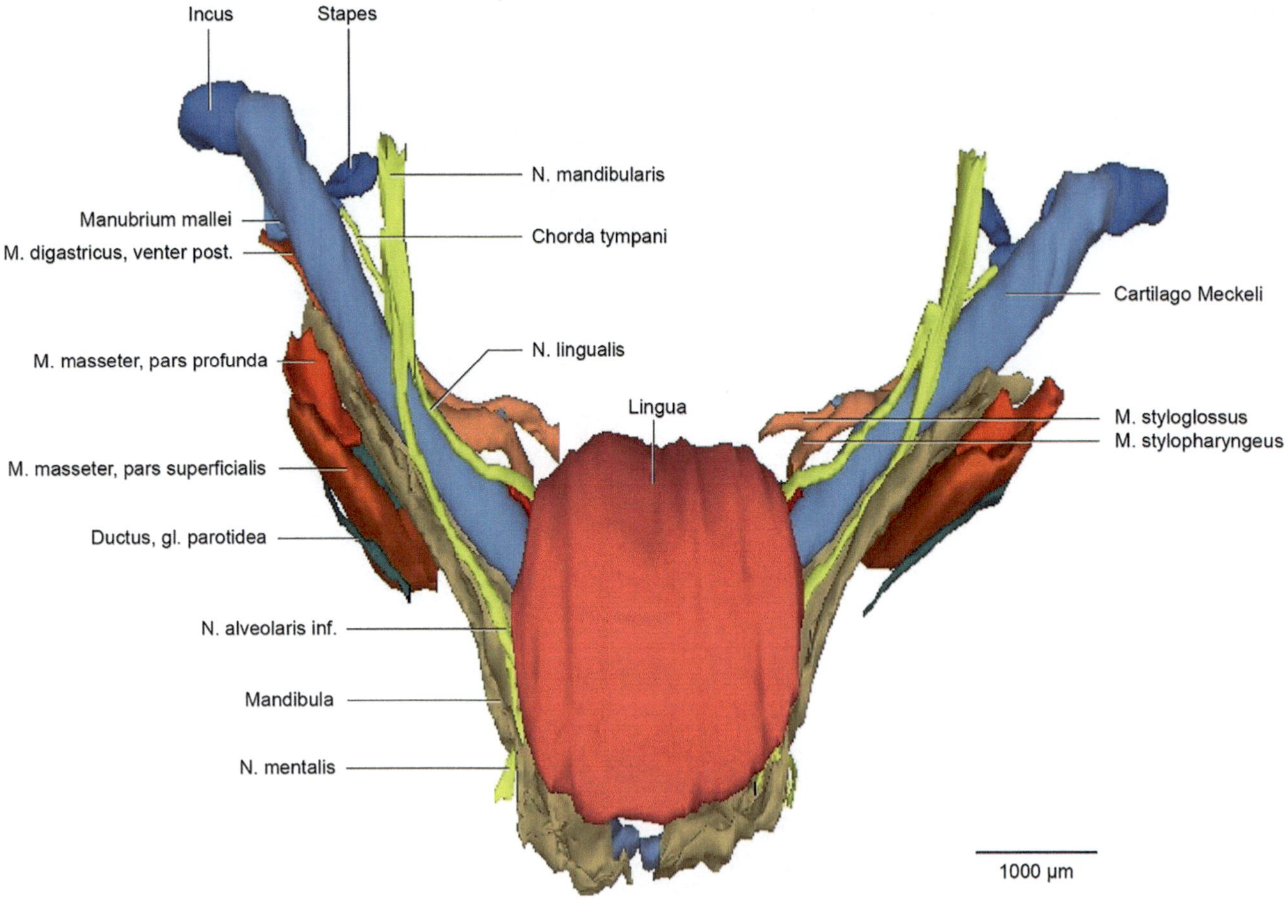

Fig. 19.24 Human embryo, 53 mm CRL, mandible, cranial view, layer 1 of preparation (*From* Radlanski and Renz [5]; *with permission*.)

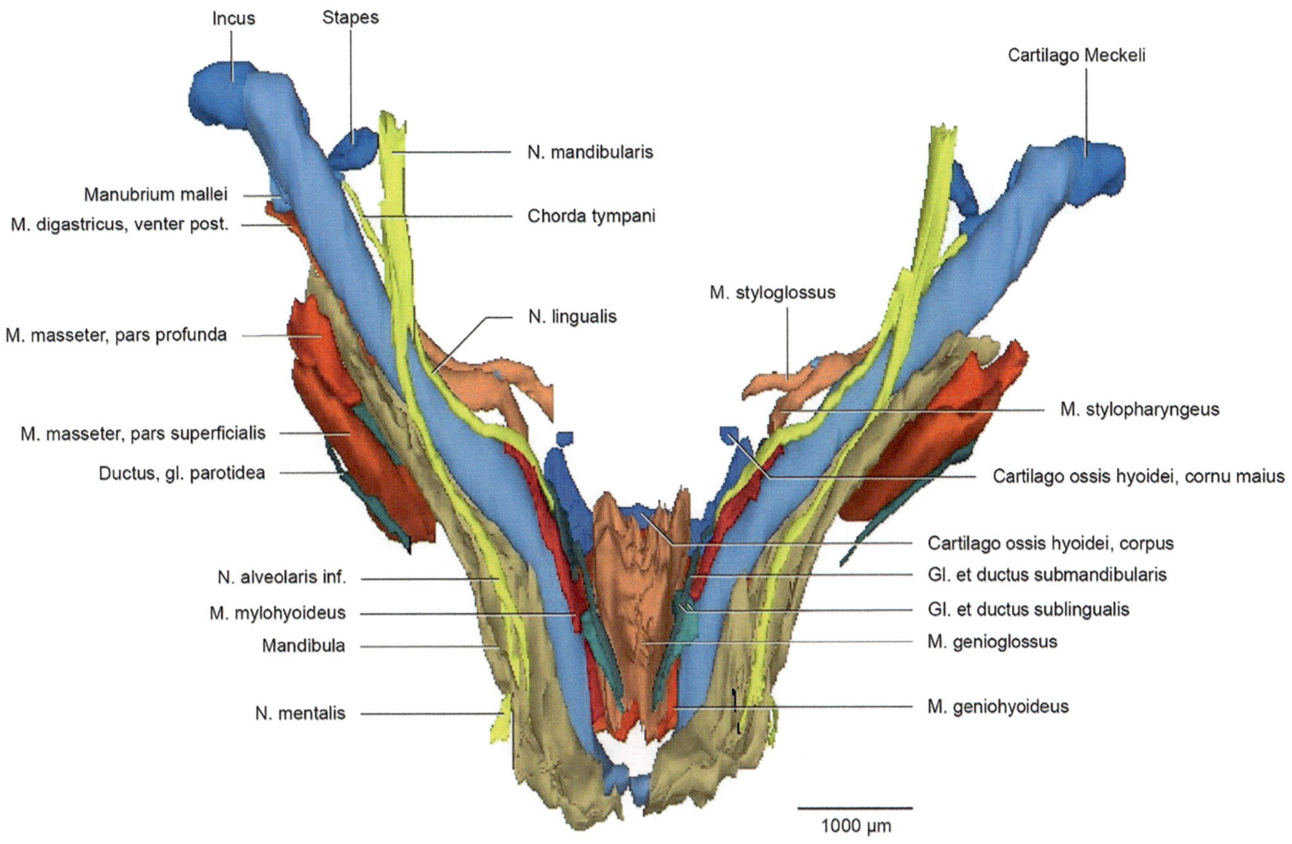

Fig. 19.25 Human embryo, 53 mm CRL, mandible, cranial view, layer 2 of preparation (*From* Radlanski and Renz [5]; *with permission.*)

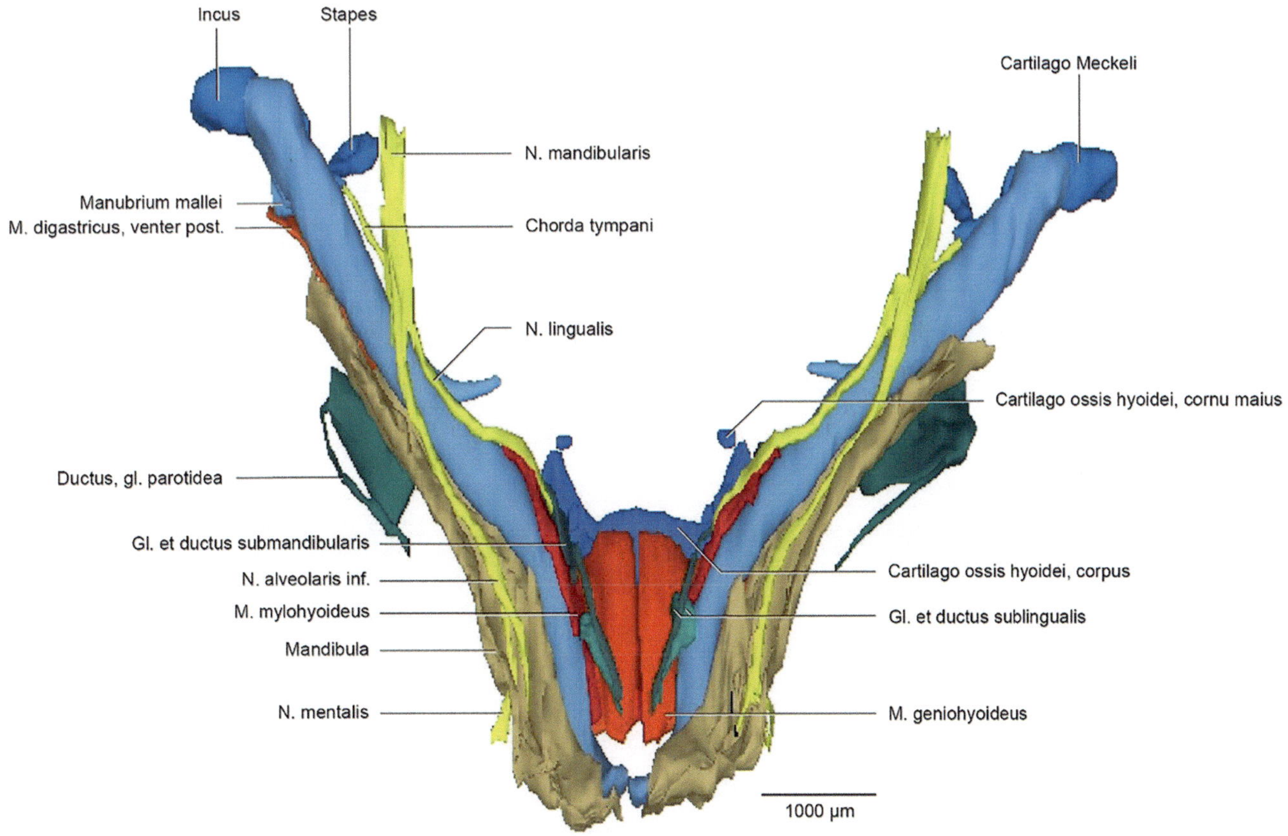

Fig. 19.26 Human embryo, 53 mm CRL, mandible, cranial view, layer 3 of preparation (*From* Radlanski and Renz [5]; *with permission.*)

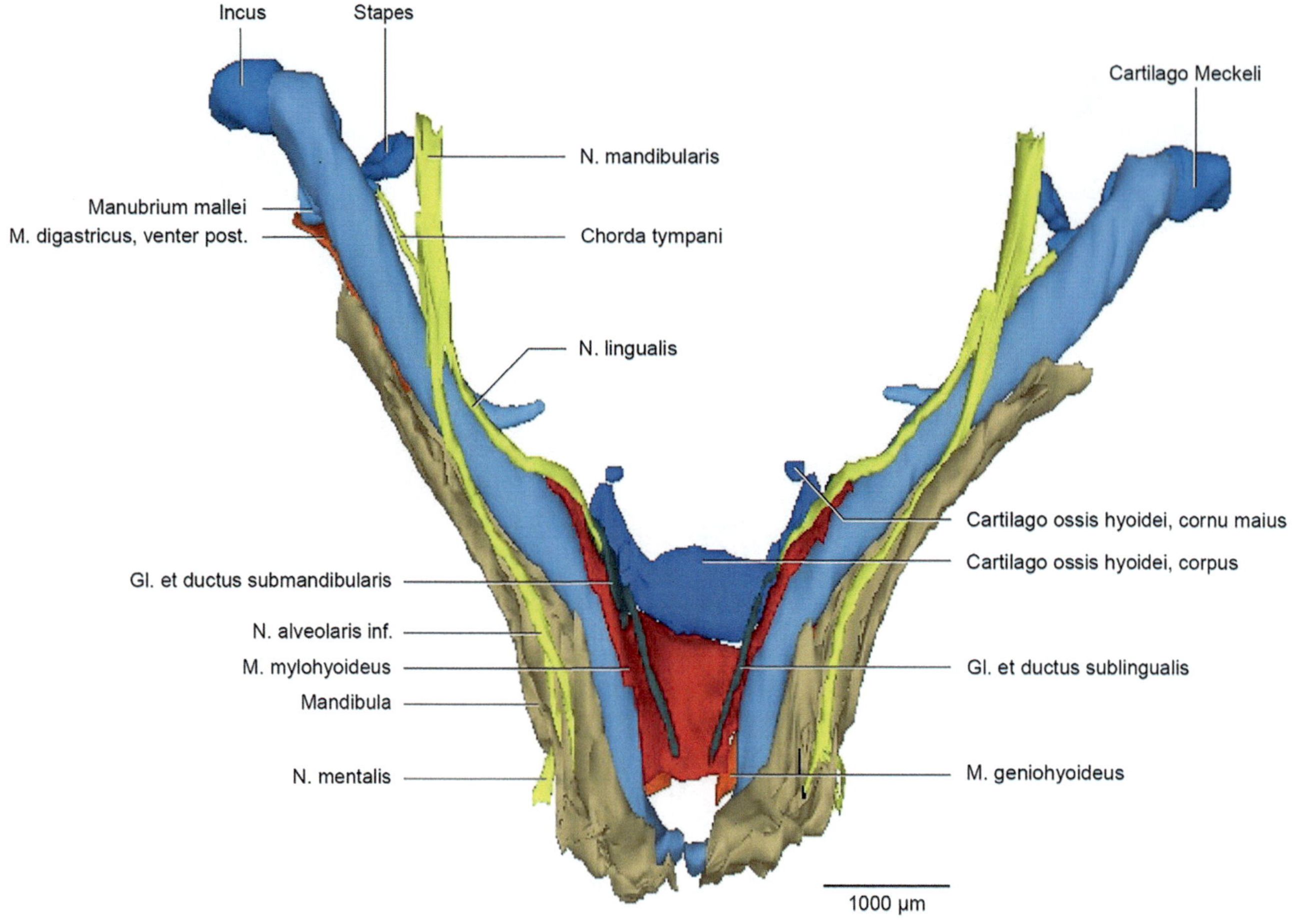

Fig. 19.27 Human embryo, 53 mm CRL, mandible, cranial view, layer 4 of preparation (*From* Radlanski and Renz [5]; *with permission.*)

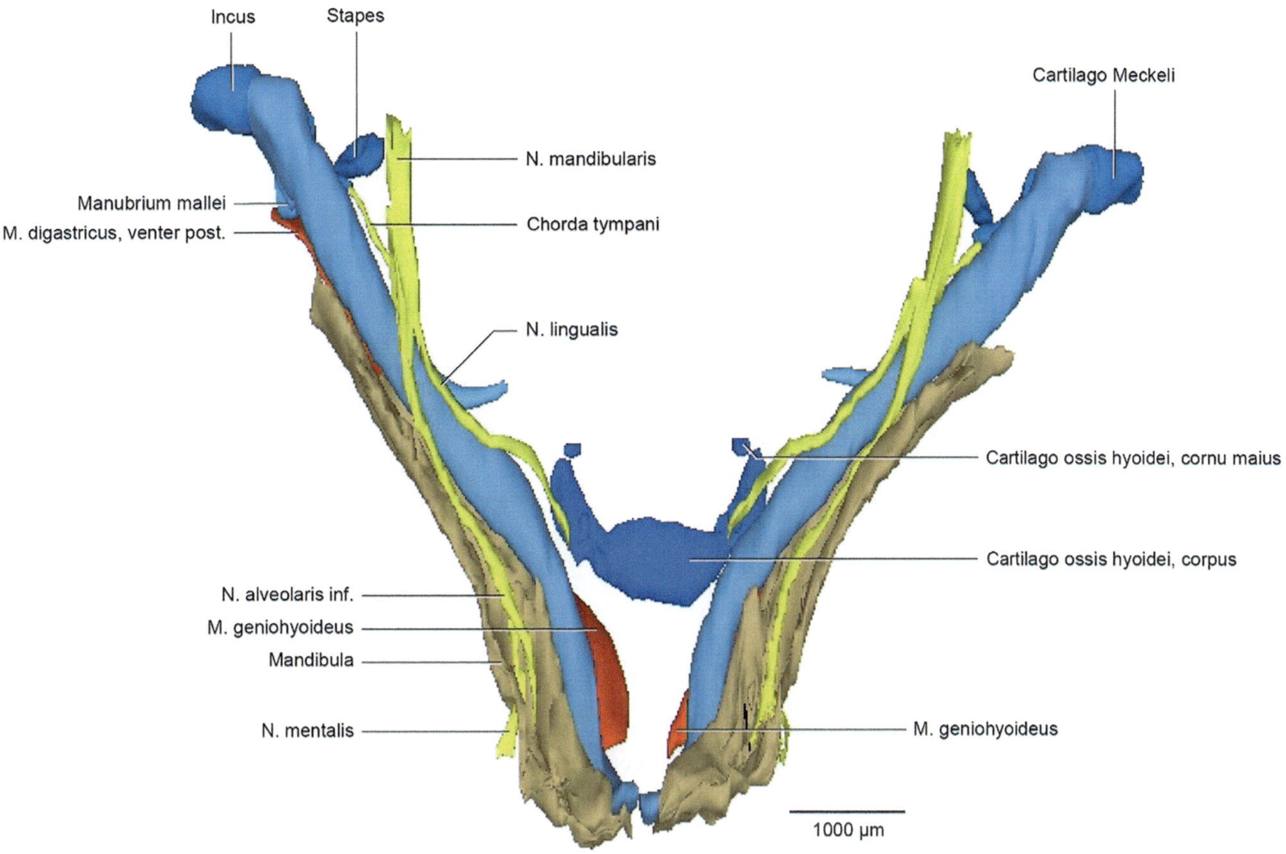

Fig. 19.28 Human embryo, 53 mm CRL, mandible, cranial view, layer 5 of preparation (*From* Radlanski and Renz [5]; *with permission.*)

Fig. 19.29 Human embryo, 53 mm CRL, mandible, cranial view, layer 6 of preparation (*From* Radlanski and Renz [5]; *with permission.*)

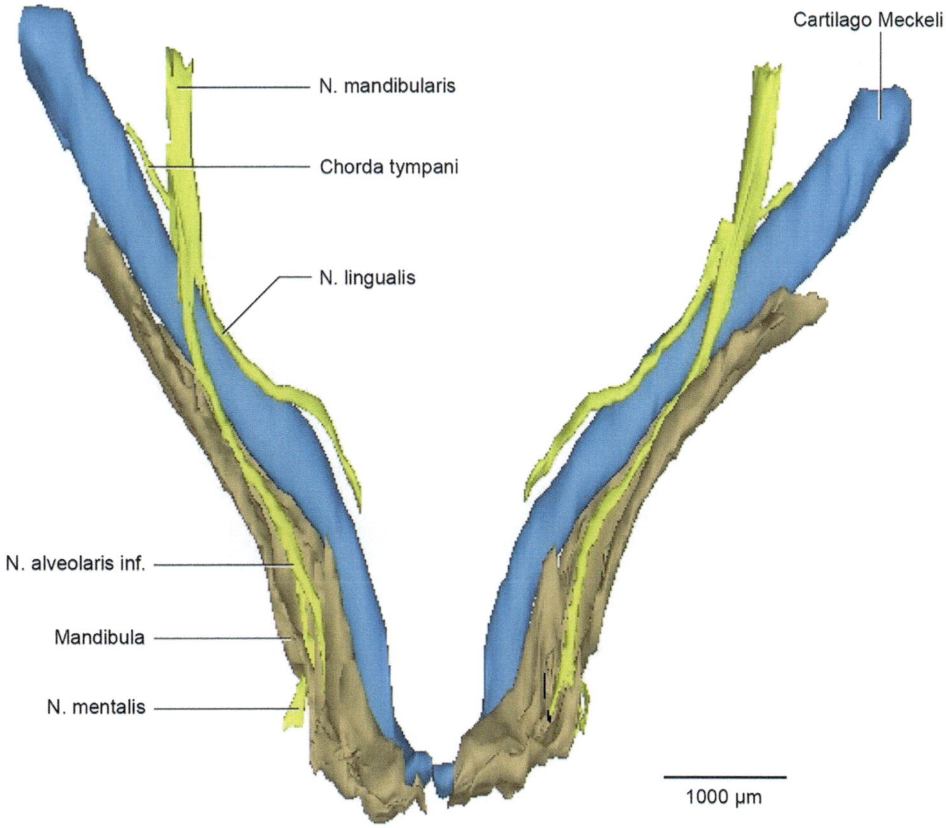

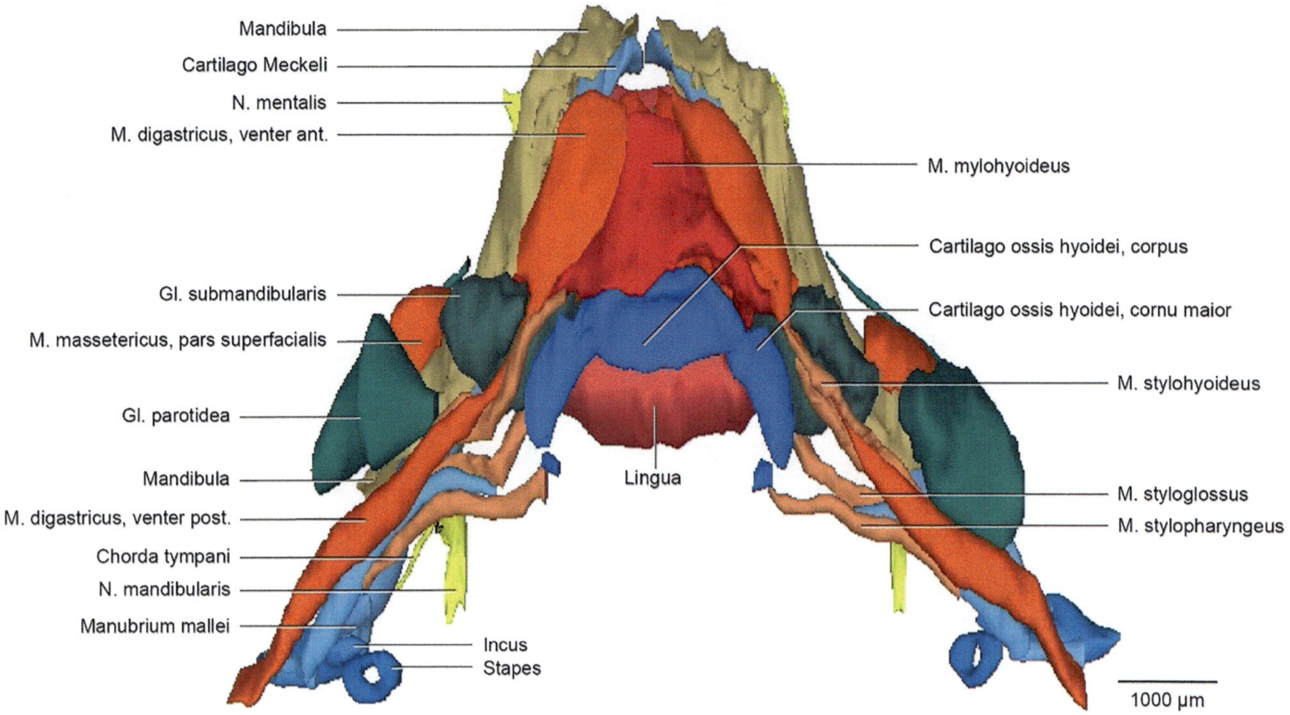

Fig. 19.30 Human embryo, 53 mm CRL, mandible, caudal view, layer 1 of preparation

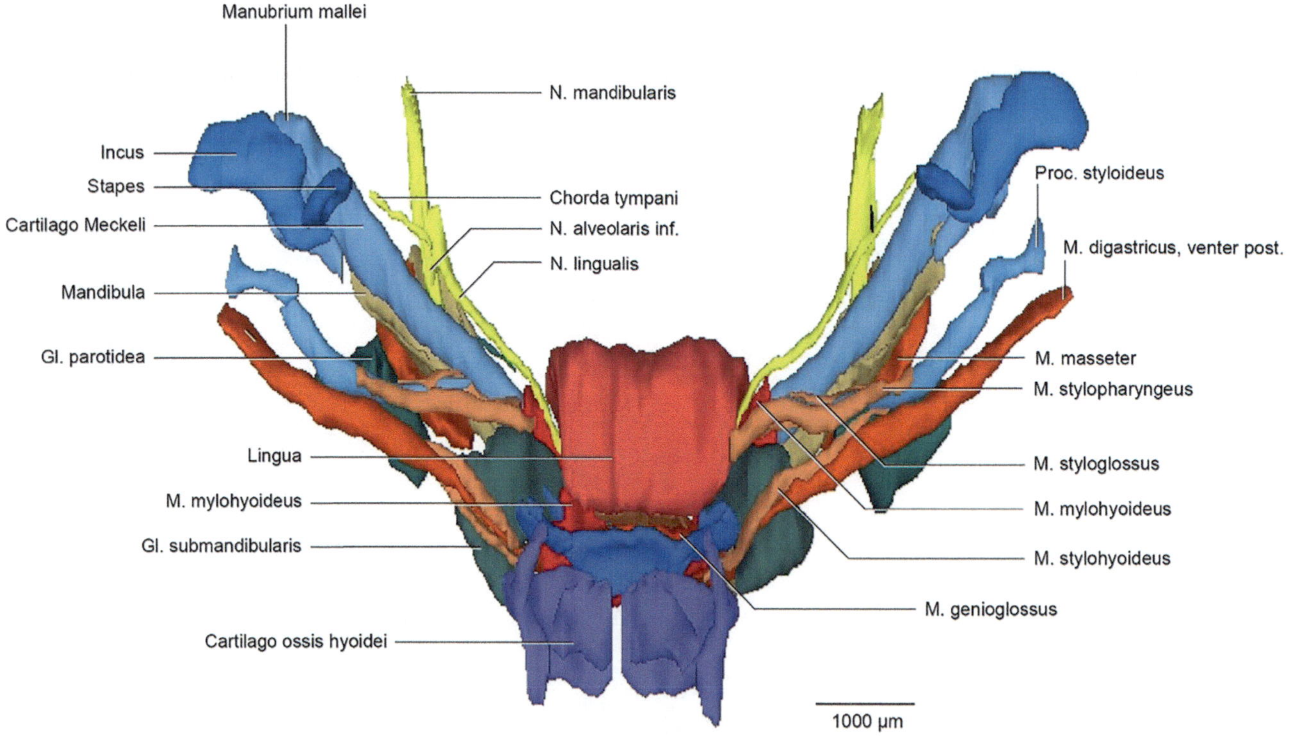

Fig. 19.31 Human embryo, 53 mm CRL, mandible, dorsal view, layer 1 of preparation

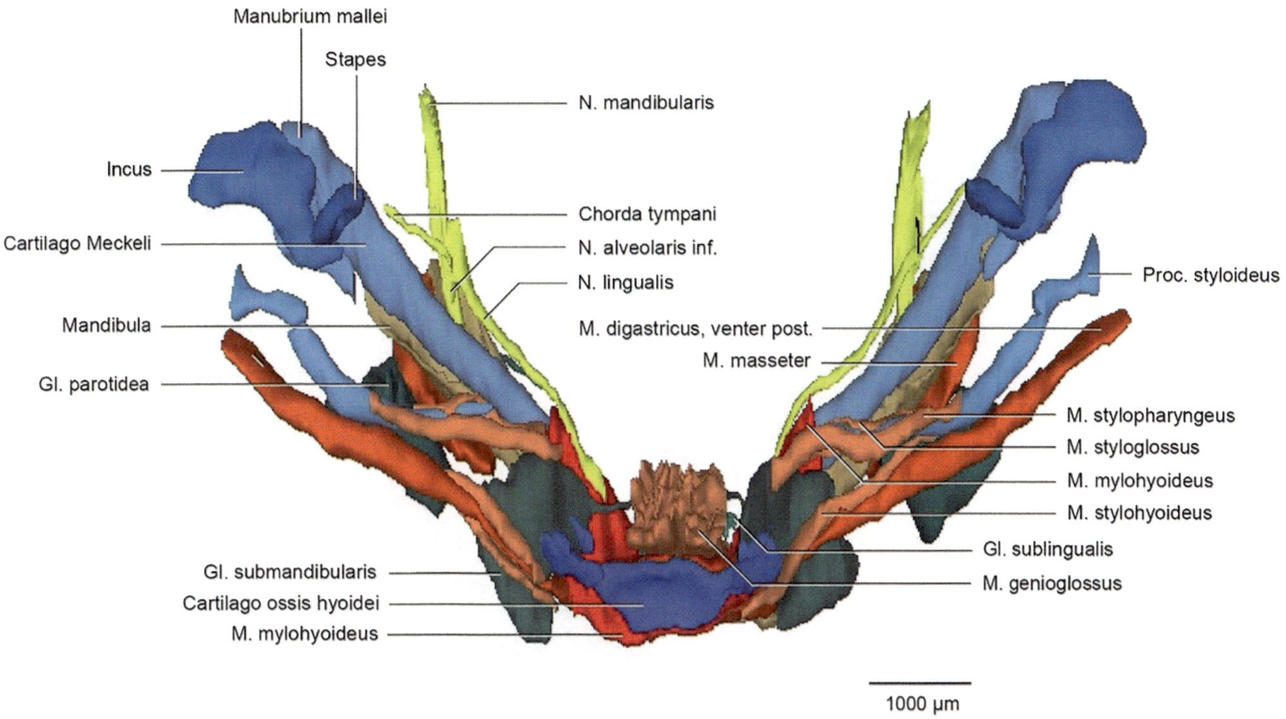

Fig. 19.32 Human embryo, 53 mm CRL, mandible, dorsal view, layer 2 of preparation

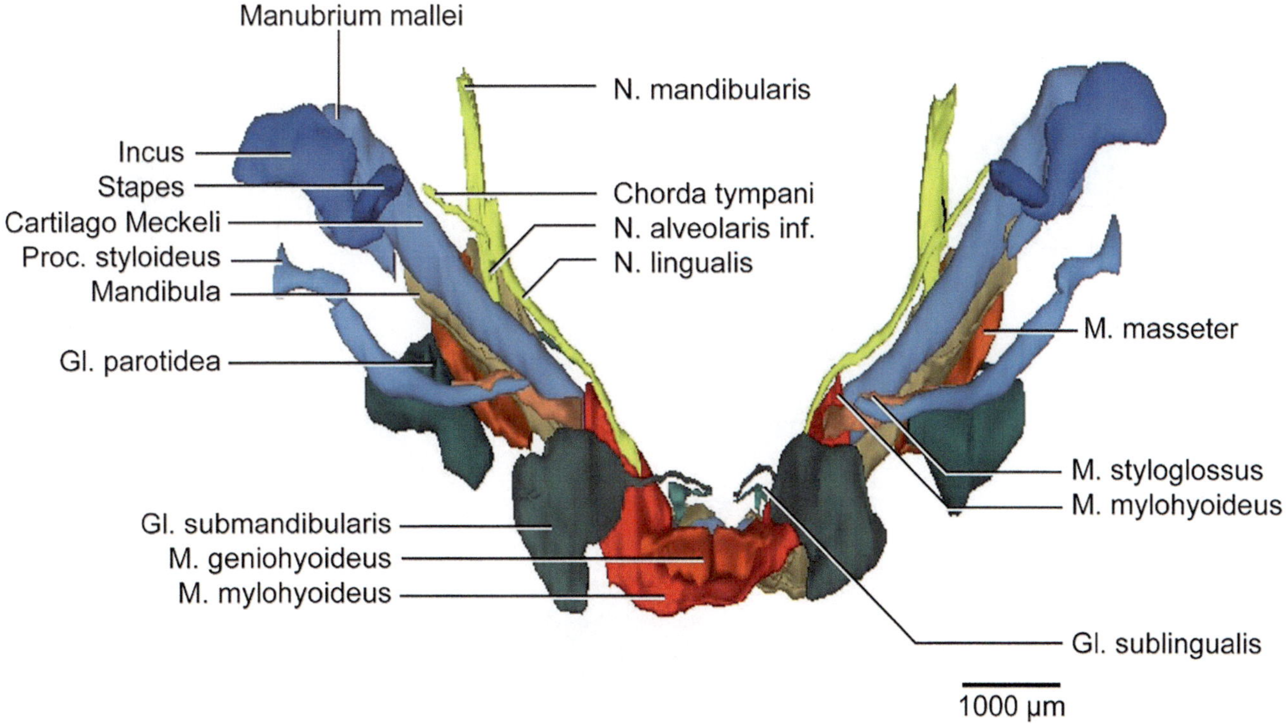

Fig. 19.33 Human embryo, 53 mm CRL, mandible, dorsal view, layer 3 of preparation

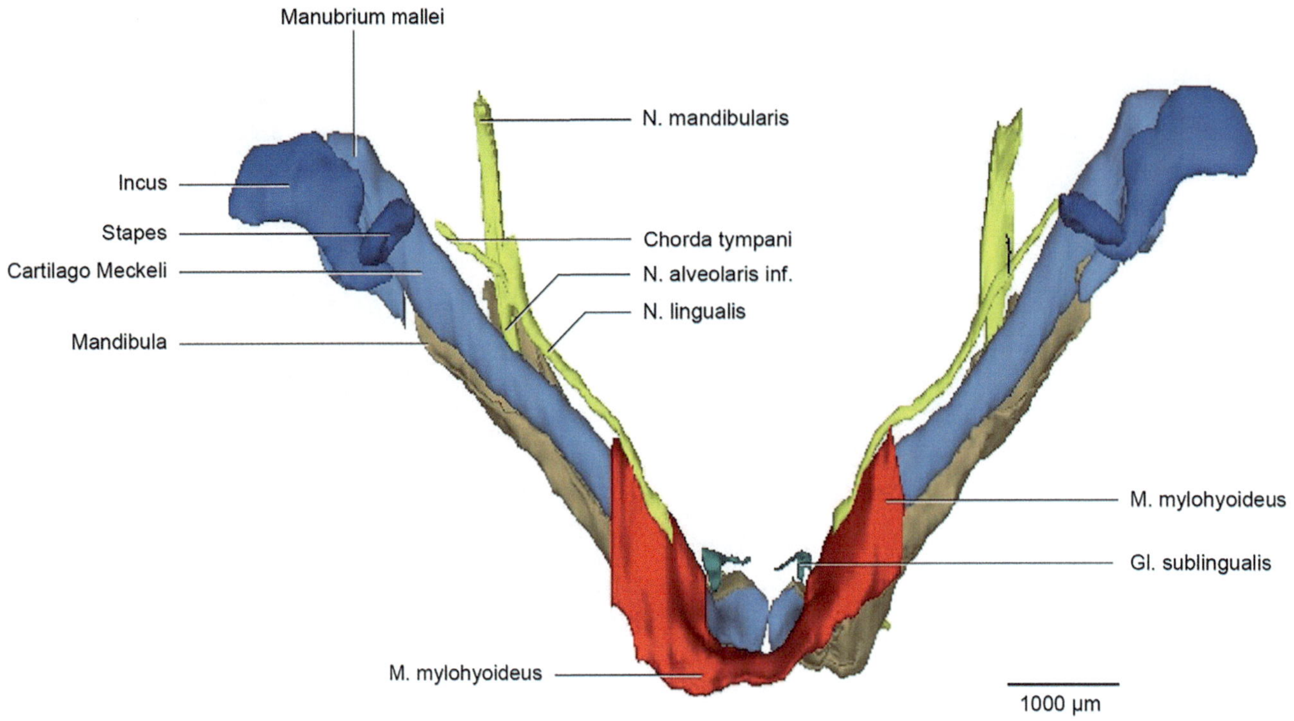

Fig. 19.34 Human embryo, 53 mm CRL, mandible, dorsal view, layer 4 of preparation

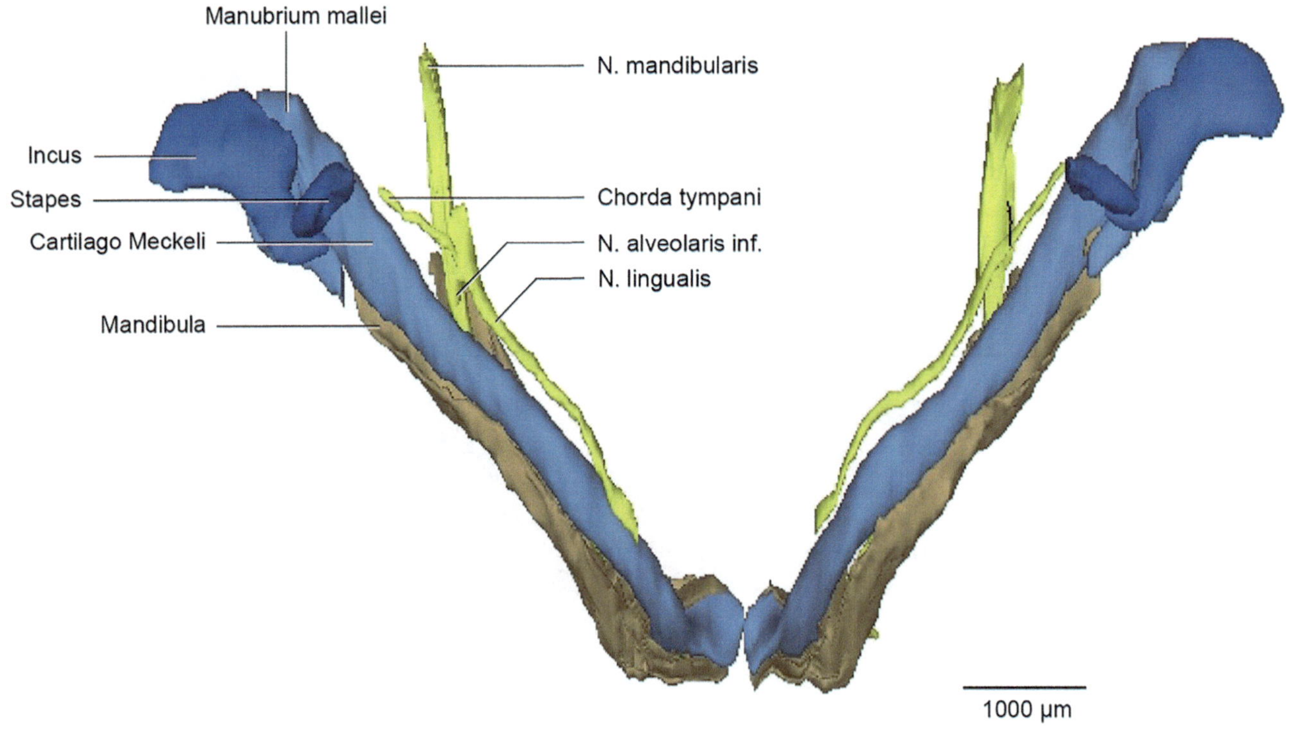

Fig. 19.35 Human embryo, 53 mm CRL, mandible, dorsal view, layer 5 of preparation

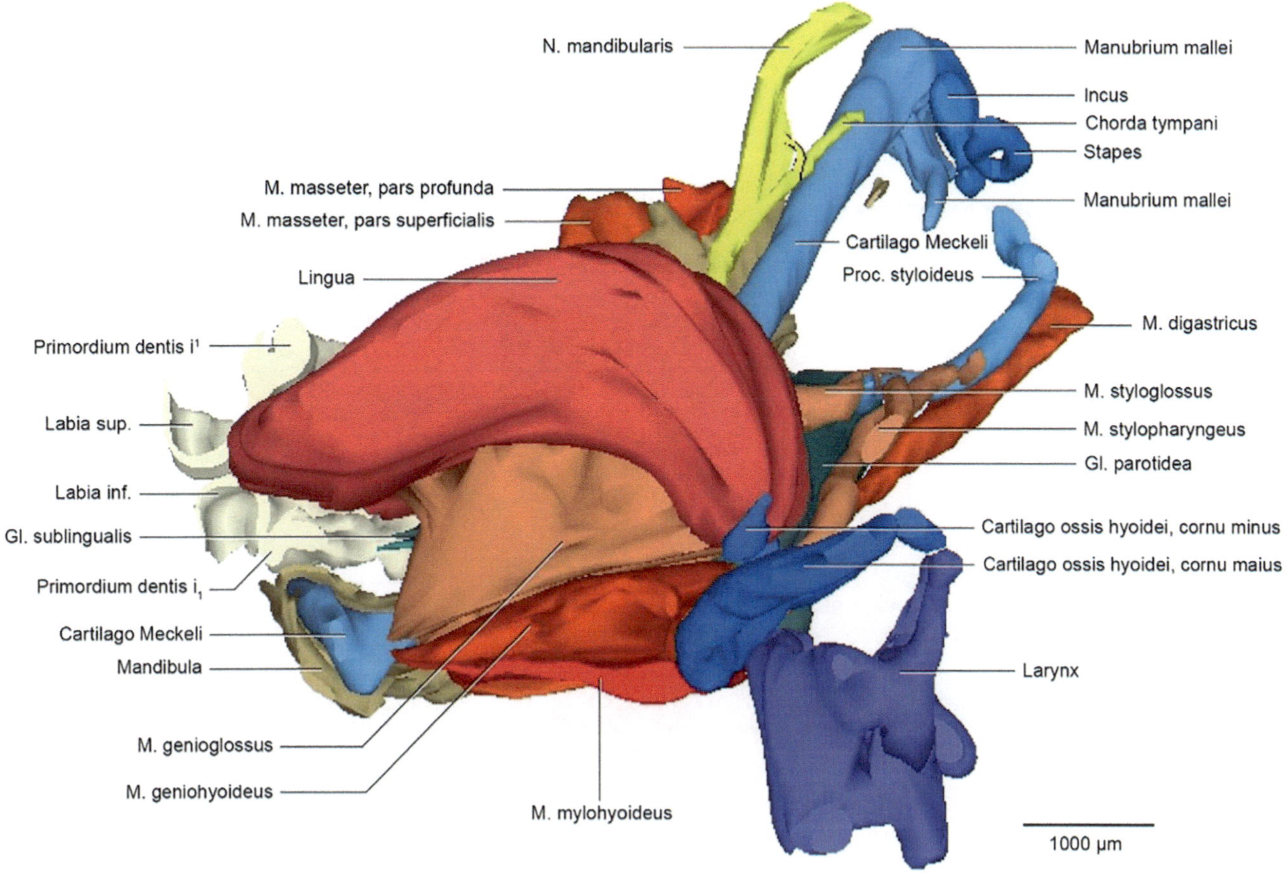

Fig. 19.36 Human embryo, 53 mm CRL, right half of mandible, medial view, layer 1 of preparation

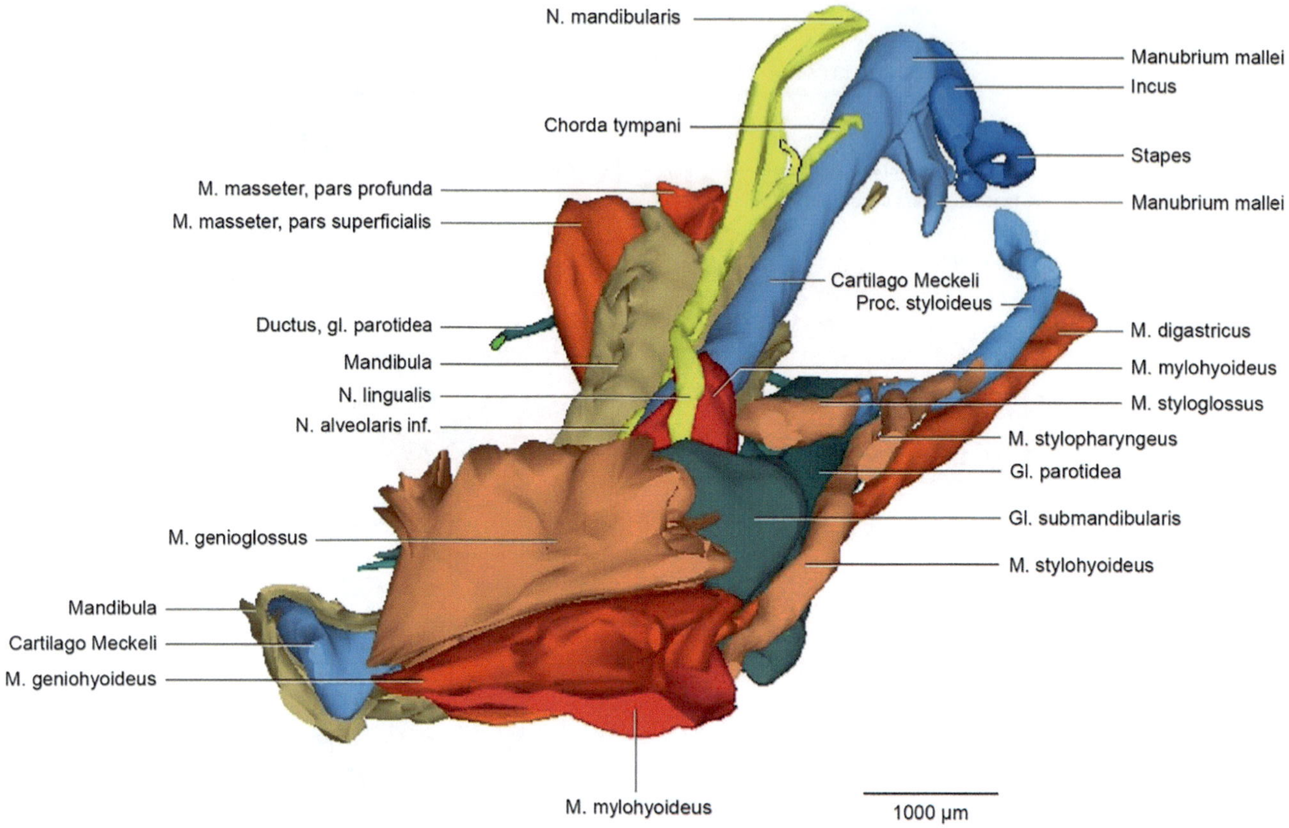

Fig. 19.37 Human embryo, 53 mm CRL, right half of mandible, medial view, layer 2 of preparation

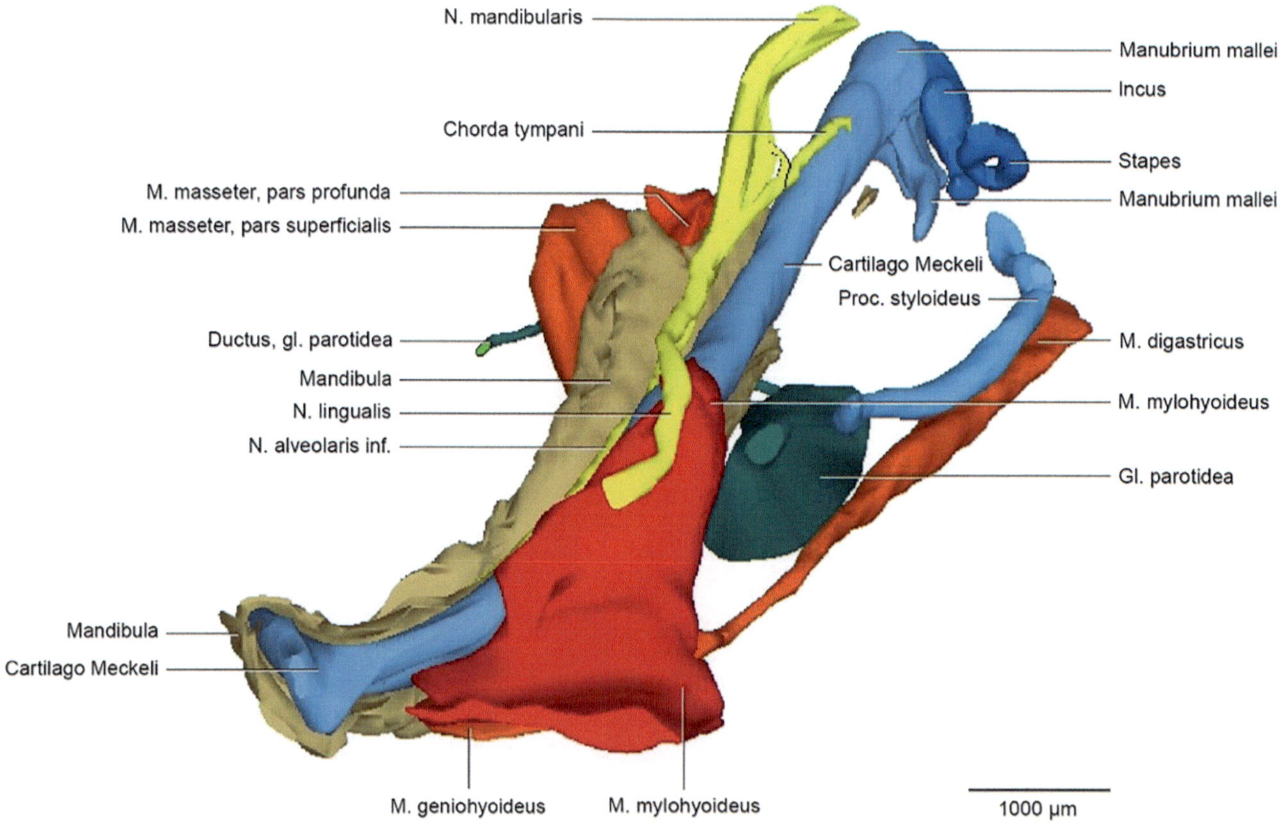

Fig. 19.38 Human embryo, 53 mm CRL, right half of mandible, medial view, layer 3 of preparation

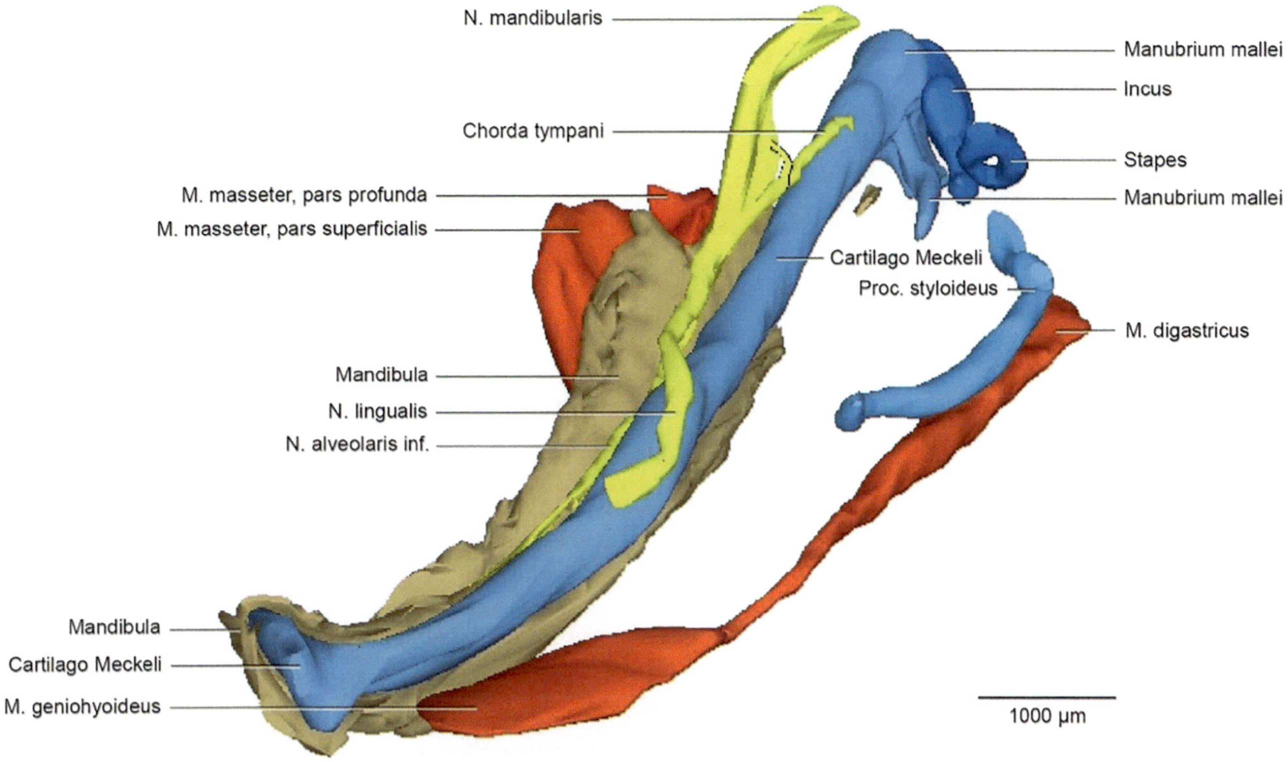

Fig. 19.39 Human embryo, 53 mm CRL, right half of mandible, medial view, layer 4 of preparation

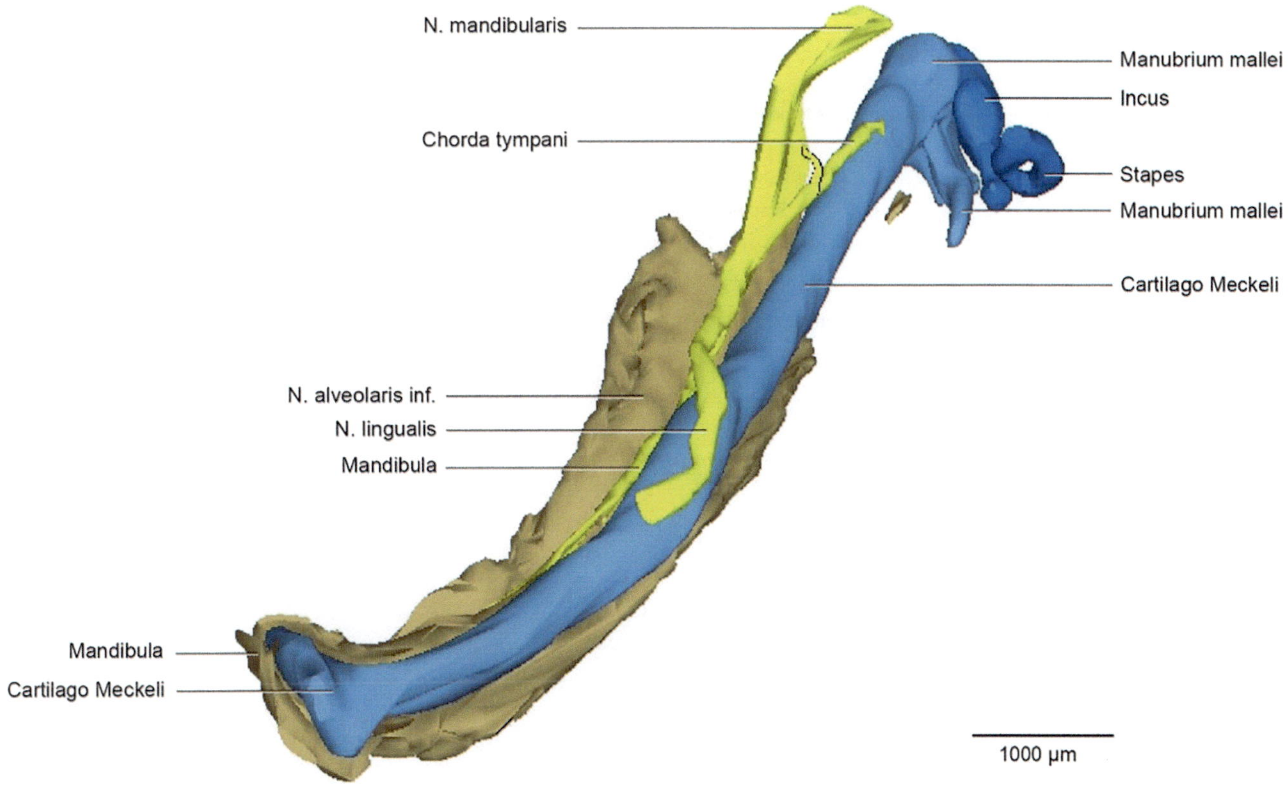

Fig. 19.40 Human embryo, 53 mm CRL, right half of mandible, medial view, layer 5 of preparation

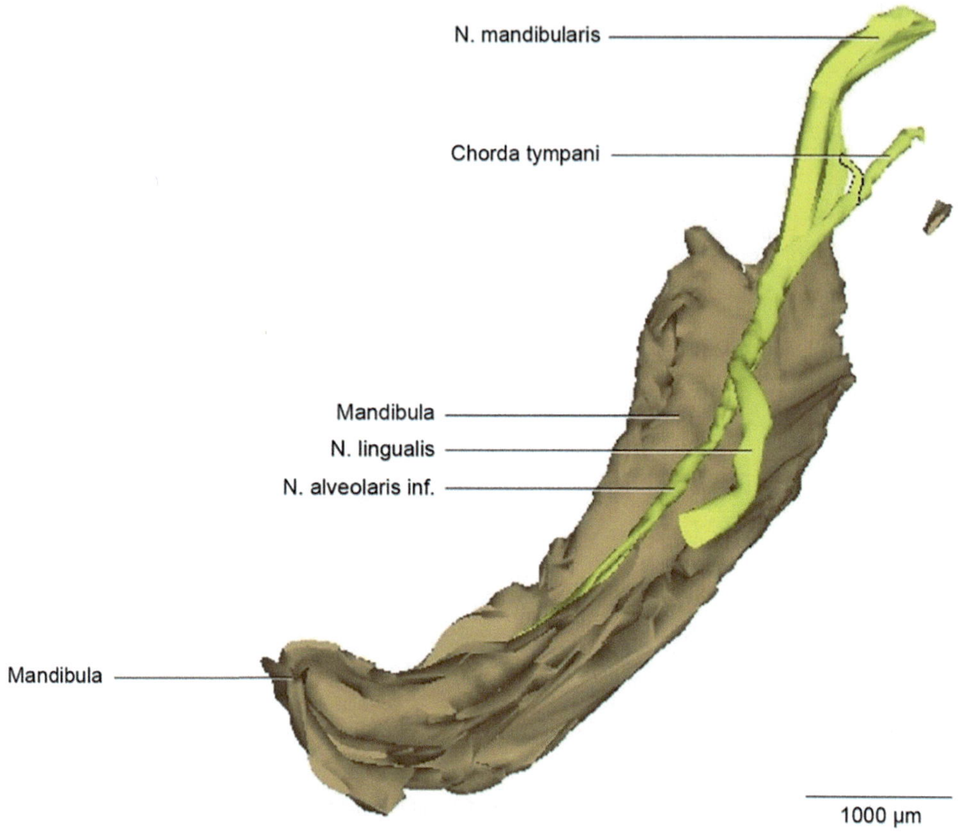

Fig. 19.41 Human embryo, 53 mm CRL, right half of mandible, medial view, layer 6 of preparation

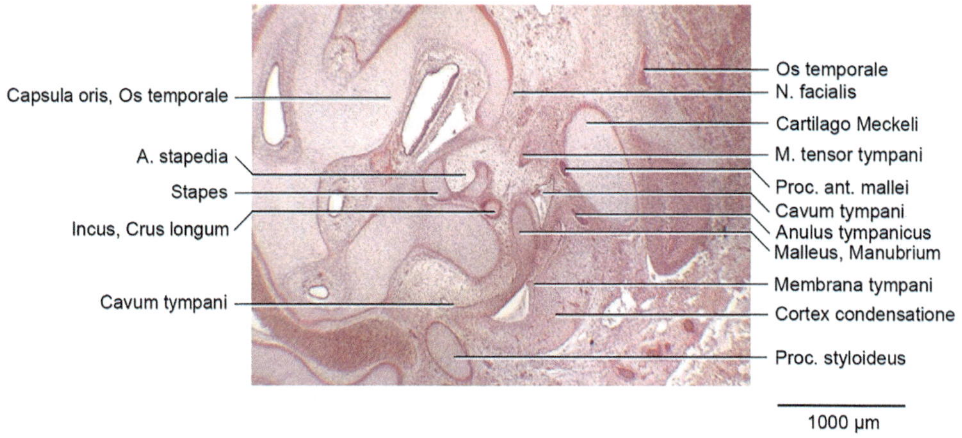

Fig. 19.42 Human embryo, 53 mm CRL, middle ear region, sagittal section. Staining: Hematoxylin-eosin

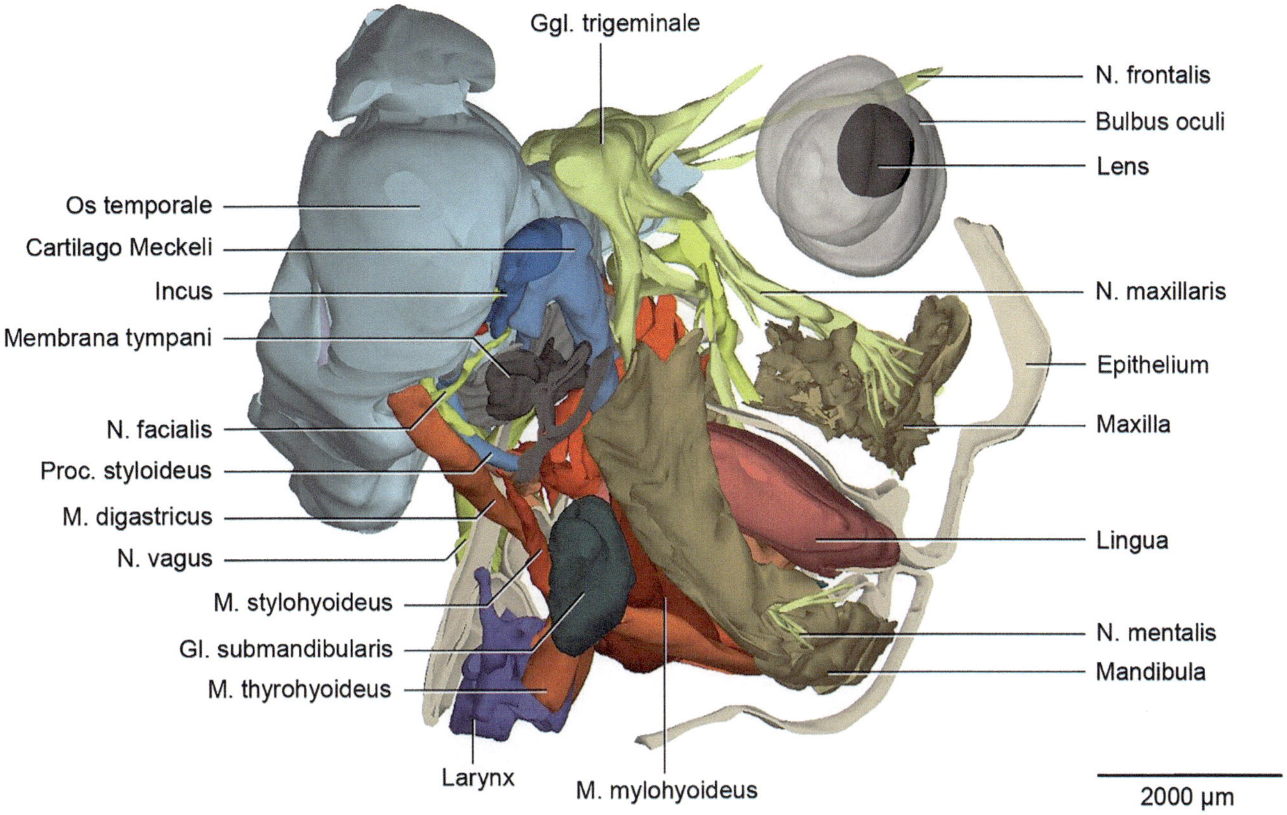

Fig. 19.43 Human embryo, 53 mm CRL, head with middle ear region, lateral view

Fig. 19.44 Human embryo, 53 mm CRL, detailed lateral view of the right auditory system. Meckel's cartilage has been cut off. Temporal bone not included in reconstruction

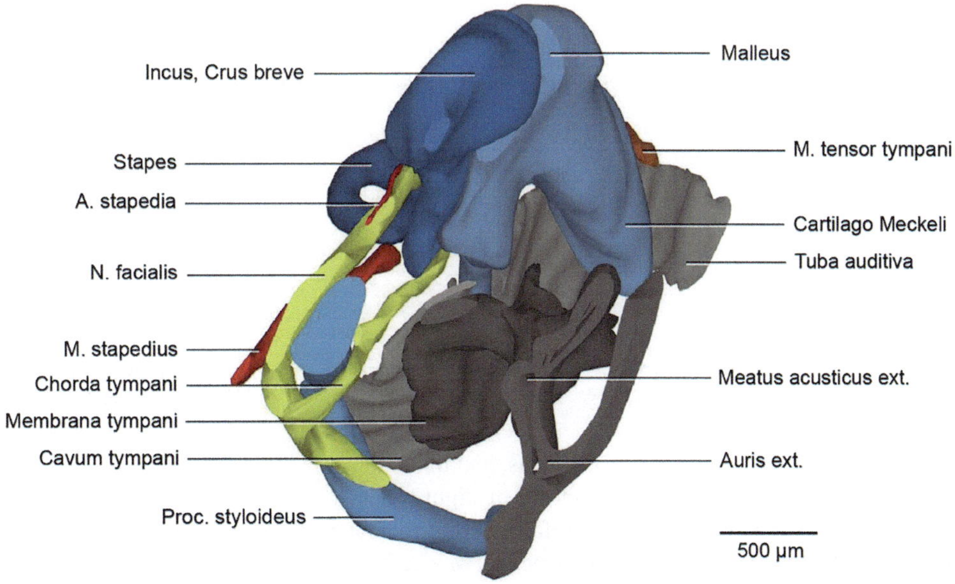

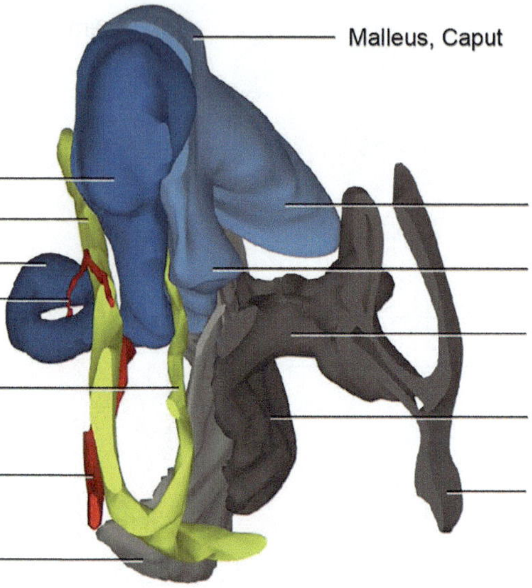

Fig. 19.45 Human embryo, 53 mm CRL, right auditory system, lateral and 45° dorsal view, styloid process removed

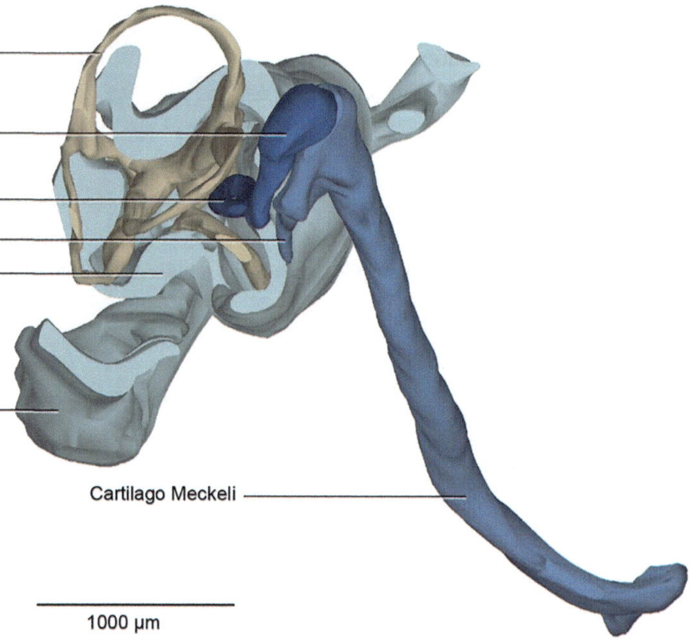

Fig. 19.46 Human embryo, 53 mm CRL, right auditory system, lateral view, with indication of vestibular part of the inner ear (semicircular canals)

Middle Ear

At this stage of development, stapes, incus, and the malleus are in good contact to each other by means of slight concavities and convexities of the articular surfaces. All of them are still embedded well in loose mesenchyme [4]. Meckel's cartilage appears straight, except for a slight curvature at its anterior third. The semicircular canals reveal their almost perfect orientation in all three planes of space.

References

1. Giambartolomei LA, Brunotto M, Gomez de Ferrais ME. Human temporomandibular joint disc: anatomy and measurements in prenatal development. Acta odontol. Latino-Americana. 2011;24:98–103.
2. Fernández Rubio EM, Radlanski RJ. Development of the human primary and secondary jaw joint. Ann Anat. 2024;251:1–8.
3. Sperber GH. Craniofacial embryology. Hamilton: BC Decker; 2001.
4. Beckert K. Beiträge zur pränatalen Morphogenese der Region des Mittelohres des Menschen—Computergestützte 3D-Rekonstruktionen und histologische Untersuchungen menschlicher Embryonen/Feten von 19 bis 150 mm CRL. (7. bis 18. Woche). Med. ed. Diss. Charité-Universitätsmedizin Berlin; 2010.
5. Radlanski RJ, Renz H. An Atlas of the human prenatal craniofacial development. Eur J Oral Sci. 2010;118:321–4.
6. Radlanski RJ. Oral structure and biology. Batavia, USA: Quintessence Publishing; 2018.

Embryo 54 mm CRL (ca. 9th wk): THE230494

20

Overall Remarks

This embryo is also very well preserved and has lent itself very well to representation in the form of three-dimensional reconstructions. We were therefore able to show the spatial positional relationships of many structures in great detail. First of all, this concerns the sublingual and prelanryngeal musculature, the cartilaginous structures, and the developmental state of the mandible. The development of the dental primordia and their interaction with the peridental bone structures in the maxilla and mandible could also be given good attention in detail at this stage. Finally, we were able to take a closer look at the development of the ear region (Figs. 20.1, 20.2, 20.3, 20.4, 20.5, 20.6, 20.7, 20.8, 20.9, 20.10, 20.11, 20.12, 20.13, 20.14, 20.15, 20.16, 20.17, 20.18, 20.19, 20.20, 20.21, 20.22, 20.23, 20.24, 20.25, 20.26, 20.27, 20.28, 20.29, 20.30, 20.31, 20.32, 20.33, 20.34, 20.35, 20.36, 20.37, 20.38, and 20.39).

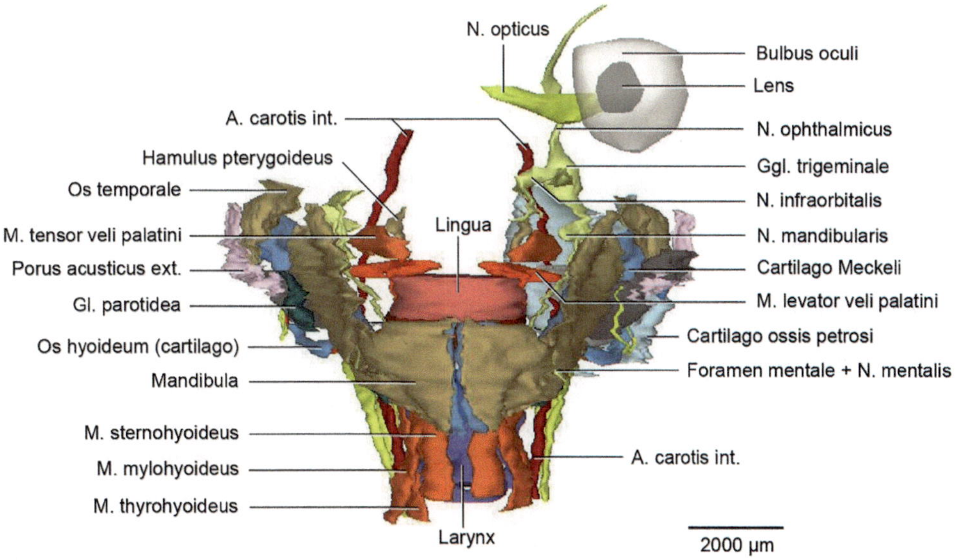

Fig. 20.1 Human embryo, 54 mm CRL, head, frontal view, skin removed, layer 1 of preparation

R. J. Radlanski, *Atlas of Human Prenatal Craniofacial Development*, https://doi.org/10.1007/978-3-031-88812-0_20

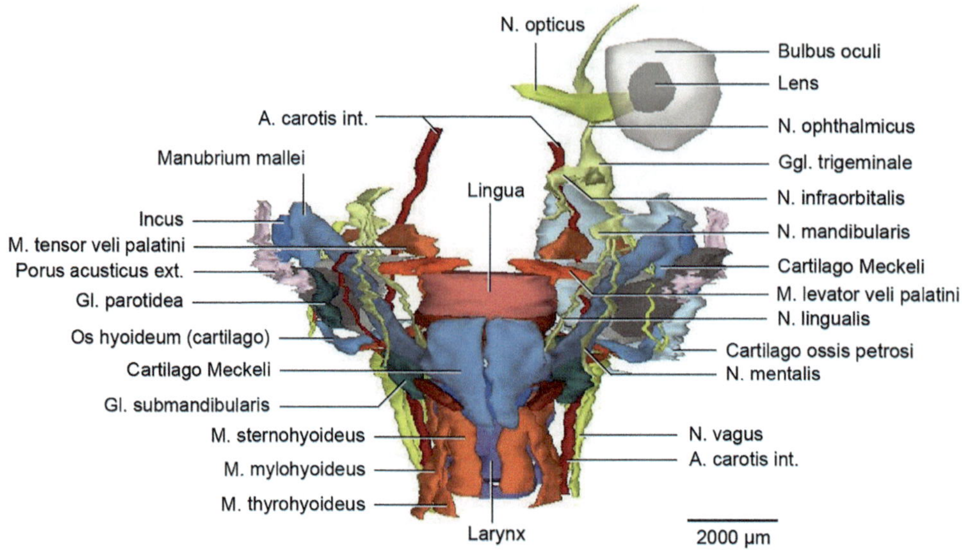

Fig. 20.2 Human embryo, 54 mm CRL, head, frontal view, layer 2 of preparation

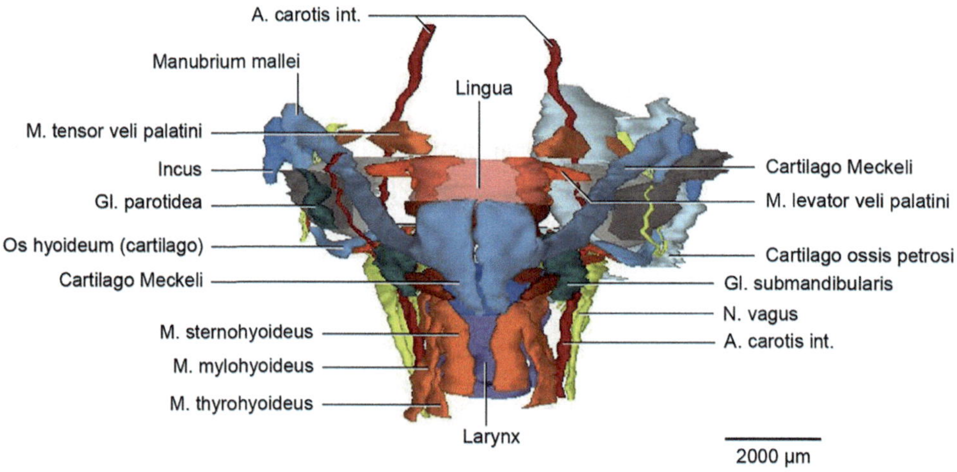

Fig. 20.3 Human embryo, 54 mm CRL, head, frontal view, layer 3 of preparation

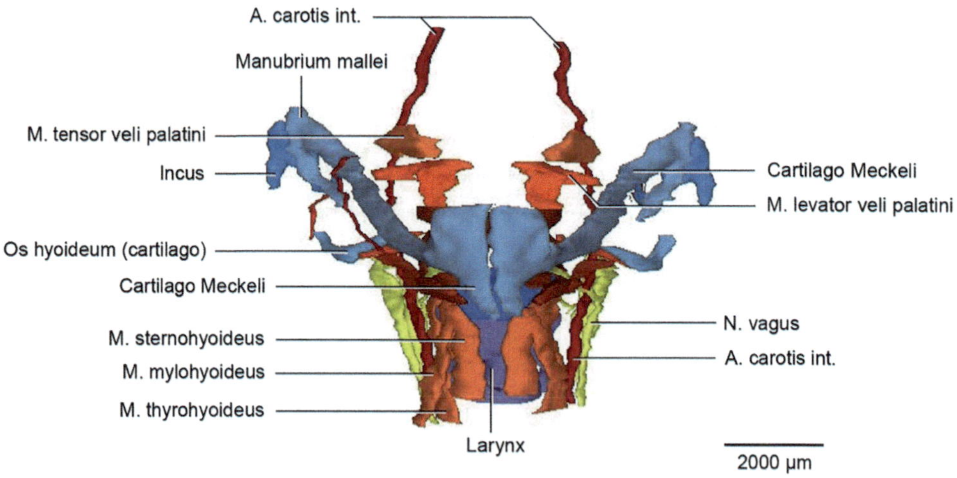

Fig. 20.4 Human embryo, 54 mm CRL, head, frontal view, cartilages, nerves, and muscles

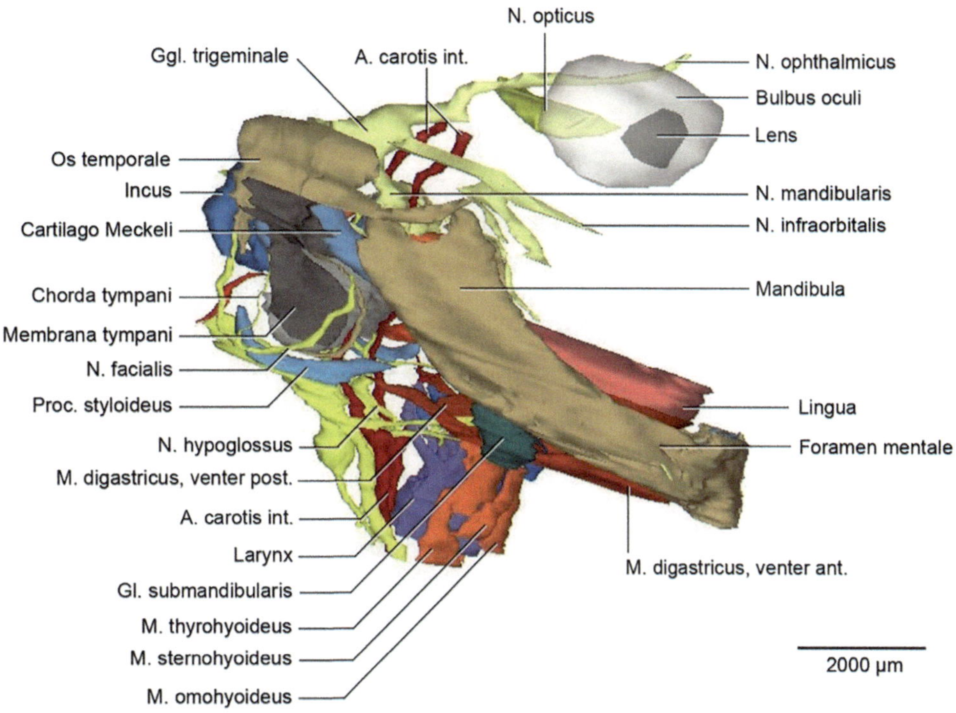

Fig. 20.5 Human embryo, 54 mm CRL, head, lateral view, skin removed, layer 1 of preparation

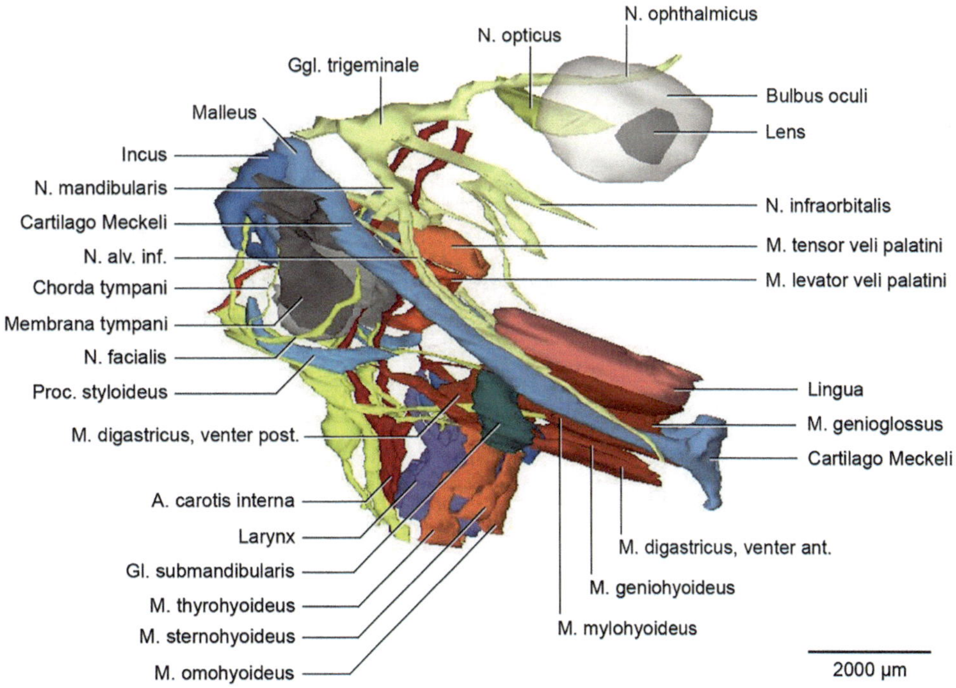

Fig. 20.6 Human embryo, 54 mm CRL, head, lateral view, layer 2 of preparation

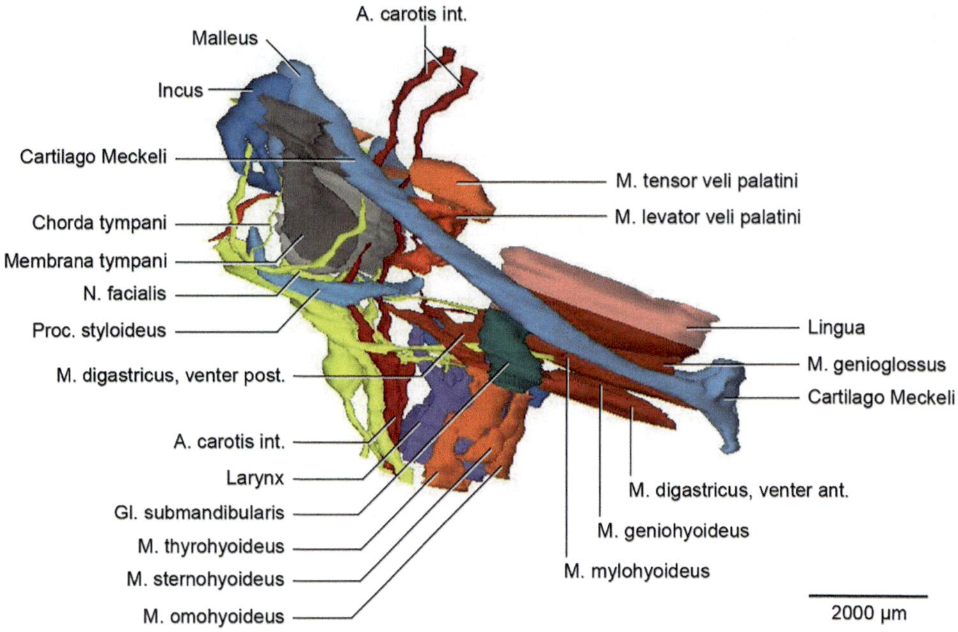

Fig. 20.7 Human embryo, 54 mm CRL, head, lateral view, layer 3 of preparation

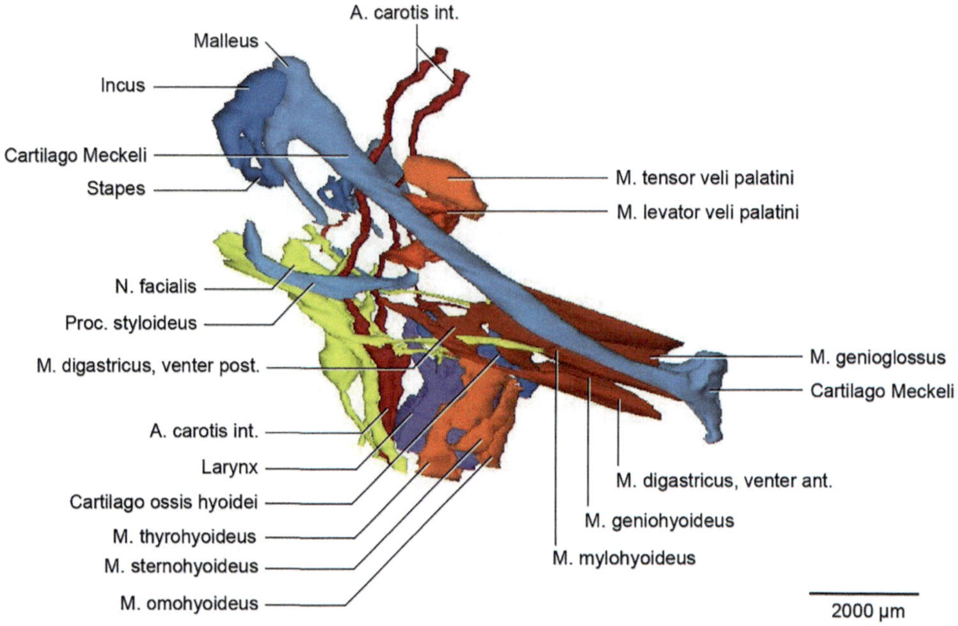

Fig. 20.8 Human embryo, 54 mm CRL, head, lateral view, layer 4 of preparation

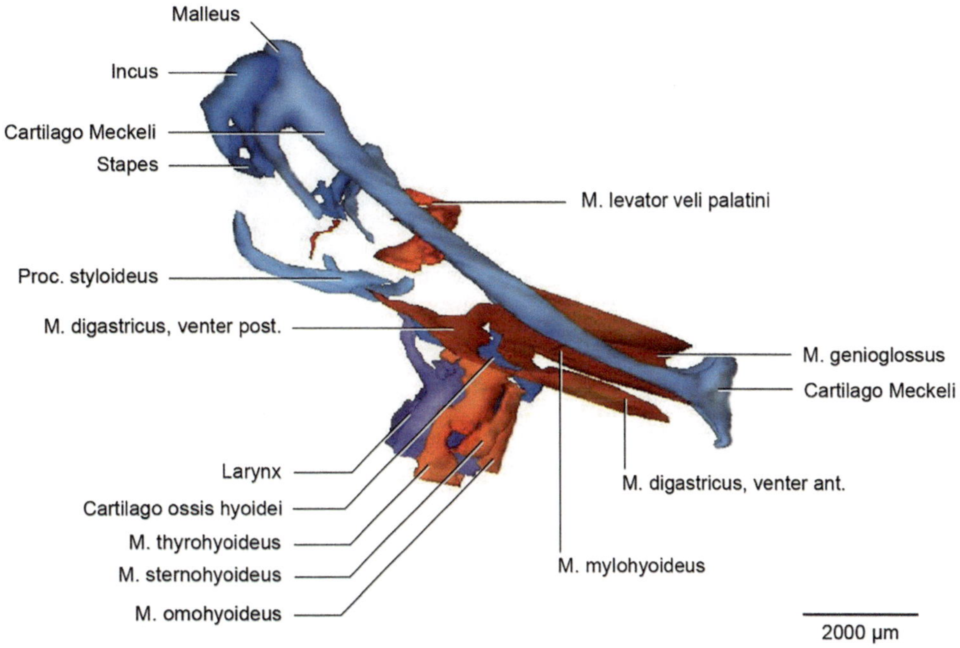

Fig. 20.9 Human embryo, 54 mm CRL, head, lateral view, layer 5 of preparation

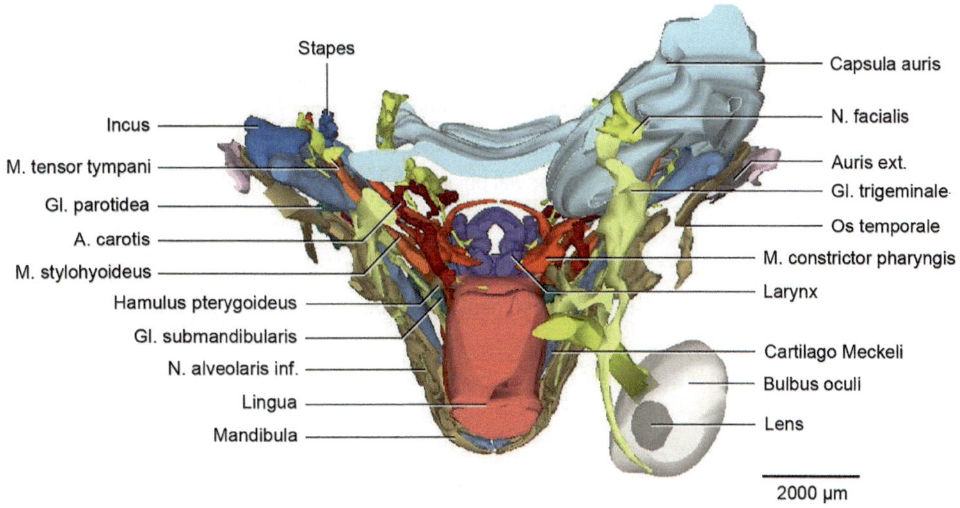

Fig. 20.10 Human embryo, 54 mm CRL, head, cranial view, layer 1 of preparation

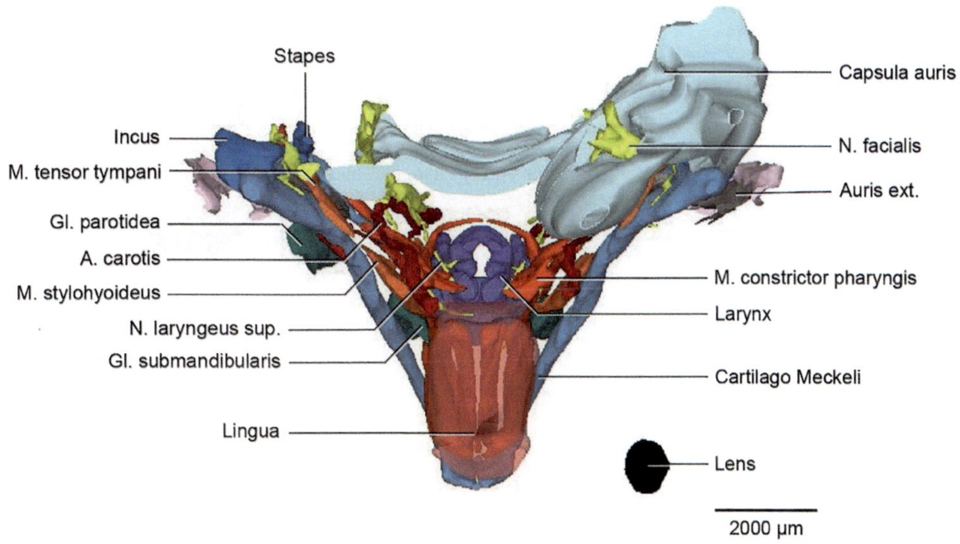

Fig. 20.11 Human embryo, 54 mm CRL, head, cranial view, layer 2 of preparation

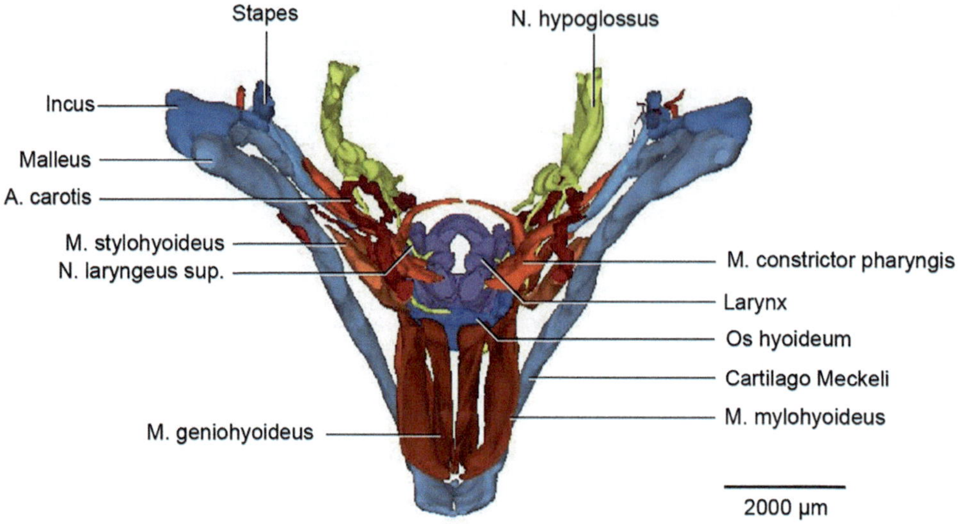

Fig. 20.12 Human embryo, 54 mm CRL, head, cranial view, layer 3 of preparation

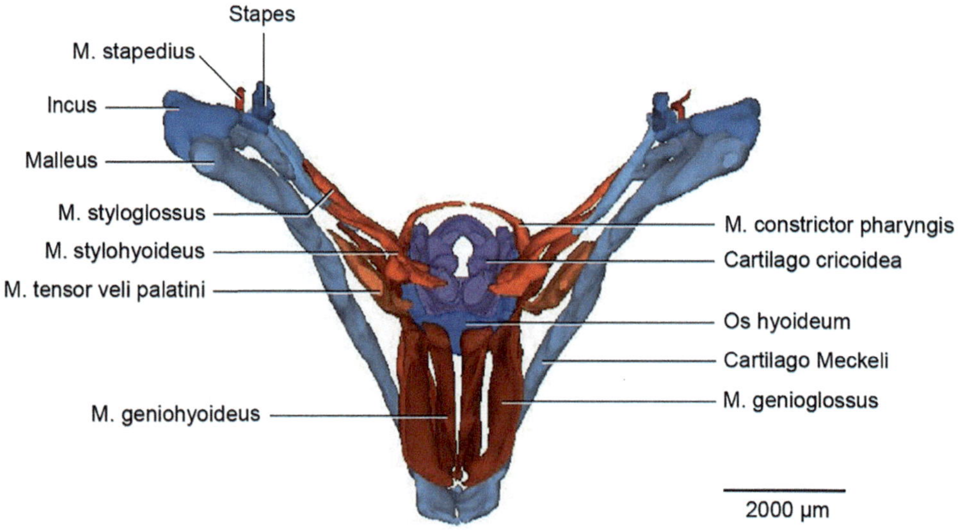

Fig. 20.13 Human embryo, 54 mm CRL, mandible, head, cranial view, layer 4 of preparation

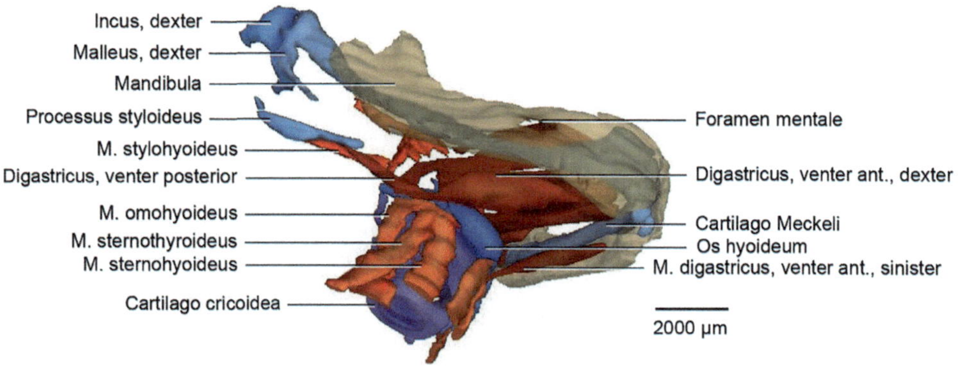

Fig. 20.14 Human embryo, 54 mm CRL, floor of the mouth, lateral view

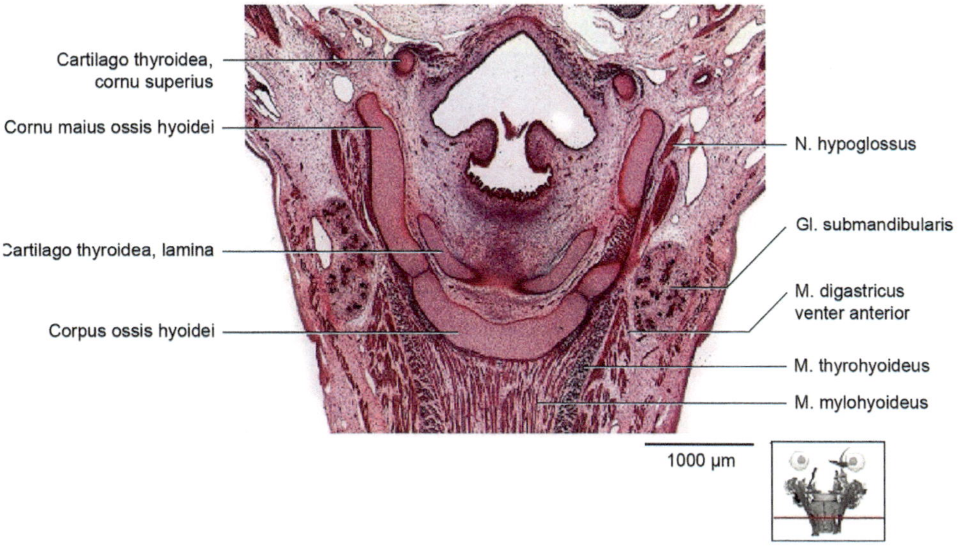

Fig. 20.15 Human embryo, 54 mm CRL, floor of the mouth, histological section (staining: Hematoxylin-eosin) at level of primordium of hyoid bone

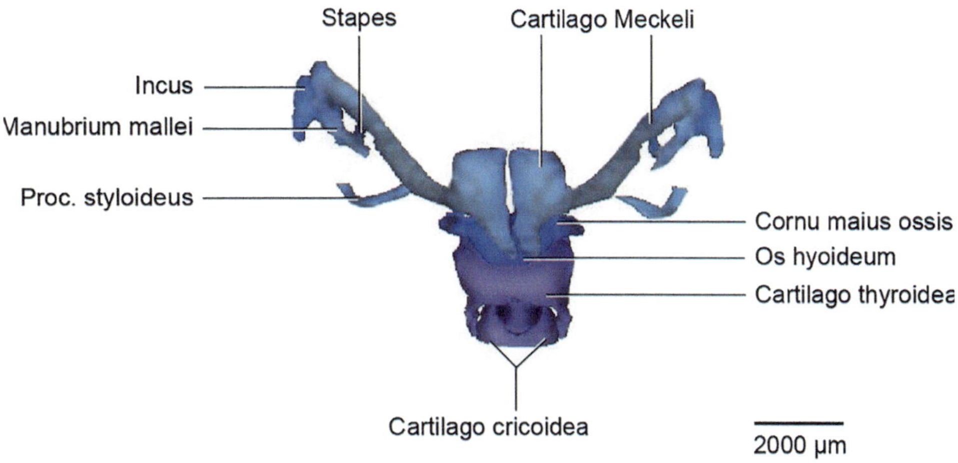

Fig. 20.16 Human embryo, 54 mm CRL, cartilaginous skeleton of 1st, 2nd, 3rd, and 4th visceral arches, anterior view

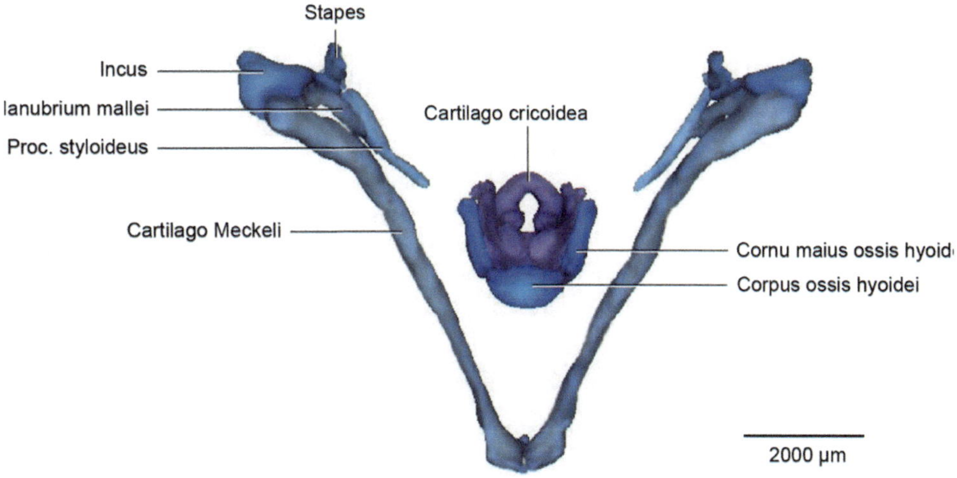

Fig. 20.17 Human embryo, 54 mm CRL, cartilaginous skeleton of 1st, 2nd, 3rd, and 4th visceral arches, cranial view

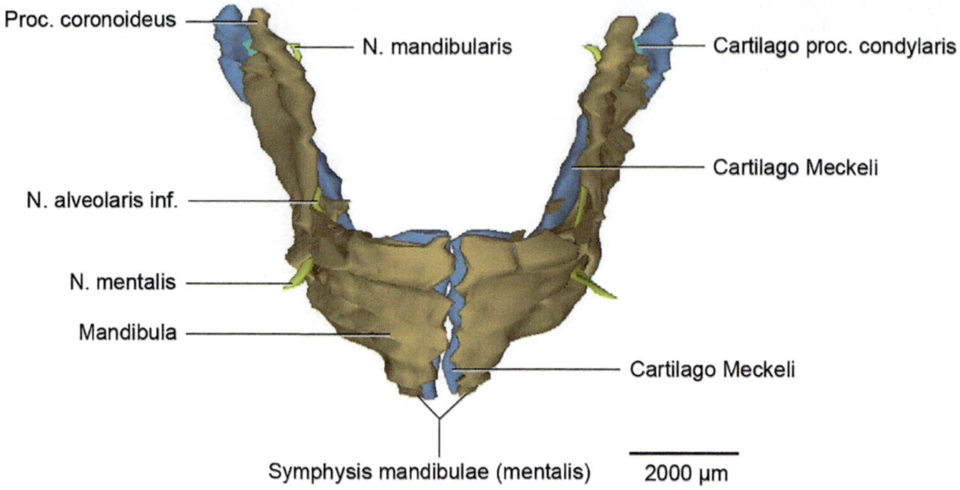

Fig. 20.18 Human embryo, 54 mm CRL, mandible, anterior view

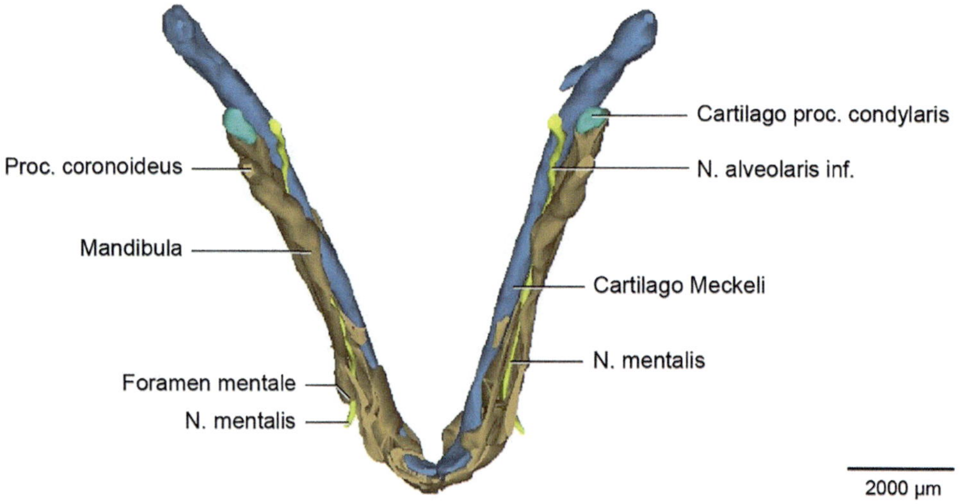

Fig. 20.19 Human embryo, 54 mm CRL, mandible, cranial view

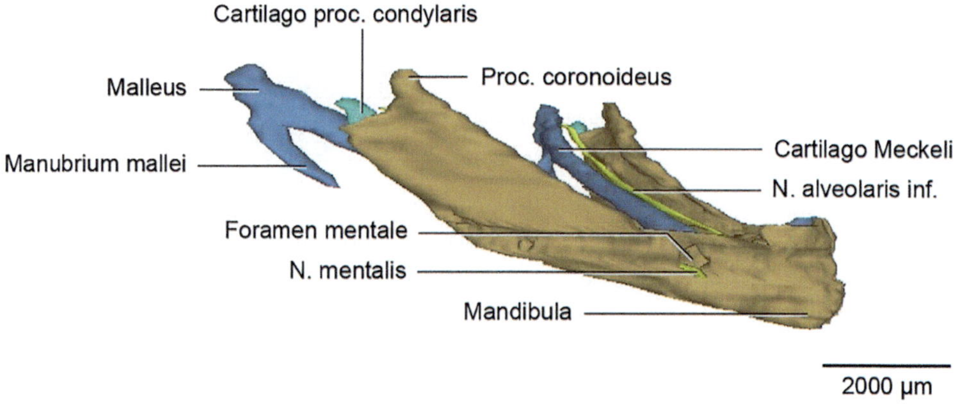

Fig. 20.20 Human embryo, 54 mm CRL, mandible, lateral view

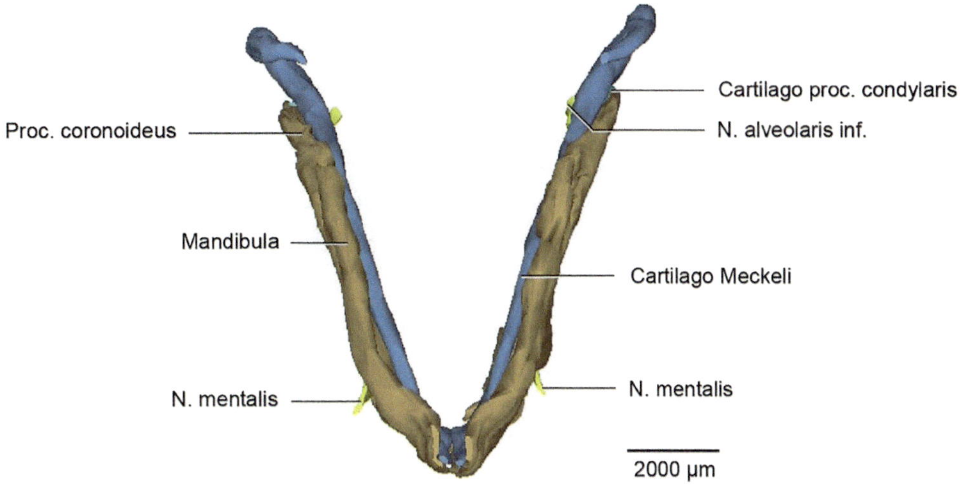

Fig. 20.21 Human embryo, 54 mm CRL, mandible, caudal view

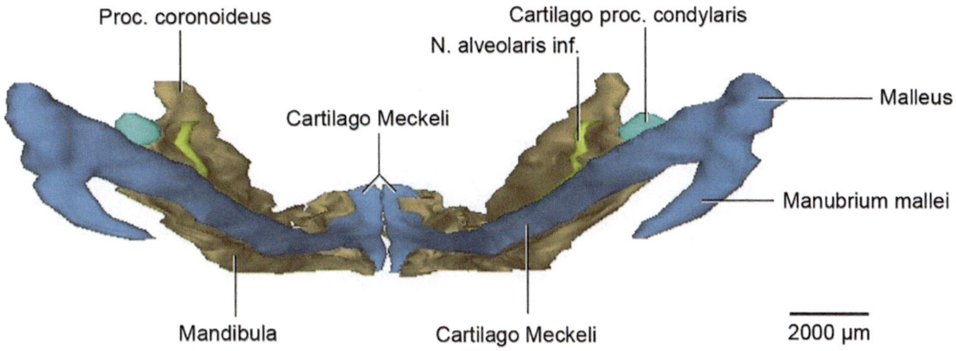

Fig. 20.22 Human embryo, 54 mm CRL, mandible, dorsal view

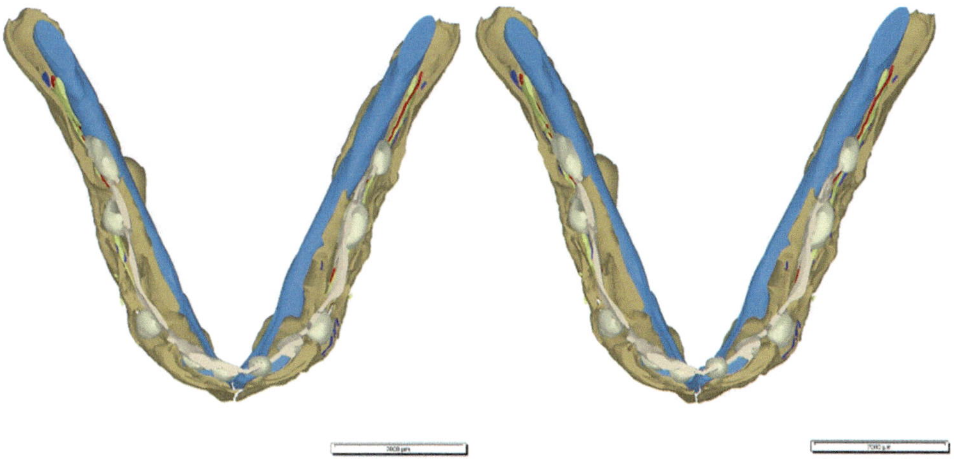

Fig. 20.23 Stereoscopical pair of images. Human embryo, 54 mm CRL, mandible (ochre) with Meckel's cartilage (blue), inferior alveolar n. (yellow), inferior alveolar a. (red), inferior alveolar v. (dark blue), and dental primordia (light gray), cranial view. Bar: 2000µm

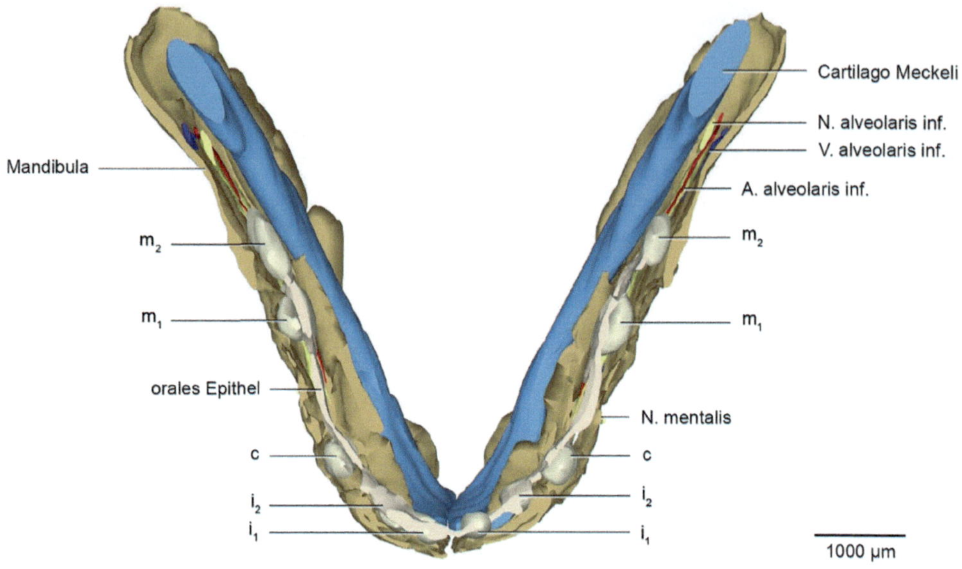

Fig. 20.24 Human embryo, 54 mm CRL, mandible with Meckel's cartilage and dental primordia, cranial view

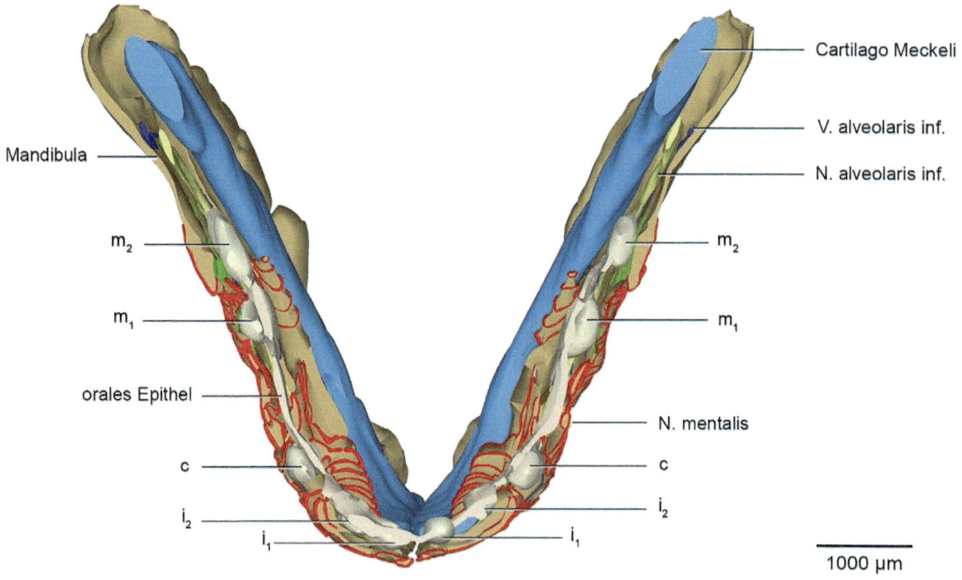

Fig. 20.25 Human embryo, 54 mm CRL, mandible with Meckel's cartilage and dental primordia, cranial view, with bone remodeling pattern superimposed (red: apposition, green: resorption)

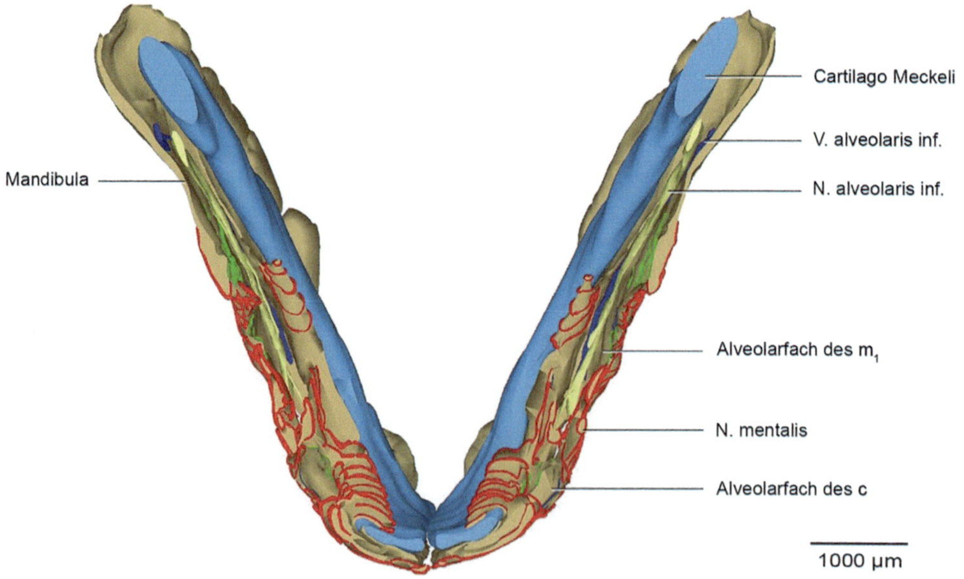

Fig. 20.26 Human embryo, 54 mm CRL, mandible with Meckel's cartilage (dental primordia removed), cranial view, with bone remodeling pattern superimposed (red: apposition, green: resorption)

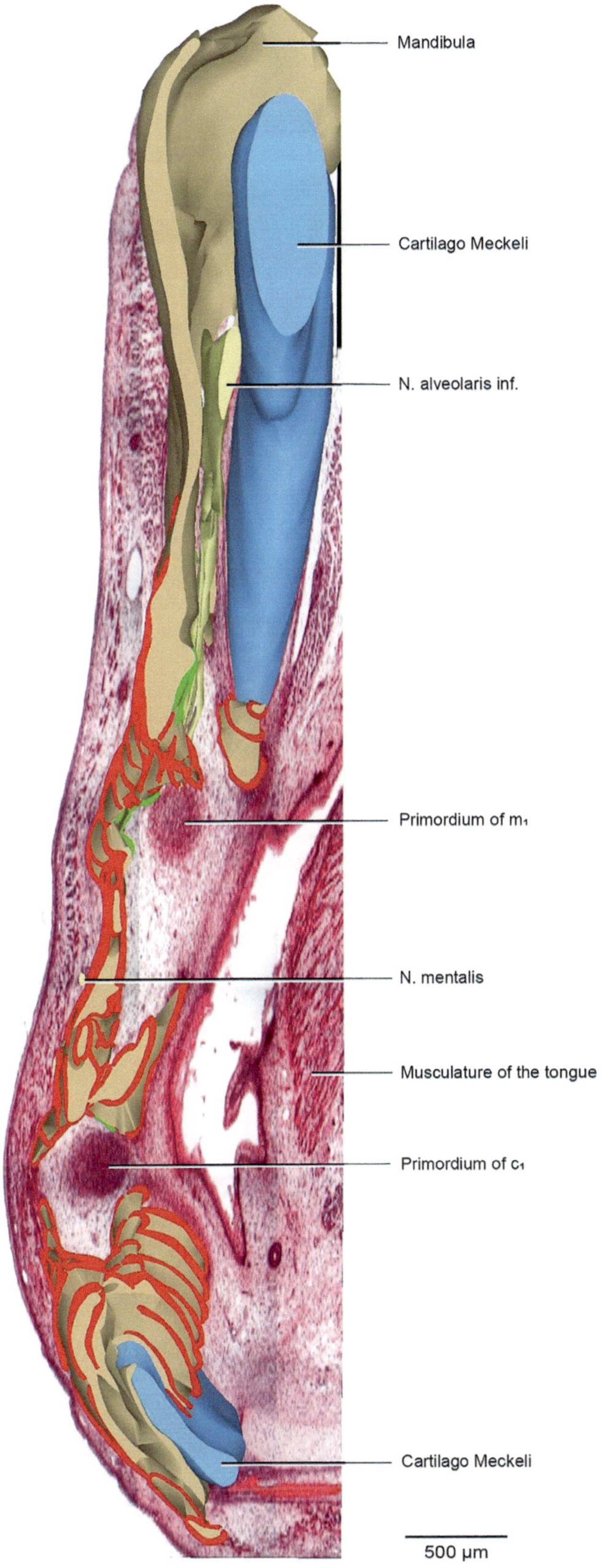

Fig. 20.27 Human embryo, 54 mm CRL, right half of mandible with Meckel's cartilage, cranial view. Superimposed on horizontal section (stained Hematoxylin-eosin). Bone remodeling pattern superimposed (red: apposition, green: resorption)

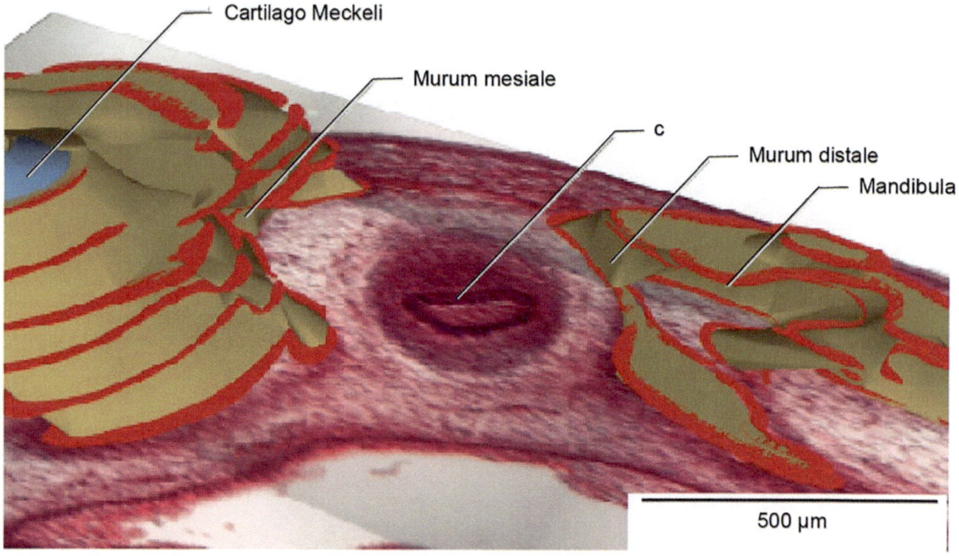

Fig. 20.28 Human embryo, 54 mm CRL, right half of mandible with Meckel's cartilage, cranial view. Region of canine primordium superimposed on horizontal section (stained Hematoxylin-eosin). Bone remodeling pattern superimposed (red: apposition)

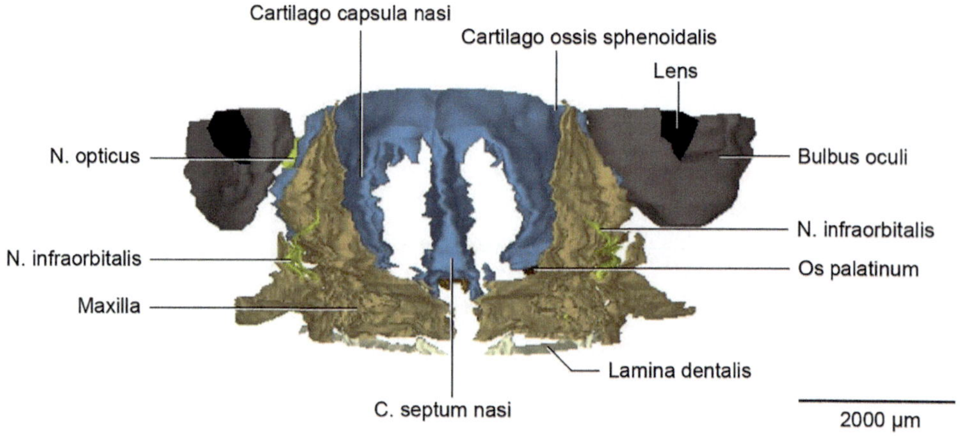

Fig. 20.29 Human embryo, 54 mm CRL, maxilla and nasal capsule, frontal view

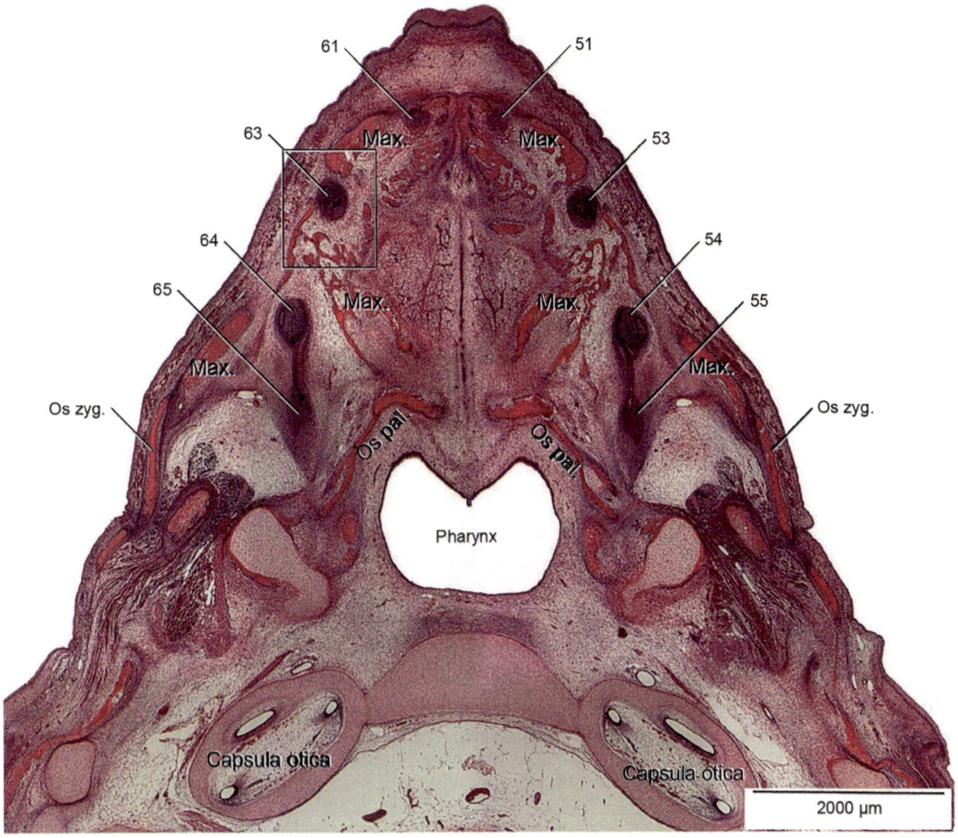

Fig. 20.30 Human embryo, 54 mm CRL, horizontal section (staining: Hematoxylin-eosin.) at level of upper dental primordia. Box: Region of Fig. 20.31

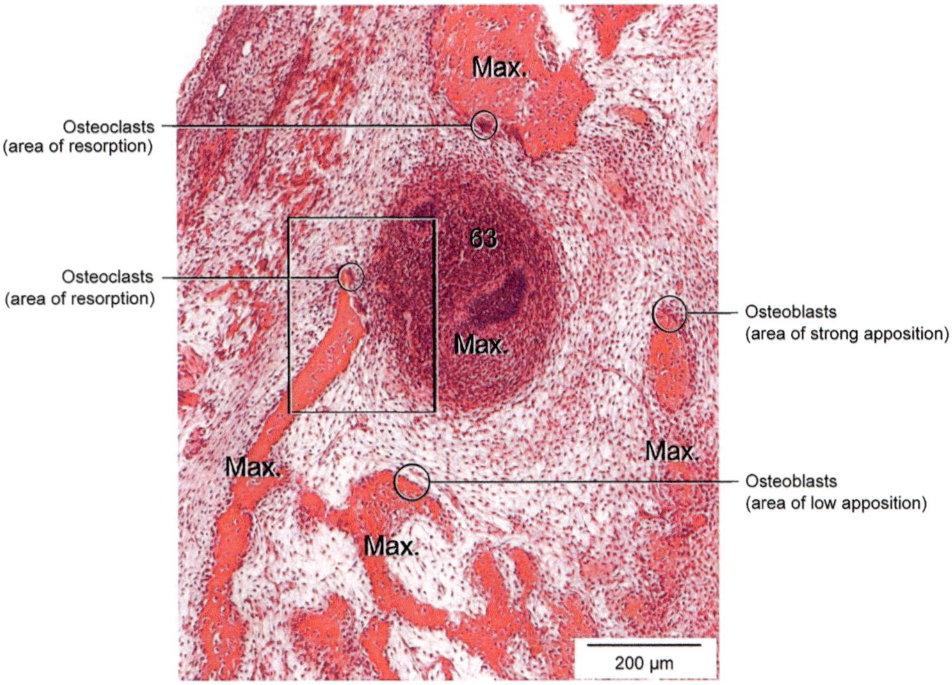

Fig. 20.31 Human embryo, 54 mm CRL, horizontal section (staining: Hematoxylin-eosin.) on level of upper deciduous canine primordium (63). Box: Region of Fig. 20.32

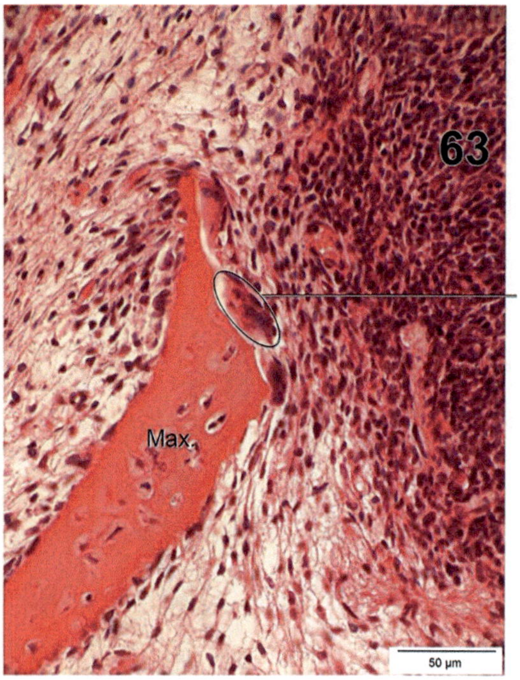

Fig. 20.32 Human embryo, 54 mm CRL, horizontal section (staining: Hematoxylin-eosin) on level of upper deciduous canine primordium (63)

Fig. 20.33 Human embryo, 54 mm CRL, horizontal section, at level of the stapedic foramen. Box: Region of Fig. 20.34

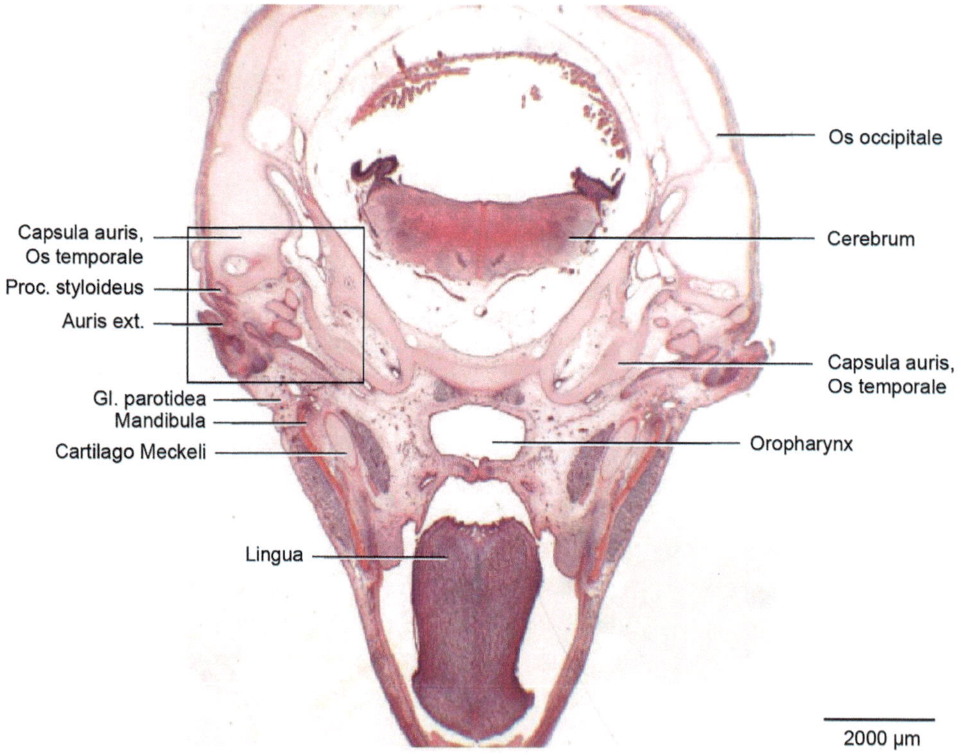

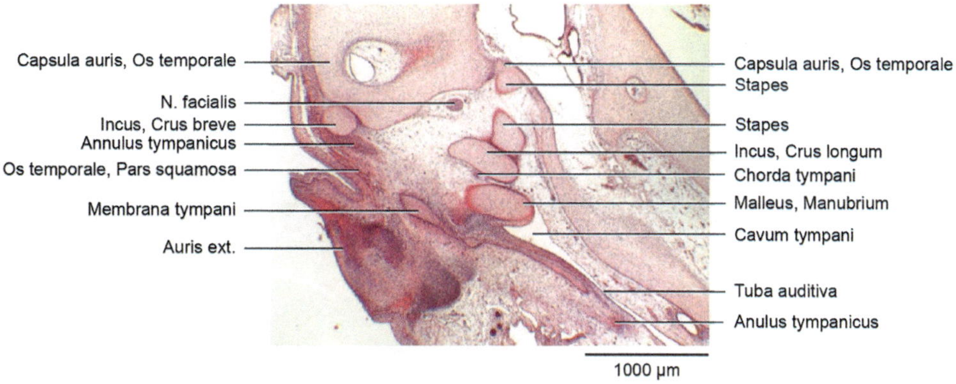

Fig. 20.34 Human embryo, 54 mm CRL, detailed view of the region marked of Fig. 20.33

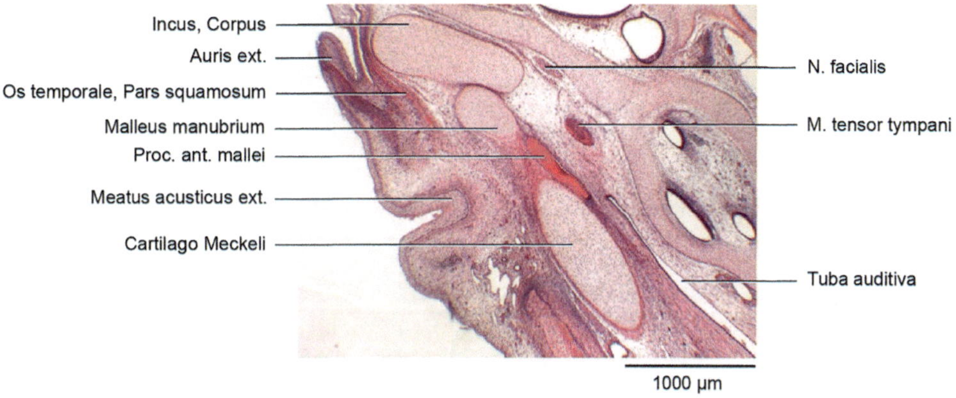

Fig. 20.35 Human embryo, 54 mm CRL, horizontal section at level of the external acoustic duct

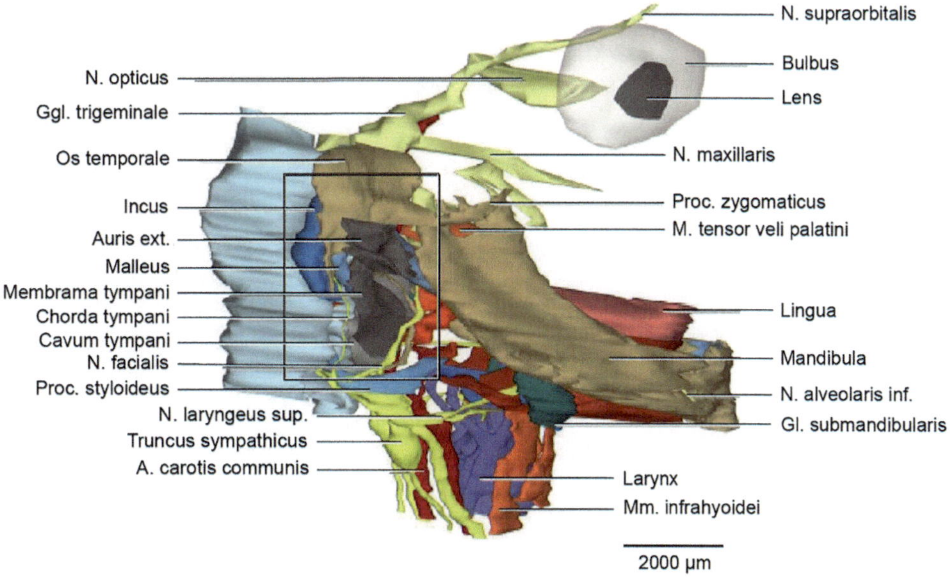

Fig. 20.36 Human embryo, 54 mm CRL, survey of otic region, lateral view. Box: Region of Fig. 20.37

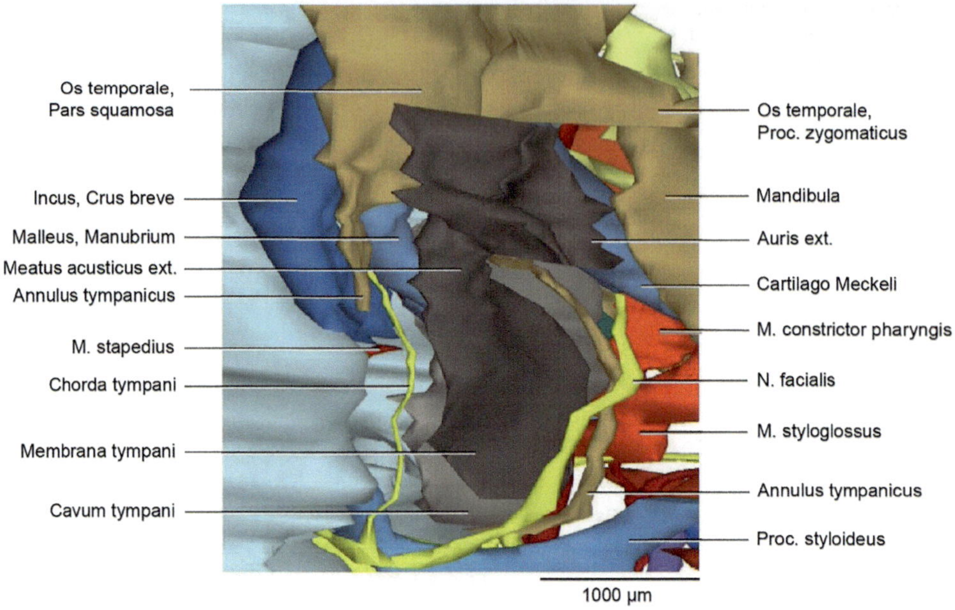

Fig. 20.37 Human embryo, 54 mm CRL, right auditory system, marked region of Fig. 20.36, lateral view

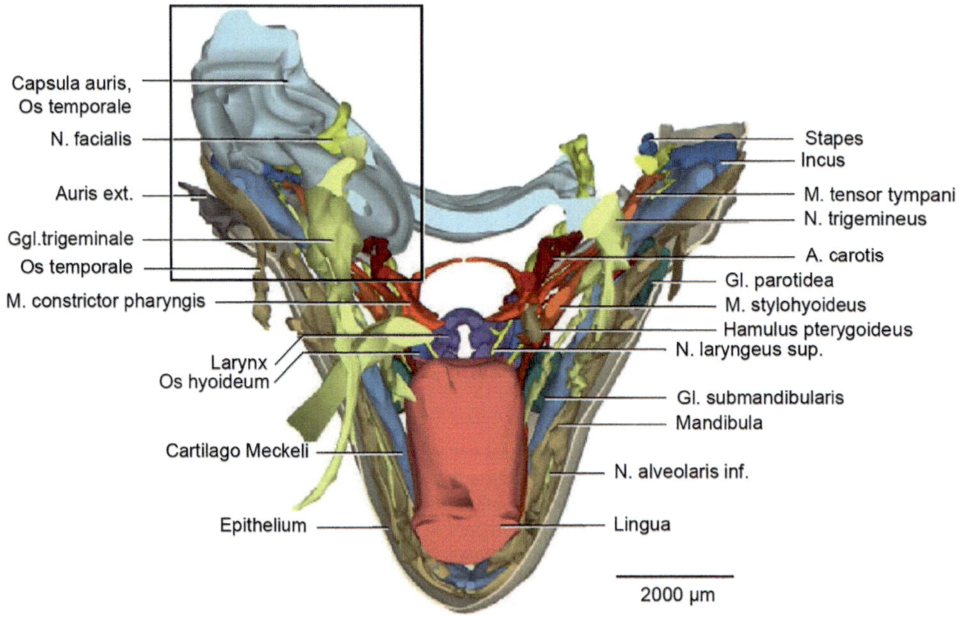

Fig. 20.38 Human embryo, 54 mm CRL, survey of otic region, cranial view. Box: Region of Fig. 20.39

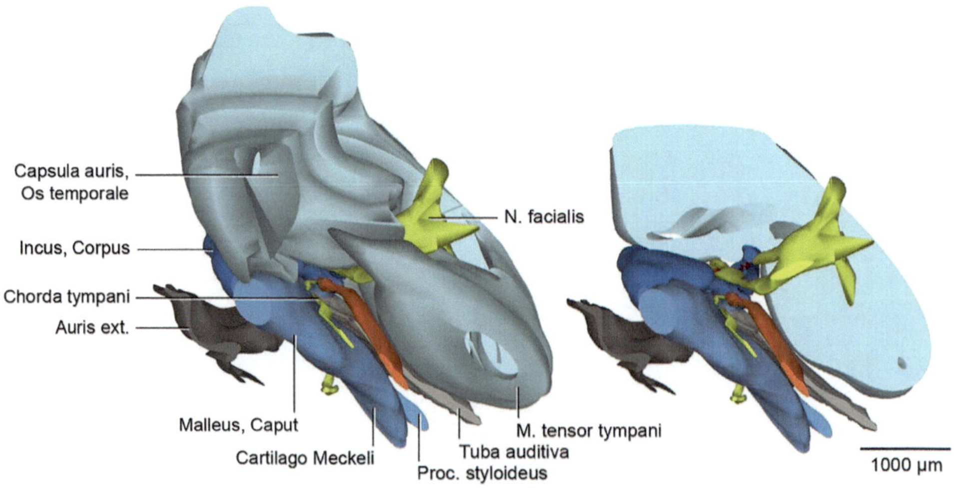

Fig. 20.39 Human embryo, 54 mm CRL, right auditory system, marked region of Fig. 20.38, cranial view, in right image, capsula auris partly removed

Further Reading

Beckert K. Beiträge zur pränatalen Morphogenese der Region des Mittelohres des Menschen – Computergestützte 3D-Rekonstruktionen und histologische Untersuchungen menschlicher Embryonen/Feten von 19 bis 150 mm CRL (7. bis 18. Woche). Med. Diss. Charité-Universitätsmedizin Berlin; 2010.

Fernández Rubio EM, Radlanski RJ. Development of the human primary and secondary jaw joint. Ann Anat. 2024;251:1–8.

Giambartolomei LA, Brunotto M. Gomez de Ferrais ME Human temporomandibular joint disc: anatomy and measurements in prenatal development. Acta Odontol Latino-Americana. 2011;24:98–103.

Germelmann AR (née Than) Beitrag zur pränatalen Morphogenese des Os hyoideum des Menschen—Computergestützte 3D-Rekonstruktionen und histologische Untersuchungen menschlicher Embryonen/Feten von 19–150 mm CRL (7. bis 18. Woche) Med. Diss. Charité-Universitätsmedizin Berlin; 2008

Hinrichsen KV. The early development of morphology and patterns of the face in the human embryo. Berlin Heidelberg New York Tokyo: Springer Verlag; 1985.

Kjaer I, Keeling JW, Fischer-Hansen B. The prenatal human cranium—normal and pathololologic development. Copenhagen: Munksgaard; 1999.

Radlanski RJ, Kjaer I, Vastardis H. Renz H Morphometrische Untersuchungen zur Fetalentwicklung der menschlichen Mandibula. Fortschr Kieferorthop. 1994;55:77–83.

Radlanski RJ, Emmerich S, Renz H. Prenatal morphogenesis of the human incisive canal. Anat Embryol. 2004;208:265–71.

Schuster FP. Zur Entwicklung des Kieferknochens im Bereich der Zahnanlagen. Histologische und morphologische Untersuchungen an der fetalen Mandibula des Menschen während des Glockenstadiums der Zahnanlagen (68–160 mm SSL). Med. Diss. Charité-Universitätsmedizin Berlin; 2013.

Tsengelsaikhan N. Zur Entwicklung des Kieferknochens im Bereich der Zahnanlagen – Histologische und morphologische Untersuchung an der embryonalen Mandibula des Menschen während der frühen Entwicklungsstadien der Zahnanlagen (19–68 mm SSL) Med. Diss. Charité-Universitätsmedizin Berlin; 2014.

Zimmermann C. Zur Entwicklung des Kieferknochens im Bereich der Zahnanlagen. Histologische und morphologische Untersuchungen an der fetalen Mandibula des Menschen während des späten Glockenstadiums der Zahnanlagen (160–270 mm SSL) Med. Diss. Charité-Universitätsmedizin Berlin; 2013.

Embryo 54 mm CRL (ca. 11th wk): SIM280489

21

Overall Remarks

This chapter focuses on the development of the dental primordia in the maxilla, which are in their cap stages, and on their positional relationships to the surrounding maxillary bone. The formation of the bony compartments, which will eventually constitute the bony alveoli for each individual dental primordium, takes place at this stage of development primarily through apposition of bone (Figs. 21.1, 21.2, 21.3, 21.4, 21.5, 21.6, 21.7, 21.8, 21.9, 21.10, and 21.11).

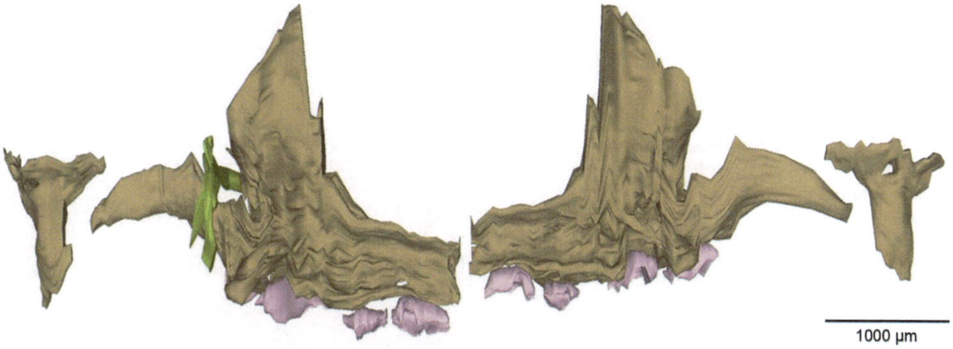

Fig. 21.1 Human embryo, 54 mm CRL, maxilla with infraorbital n., and dental primordia, frontal view

Fig. 21.2 Human embryo, 54 mm CRL, maxilla with dental primordia, caudal view

R. J. Radlanski, *Atlas of Human Prenatal Craniofacial Development*, https://doi.org/10.1007/978-3-031-88812-0_21

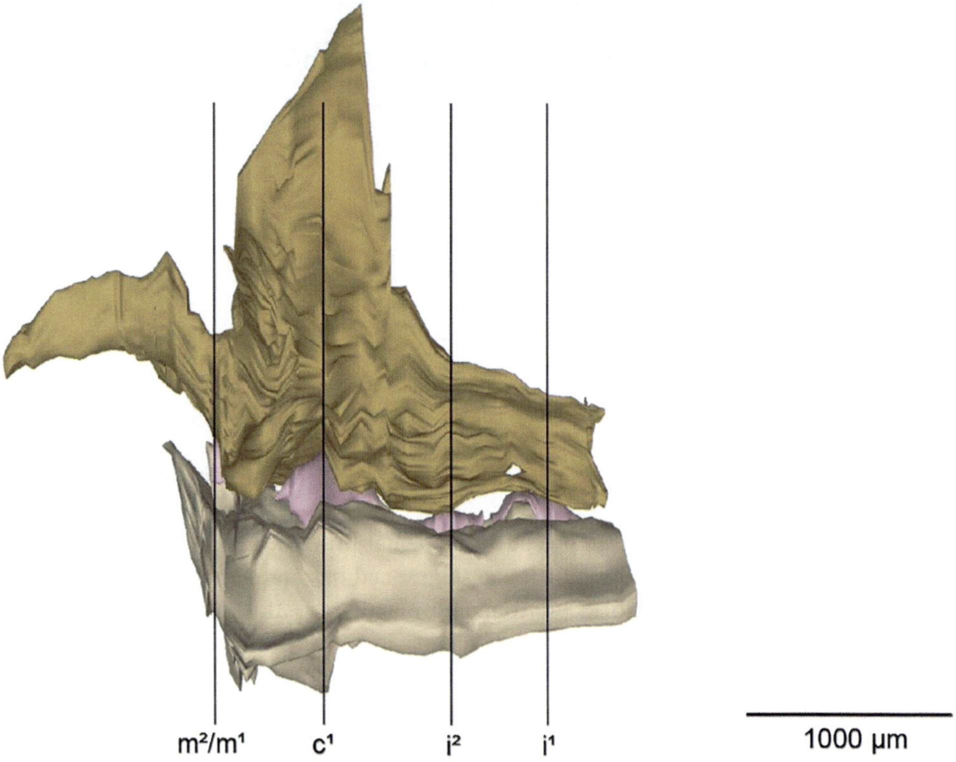

Fig. 21.3 Human embryo, 54 mm CRL, maxilla with dental primordia (pink) and dental lamina (gray), right lateral view. Section planes of the following figures through the centers of dental primordia superimposed

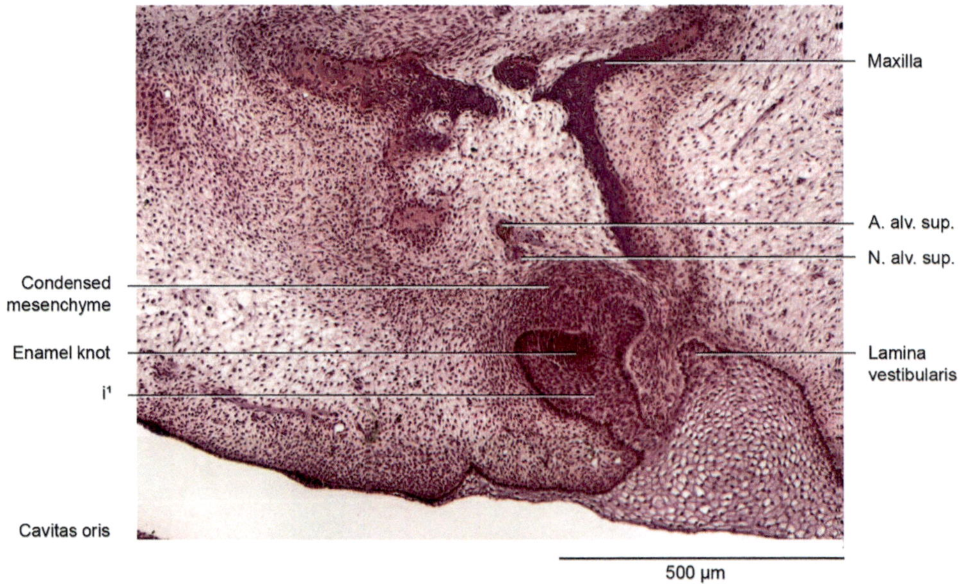

Fig. 21.4 Human embryo, 54 mm CRL, vertical section at level of dental primordium i^1 (staining: hematoxylin-eosin)

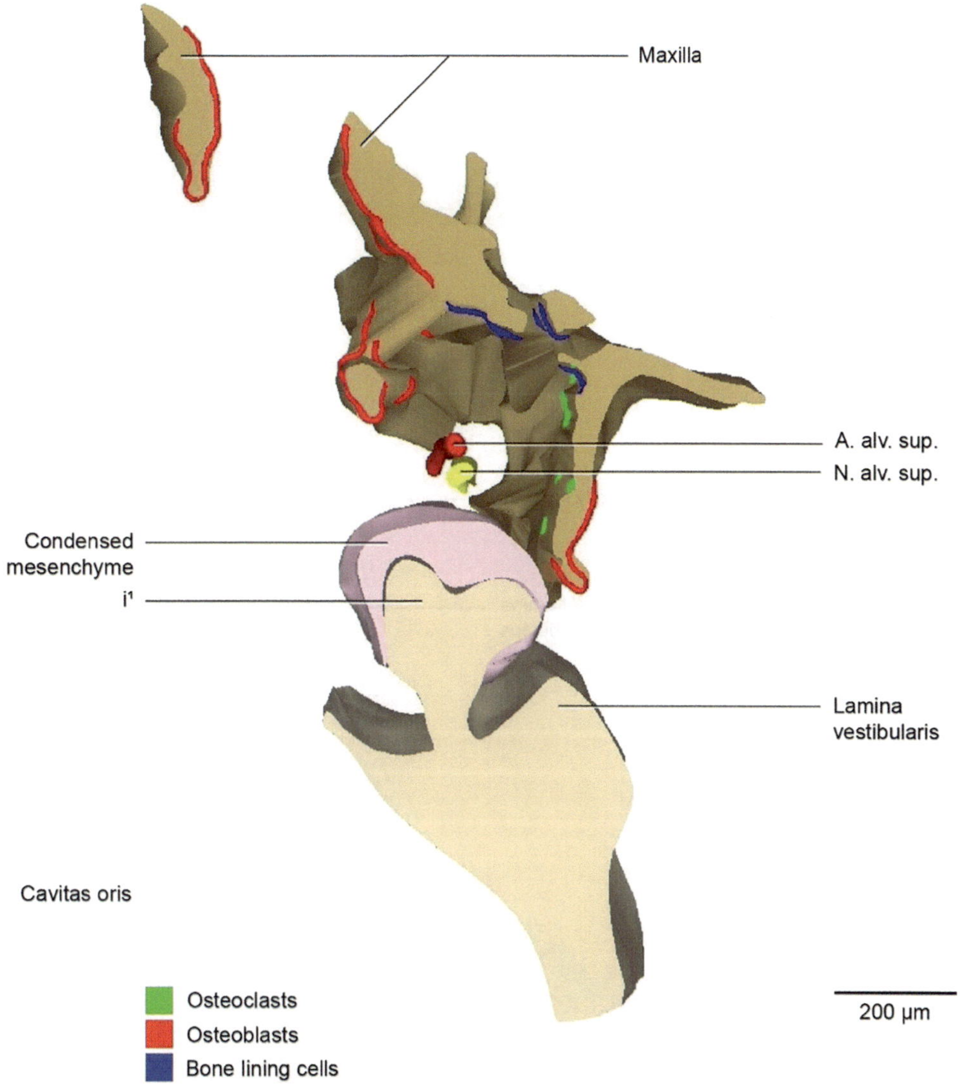

Fig. 21.5 Human embryo, 54 mm CRL, maxilla and dental primordium i^1, sectioned medially, medial view, with bone remodeling pattern superimposed

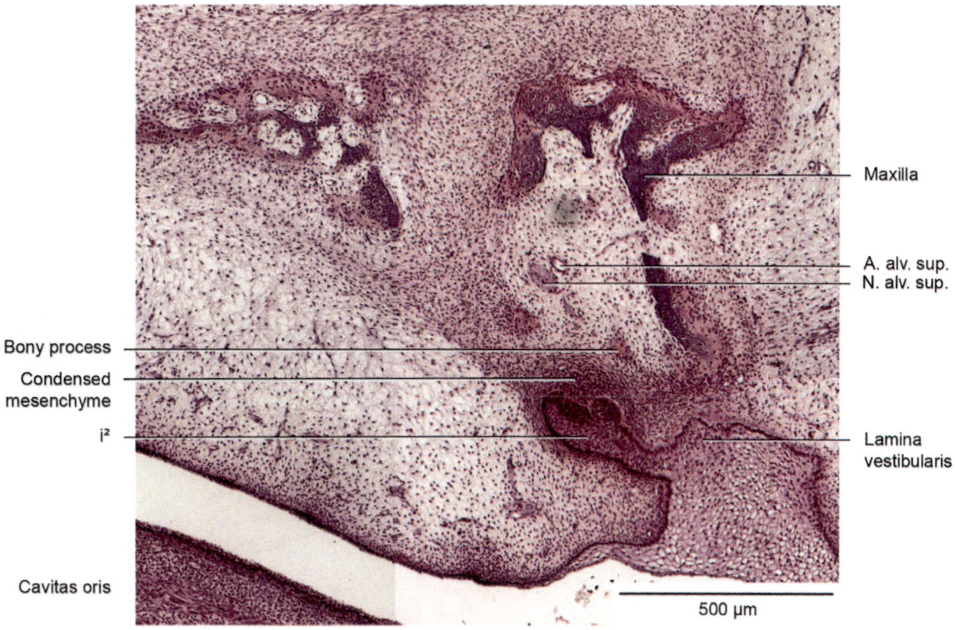

Fig. 21.6 Human embryo, 54 mm CRL, vertical section at level of dental primordium i^2 (staining: hematoxylin-eosin)

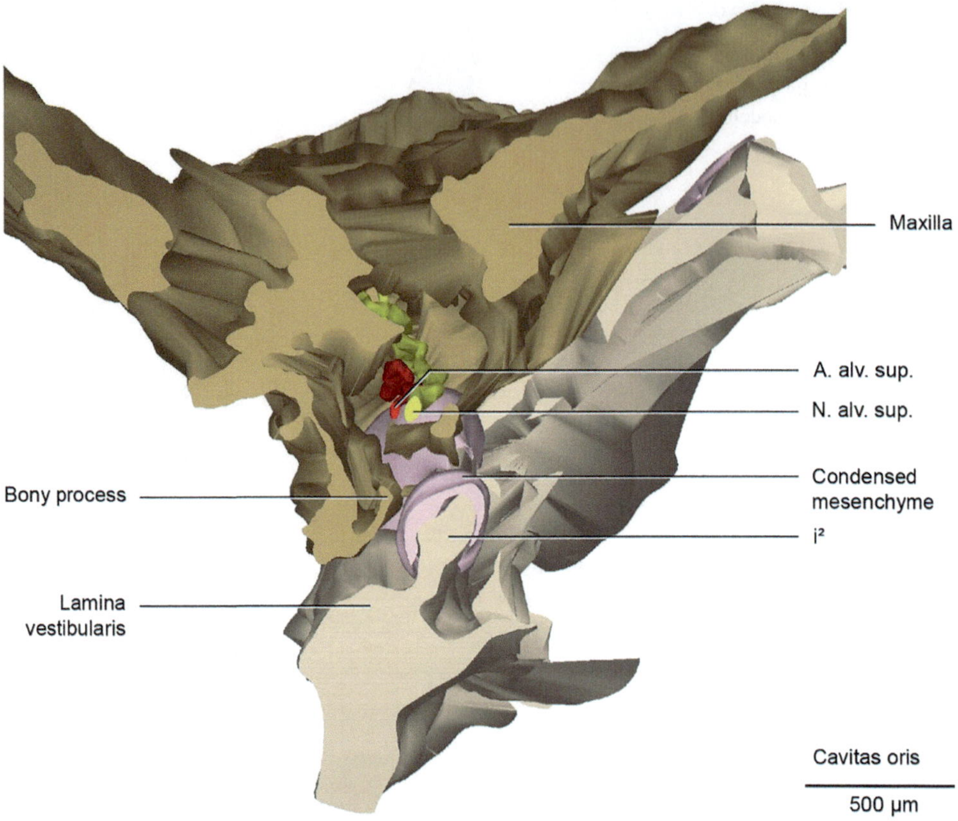

Fig. 21.7 Human embryo, 54 mm CRL, maxilla and dental primordium i^2, sectioned medially, medial view, with bone remodeling pattern superimposed

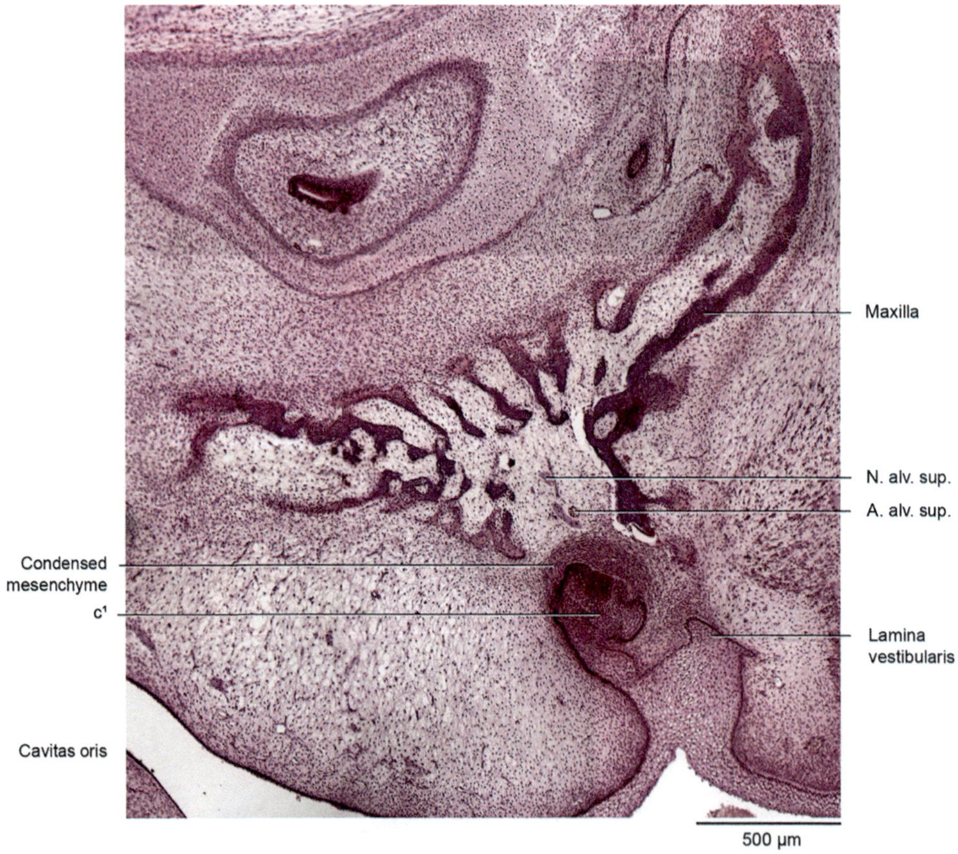

Fig. 21.8 Human embryo, 54 mm CRL, vertical section at level of dental primordium c^1 (staining: hematoxylin-eosin)

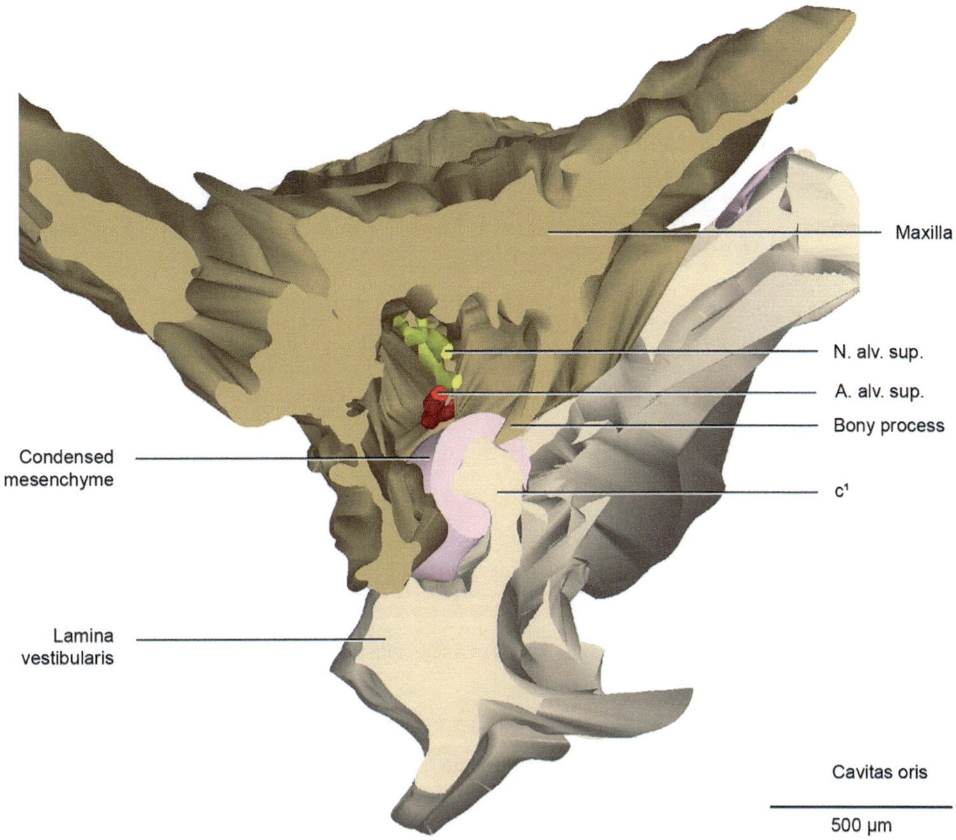

Fig. 21.9 Human embryo, 54 mm CRL, maxilla and dental primordium c^1, sectioned medially, medial view, with bone remodeling pattern superimposed

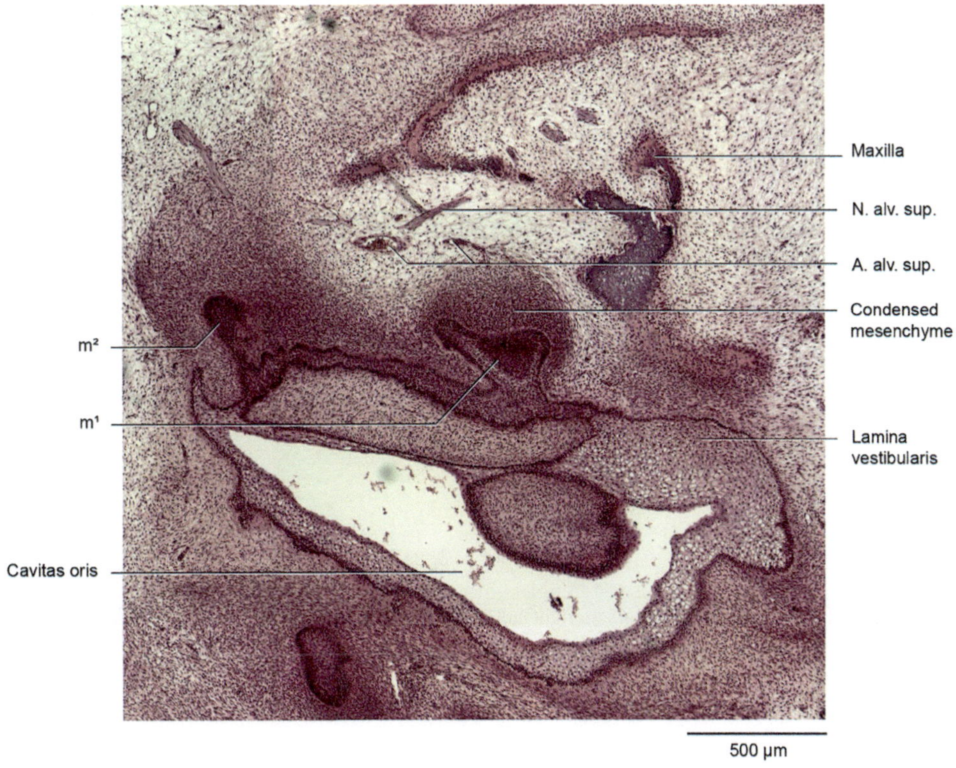

Fig. 21.10 Human embryo, 54 mm CRL, vertical section at level of dental primordia m^1 and m^2 (staining: hematoxylin-eosin)

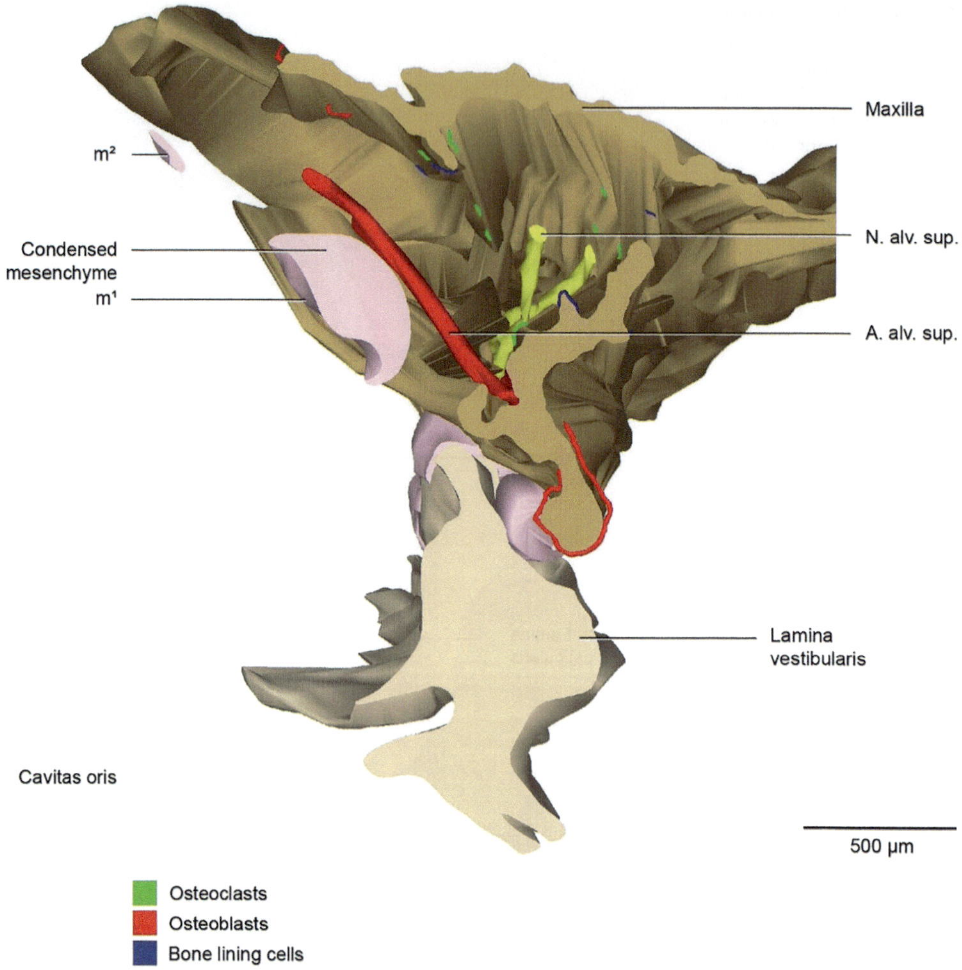

Fig. 21.11 Human embryo, 54 mm CRL, maxilla and dental primordia m[1] and m[2], sectioned medially, medial view, with bone remodeling pattern superimposed

Further Reading

Kalinke U. Embryologische Untersuchungen zur Entwicklung des Alveolarknochens im Oberkiefer beim Menschen zur Zeit der vorgeburtlichen Entwicklungsstadien von 19–68 mm SSL. Med. Diss. Charité-Universitätsmedizin Berlin; 2014.

Radlanski RJ. Contributions to the development of human deciduous tooth primordia. Chicago: Quintessence Publishing Co., Inc.; 1993.

Radlanski RJ. Oral structure and biology. Batavia: Quintessence Publishing; 2018.

Fetus 60 mm CRL (ca. 15th wk): URS250889

22

Dental Primordia

The dental primordia for the first dentition have reached their early bell stages. The folding of the inner enamel epithelium already indicates the type of tooth that is developing. The stratum reticulare is well filled with fluid, which is probably available to nourish the ameloblasts. At this stage of development, however, the secretion of enamel matrix, from which the enamel later develops, has not yet begun. The mesenchyme, which is surrounded by the young tooth bell, is clearly compacted. This will later form the dental pulp (Figs. 22.1 and 22.2).

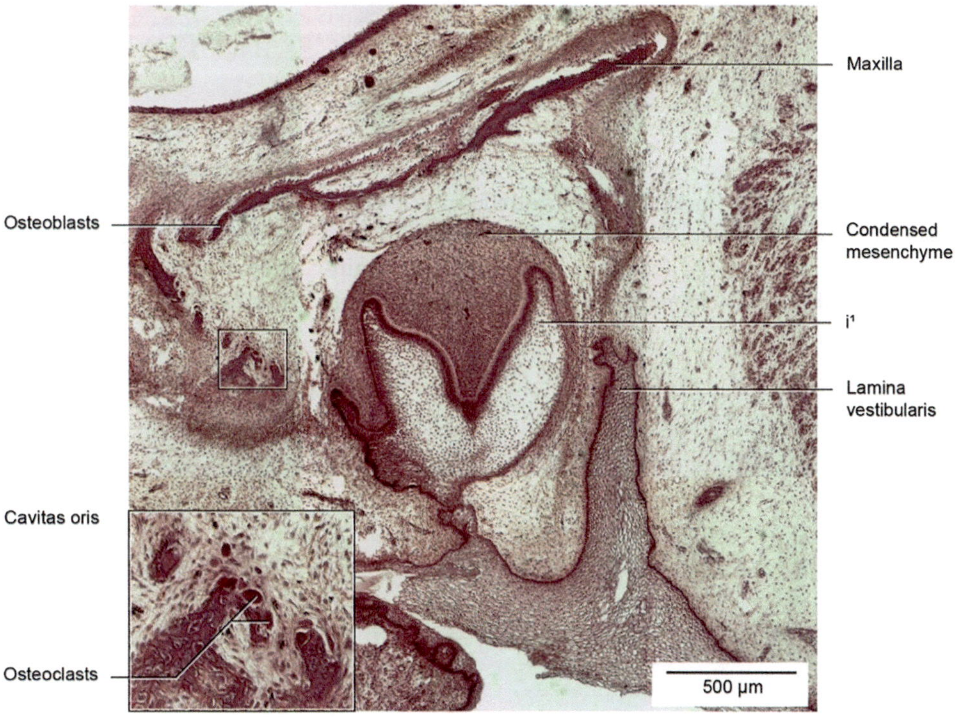

Fig. 22.1 Human fetus 60 mm CRL (15th week), vertical section at level of primordium i^1 (staining: hematoxylin-eosin)

R. J. Radlanski, *Atlas of Human Prenatal Craniofacial Development*, https://doi.org/10.1007/978-3-031-88812-0_22

Fig. 22.2 Human fetus 60 mm CRL (15th week), vertical section at level of primordium i² (staining: hematoxylin-eosin)

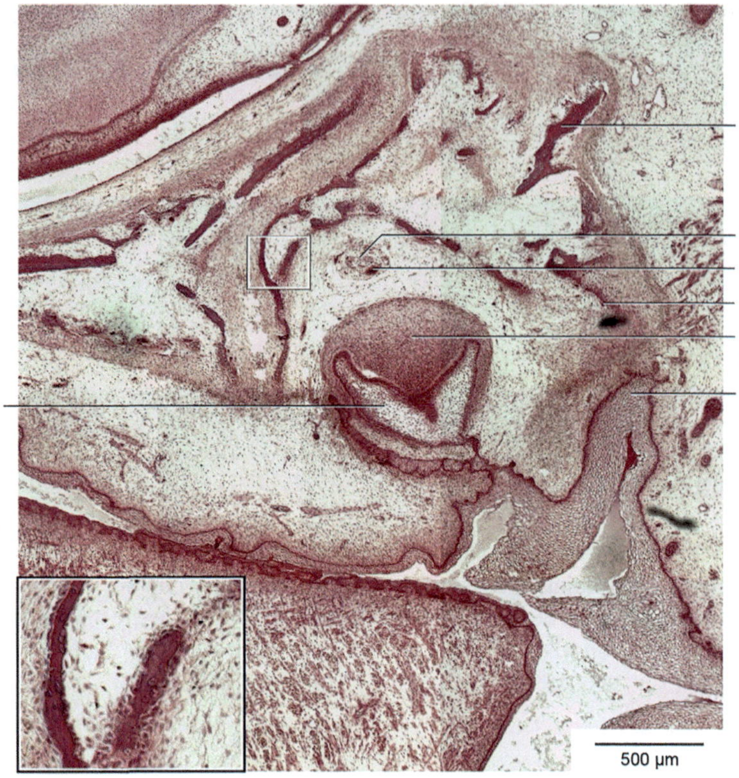

Peridental Bone

The entire dental primordium is surrounded by a dense mesenchyme, identified as the dental follicle (which is actually also a part of the dental primordium). The remaining peridental mesenchyme is loosely woven. It contains the early deposits of the peridental bone. Where the distance between the dental primordium and the bone is relatively small, the bone, which is otherwise characterized by apposition, shows signs of resorption: dental primordium and bone are interacting with each other (Figs. 22.3 and 22.4).

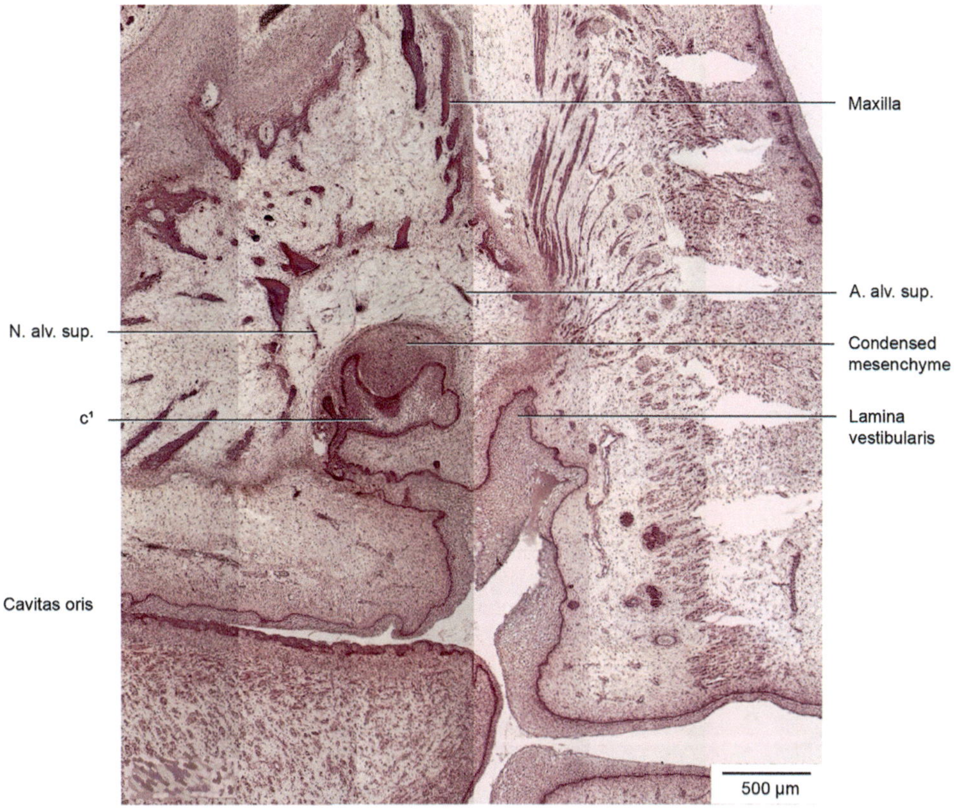

Fig. 22.3 Human fetus 60 mm CRL (15th week), vertical section at level of the upper deciduous canine primordium (staining: hematoxylin-eosin)

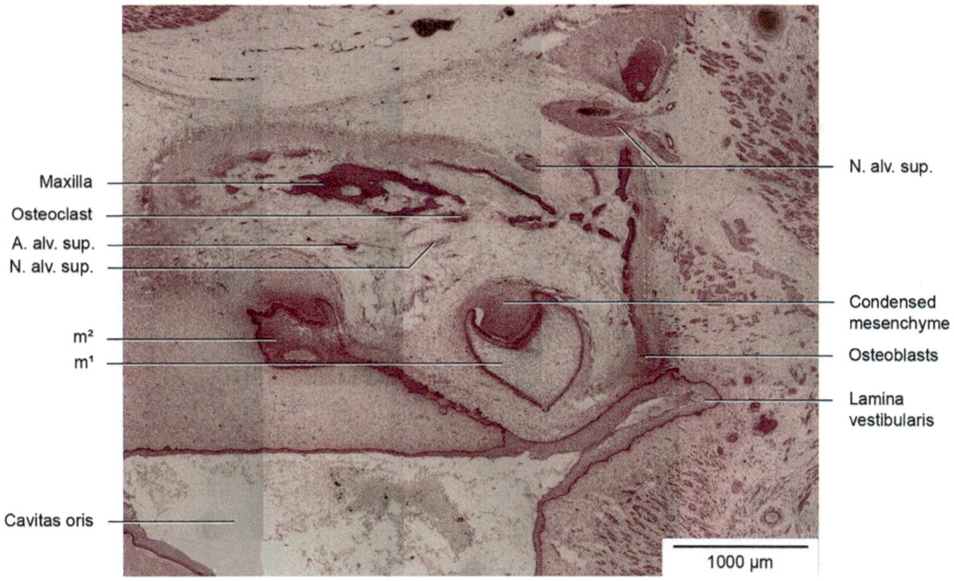

Fig. 22.4 Human fetus 60 mm CRL (15th week), vertical section at level of primordia m^1 and m^2 (staining: hematoxylin-eosin)

Further Reading

Radlanski RJ. Contributions to the development of human deciduous tooth primordia. Chicago: Quintessence Publishing Co., Inc.; 1993.

Radlanski RJ. Oral structure and biology. Batavia: Quintessence Publishing; 2018.

Embryo 66 mm CRL (ca. 10th wk): GIS210789

23

Overall Remarks

Although somewhat taller, this embryo is a bit younger and slightly less developed than the previous one in Chap. 22. This affects the anatomically ascertainable maturity of the mandible and also the developmental state of the dental anatomy. The primordia are only in the late cap stage here (Figs. 23.1, 23.2, 23.3, 23.4, 23.5, 23.6, 23.7, and 23.8).

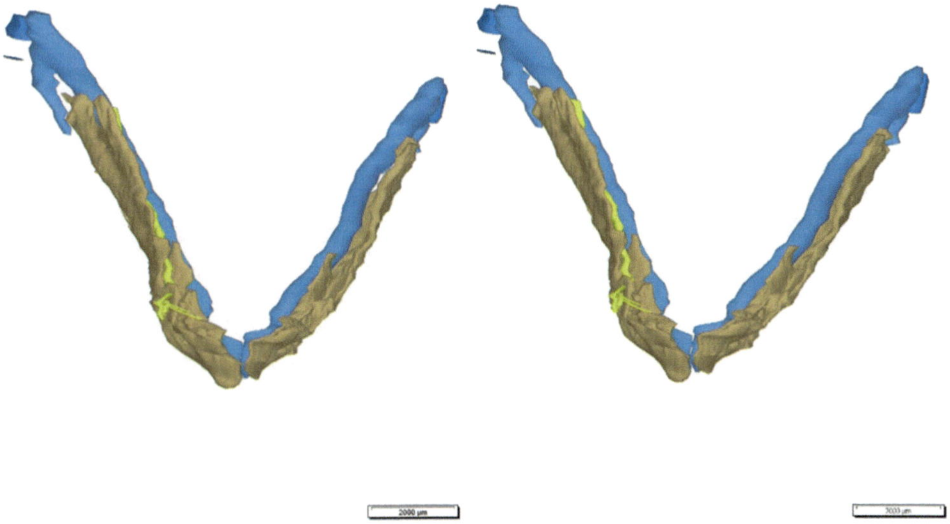

Fig. 23.1 Stereoscopical pair of images. Human embryo 66 mm CRL (ca. 10th week), mandible (ochre), Meckel's cartilage (blue), and right inferior alveolar nerve (yellow), cranial view. Bar: 2000 μm

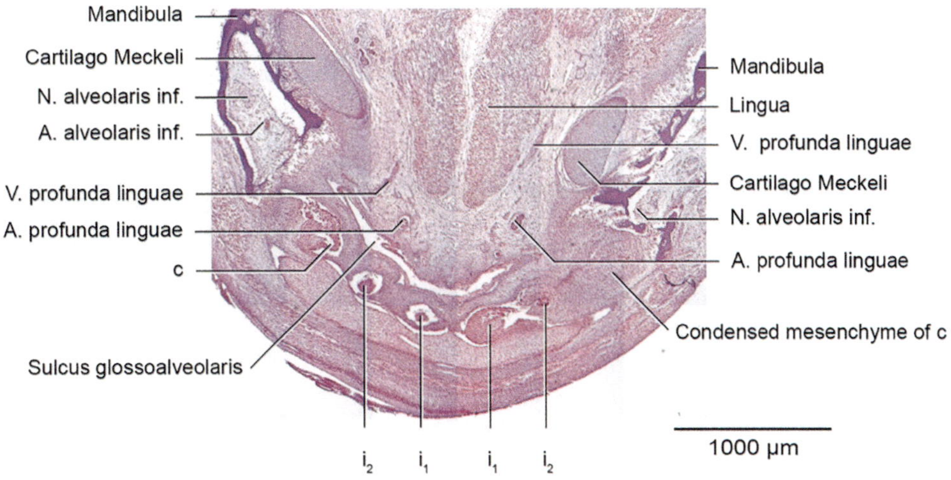

Fig. 23.2 Human embryo, 66 mm CRL, mandible, horizontal section (staining: hematoxylin-eosin) at level of i_1, i_2, and c

R. J. Radlanski, *Atlas of Human Prenatal Craniofacial Development*, https://doi.org/10.1007/978-3-031-88812-0_23

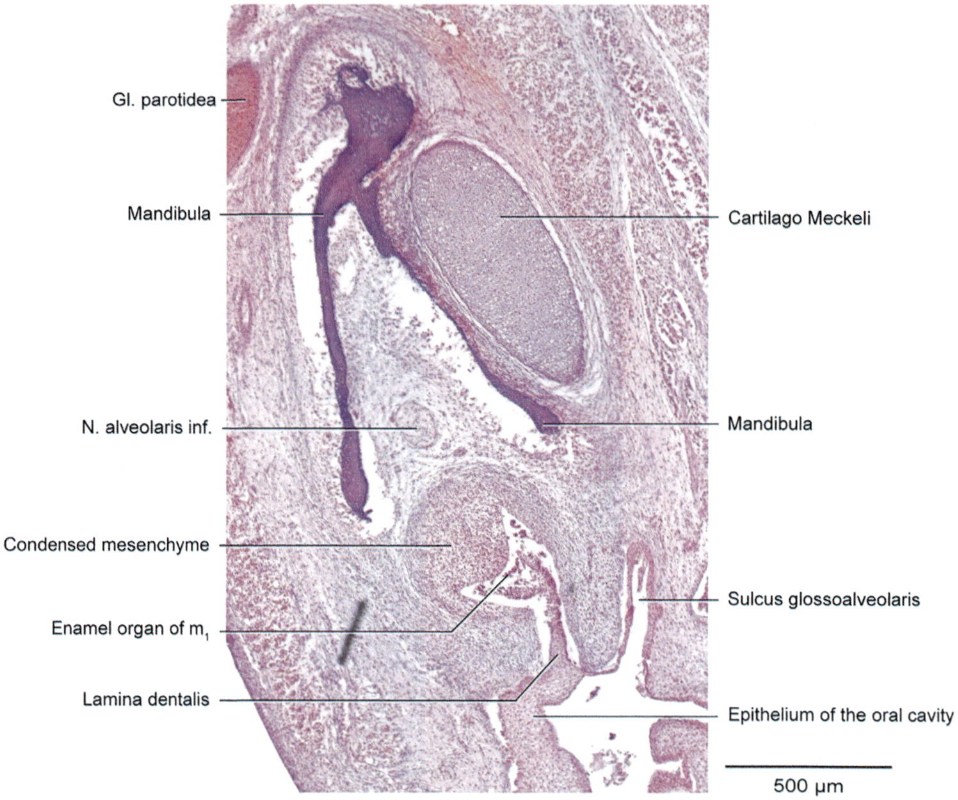

Fig. 23.3 Human embryo, 66 mm CRL, mandible, horizontal section (staining: hematoxylin-eosin) at level of m_1

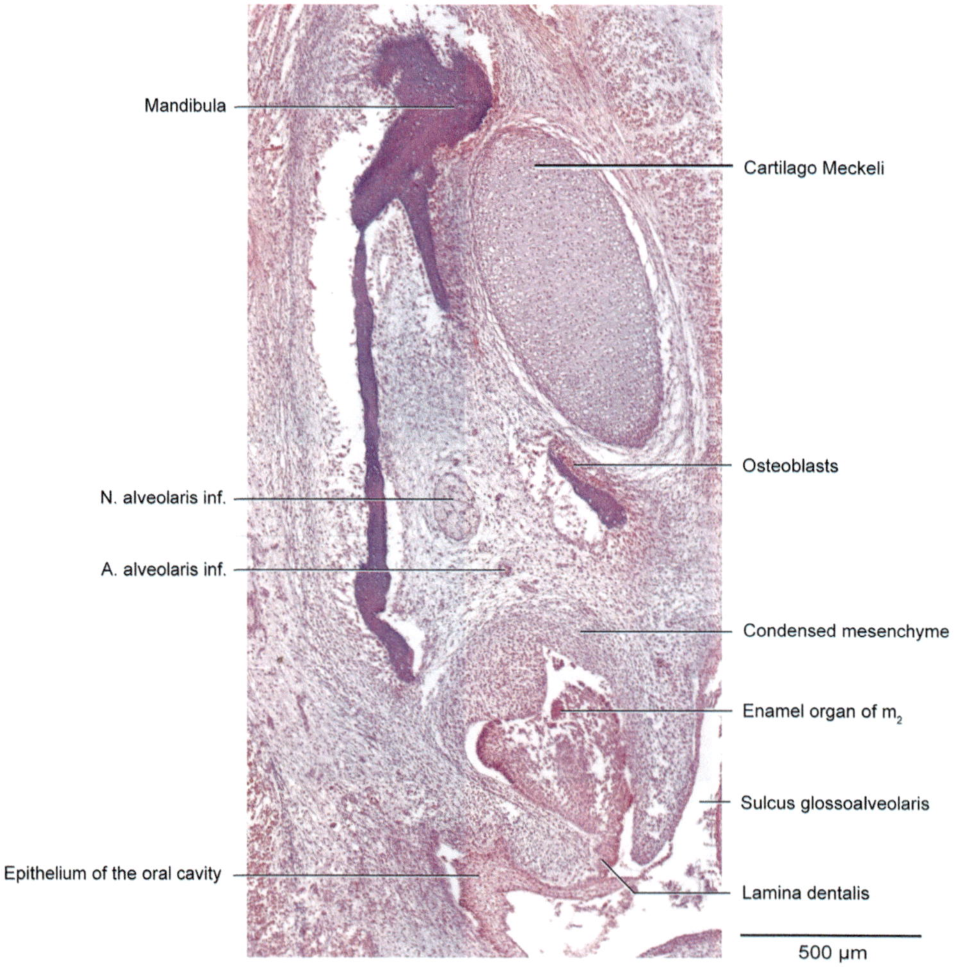

Fig. 23.4 Human embryo, 66 mm CRL, mandible, horizontal section (staining: hematoxylin-eosin) at level of m_2

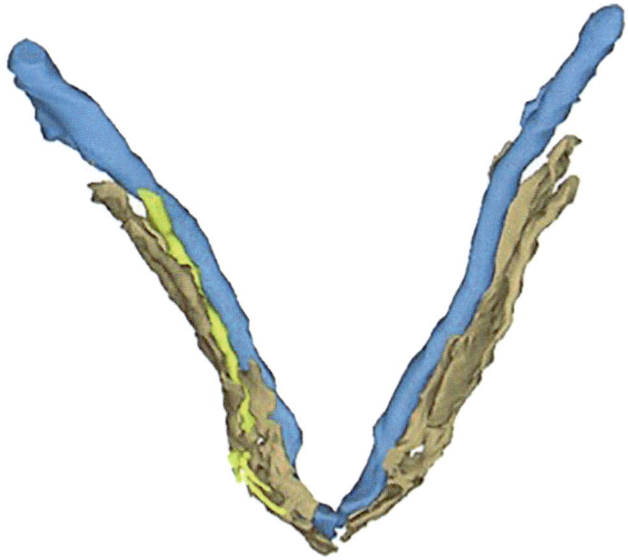

Fig. 23.5 Human embryo, 66 mm CRL, mandible, cranial view

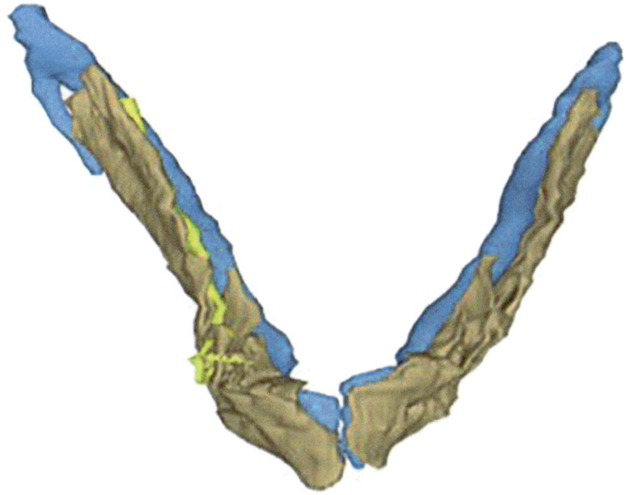

Fig. 23.6 Human embryo, 66 mm CRL, mandible, frontal view

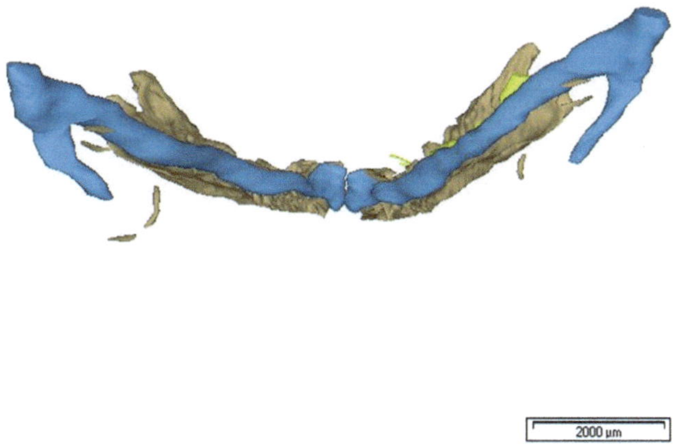

Fig. 23.7 Human embryo, 66 mm CRL, mandible, dorsal view

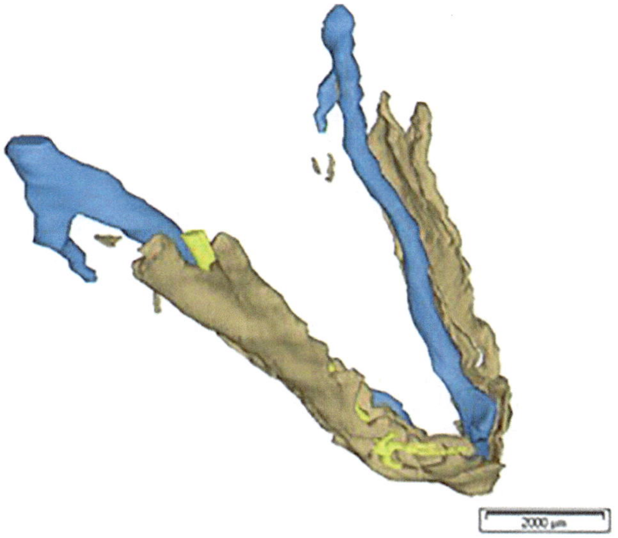

Fig. 23.8 Human embryo, 66 mm CRL, mandible, 45° lateral view

Further Reading

Radlanski RJ. Oral structure and biology. Batavia: Quintessence Publishing; 2018.

Radlanski RJ, Renz H, Klarkowski MC. Prenatal development of the human mandible. 3D reconstructions, morphometry and bone remodelling pattern, sizes 12-117 mm CRL. Anat Embryol (Berl). 2003;207:221–32.

Embryo 68 mm CRL (ca. 10th wk): ART270694

24

Overall Remarks

This embryo now shows significantly further developed dental primordia than the previous one in Chap. 23. This fact can also be seen as an indication that the respective maturity of individual organs can be very heterochronous in comparison with each other. In the reconstructions of this specimen, we were again able to incorporate the bone remodeling patterns very nicely into the illustrations. At the first glance, bone remodeling dominates in the mandible. However, there are also regions with bone resorption. These are preferably located on the inner sides of the mandible and, above all, on the inner walls of the bony crypts of the dental primordia. The same applies to the bone around the dental primordia in the maxillary region (Figs. 24.1, 24.2, 24.3, 24.4, 24.5, and 24.6).

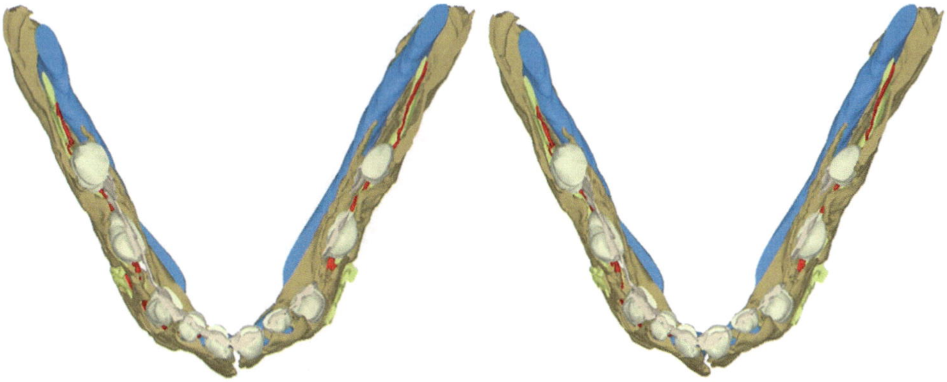

Fig. 24.1 Stereoscopical pair of images. Human embryo 68 mm CRL (ca. 10th week), mandible (ochre), Meckel's cartilage (blue), inferior alveolar nerve (yellow), inferior alveolar artery (red), and dental primordia (light gray), cranial view

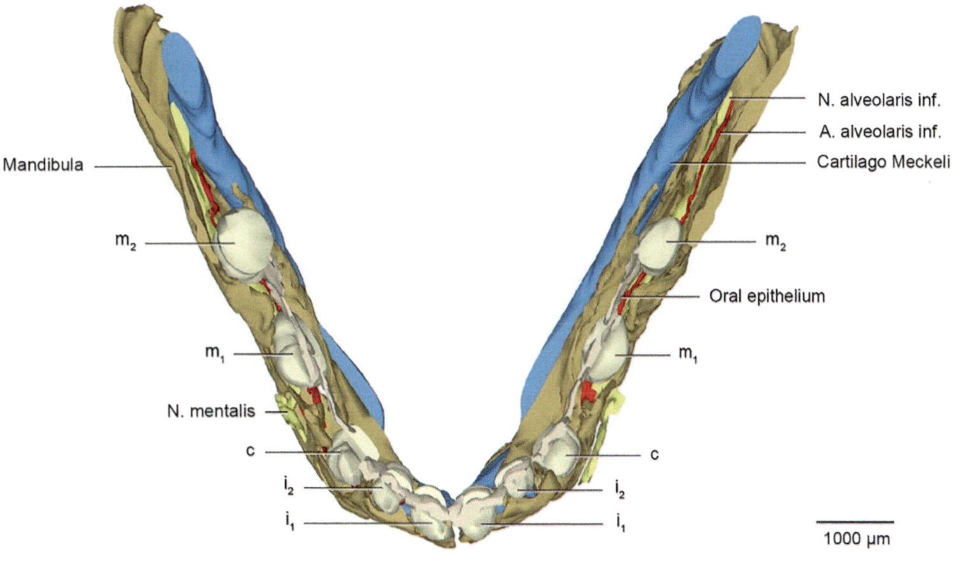

Fig. 24.2 Human embryo, 68 mm CRL, mandible, cranial view, partial 3D reconstruction of the mandible with Meckel's cartilage, alveolar inf. n., mental n., oral epithelium, and dental primordia

R. J. Radlanski, *Atlas of Human Prenatal Craniofacial Development*, https://doi.org/10.1007/978-3-031-88812-0_24

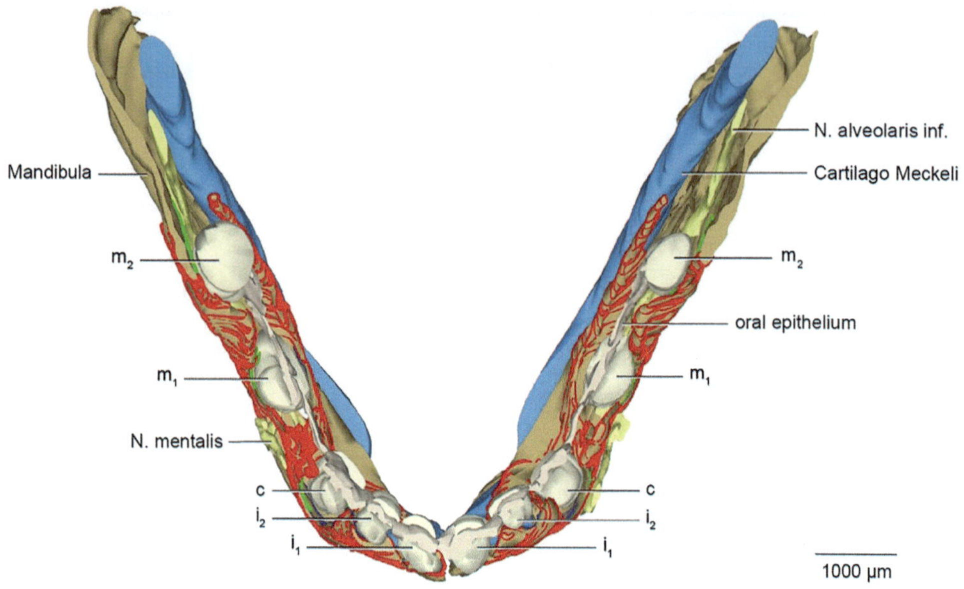

Fig. 24.3 Human embryo, 68 mm CRL, mandible, cranial view, partial 3D reconstruction of the mandible with Meckel's cartilage, alveolar inf. n., mental n., oral epithelium, and dental primordia. Bone remodeling pattern superimposed. Red: apposition, green: resorption

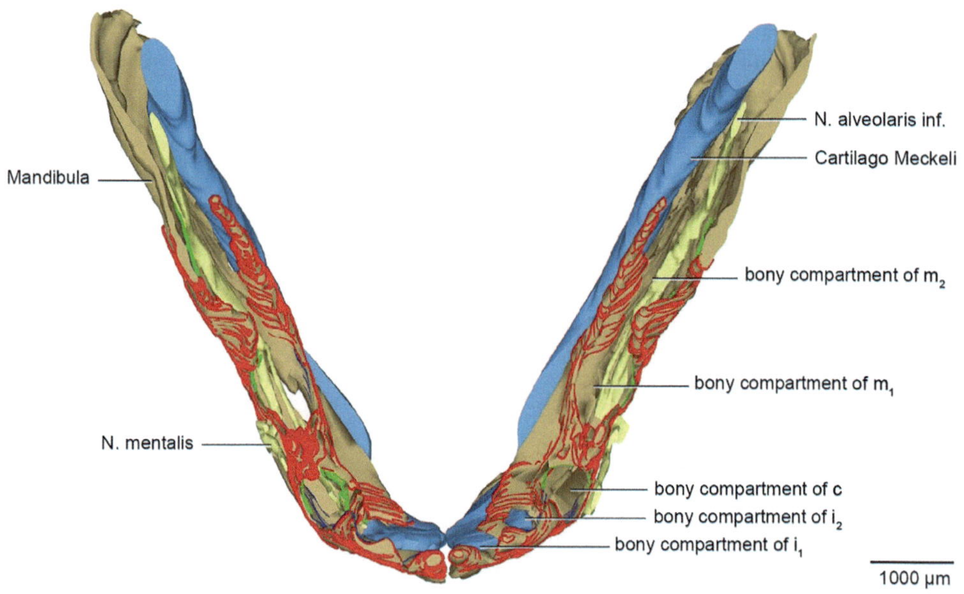

Fig. 24.4 Human embryo, 68 mm CRL, mandible, cranial view, partial 3D reconstruction of the mandible with Meckel's cartilage, alveolar inf. n., mental n. Oral epithelium, and dental primordia removed. Bone remodeling pattern superimposed. Red: apposition, green: resorption

Fig. 24.5 Human embryo, 68 mm CRL, right half of mandible, cranial view. Histological section (staining: hematoxylin-eosin) with partially reconstructed structures and bone remodeling pattern superimposed. Red: apposition, green: resorption

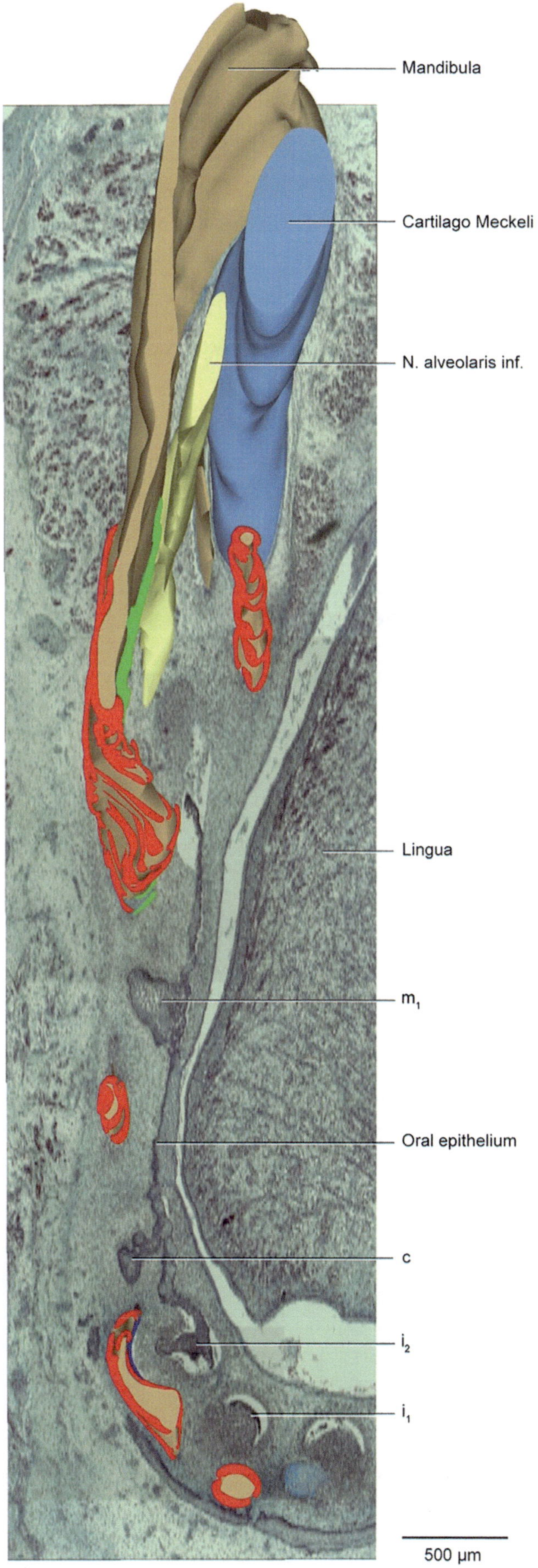

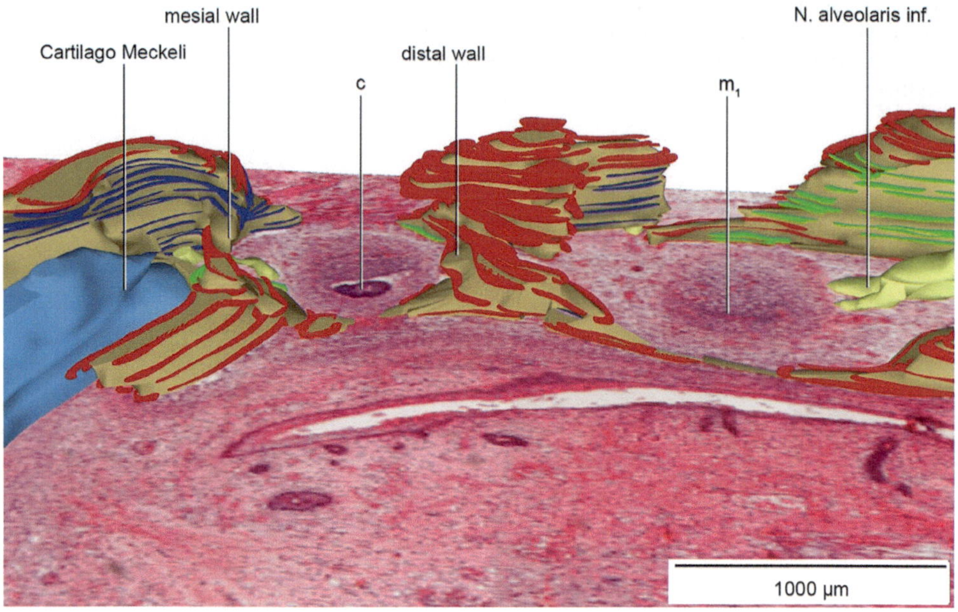

Fig. 24.6 Human embryo, 68 mm CRL, right half of mandible, cranial and 45° medial view. Histological section (staining: hematoxylin-eosin) with partially reconstructed structures and bone remodeling pattern superimposed next to crypts of dental primordia c and m^1. Red: apposition, green: resorption, blue: lining cells

Bone Remodeling

At this stage of development, peridental bone remodeling is again exemplary and typical: Apposition prevails at the margins of the alveolar process bone, whereas the inner sides of the dental crypts are characterized by bone resorption, especially when the dental primordium is expanding. Lining cells represent a temporarily inactive bone surface. We find these, for example, in the area adjacent to Meckel's cartilage or on the inner surfaces of the dental crypts that are further away from the dental follicle, which means, they are unaffected by growth expansion of the follicle (Figs. 24.7, 24.8, 24.9, 24.10, 24.11, 24.12, 24.13, 24.14, 24.15, 24.16, 24.17, 24.18, 24.19, and 24.20).

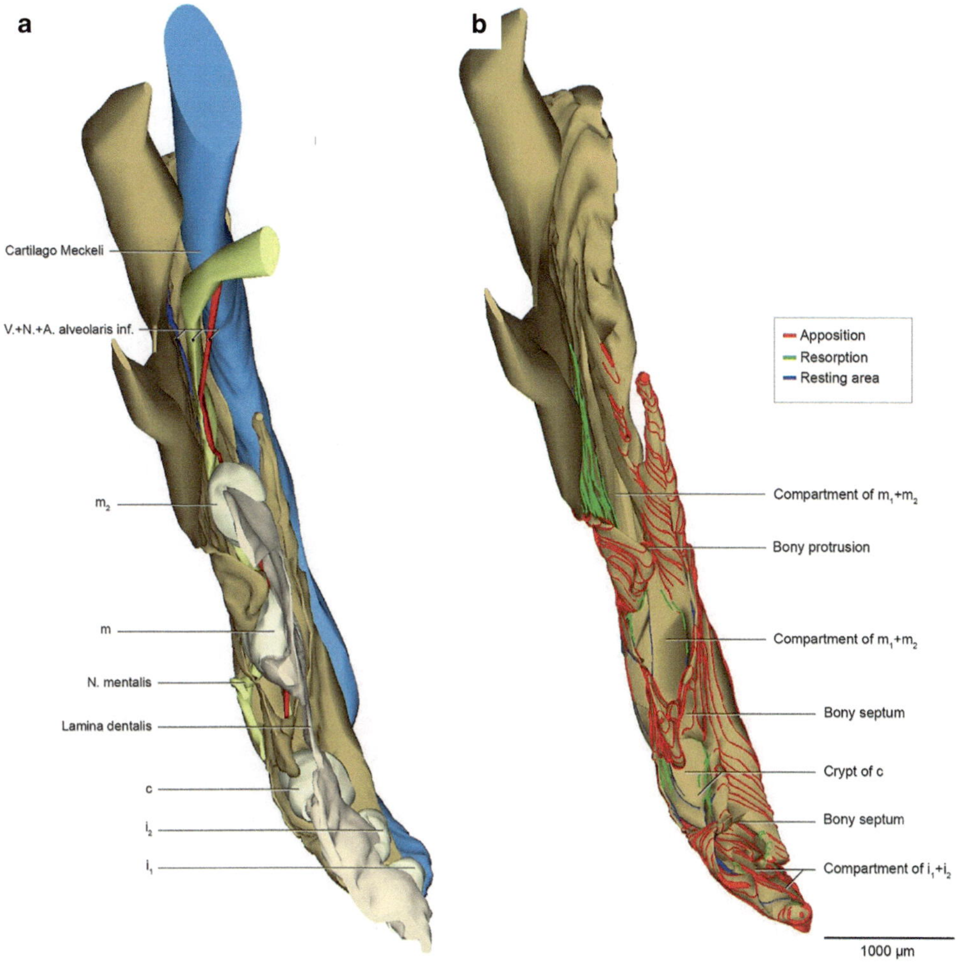

Fig. 24.7 (**a**) Human embryo, 68 mm CRL, partial 3D-reconstruction of the right half of the mandible and surrounding structures, cranial view and 45° medial view. (**b**) The same partial 3D-reconstruction dental primordia removed and bone remodeling pattern superimposed next to crypts of dental primordia. Red: apposition, green: resorption, blue: lining cells (resting area)

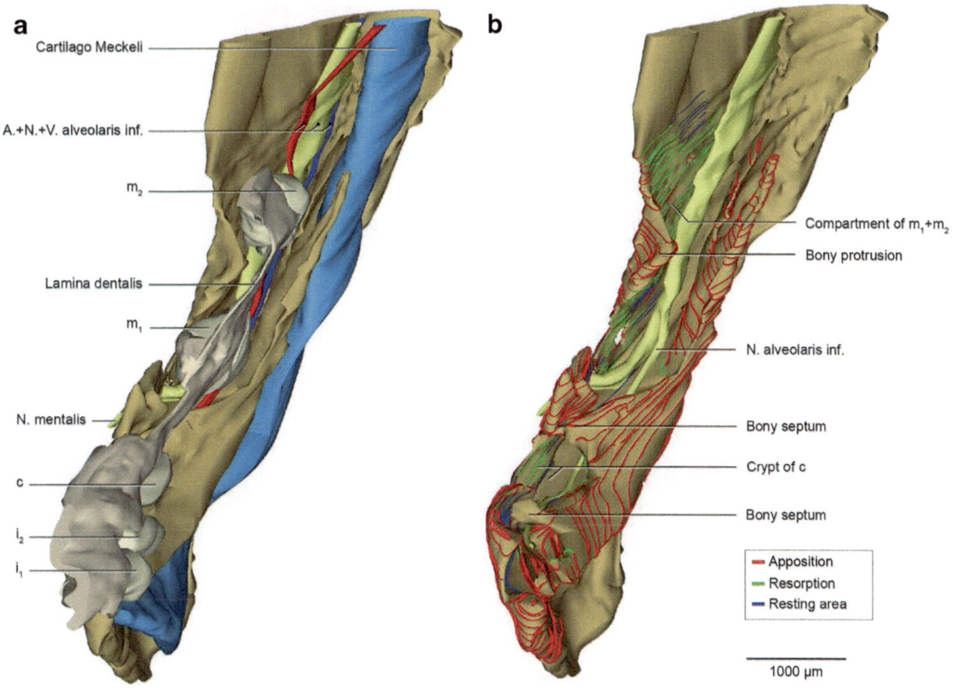

Fig. 24.8 (**a**) Human embryo, 68 mm CRL, partial 3D-reconstruction of the right half of the mandible and surrounding structures, cranial view and 45° medial view. (**b**) The same partial 3D-reconstruction dental primordia removed and bone remodeling pattern superimposed next to crypts of dental primordia. Red: apposition, green: resorption, blue: lining cells (resting area)

Fig. 24.9 (**a**) Human embryo, 68 mm CRL, horizontal section through the mandibular region of the right deciduous canine primordium (staining: hematoxylin-eosin). (**b**) 3D partial reconstruction of the right deciduous canine primordium of the mandible and the surrounding structures, histological section (staining: hematoxylin-eosin) in the background (appears blurry due to software issues). Cranial view. The dental primordium is sectioned horizontally. Bone remodeling pattern superimposed

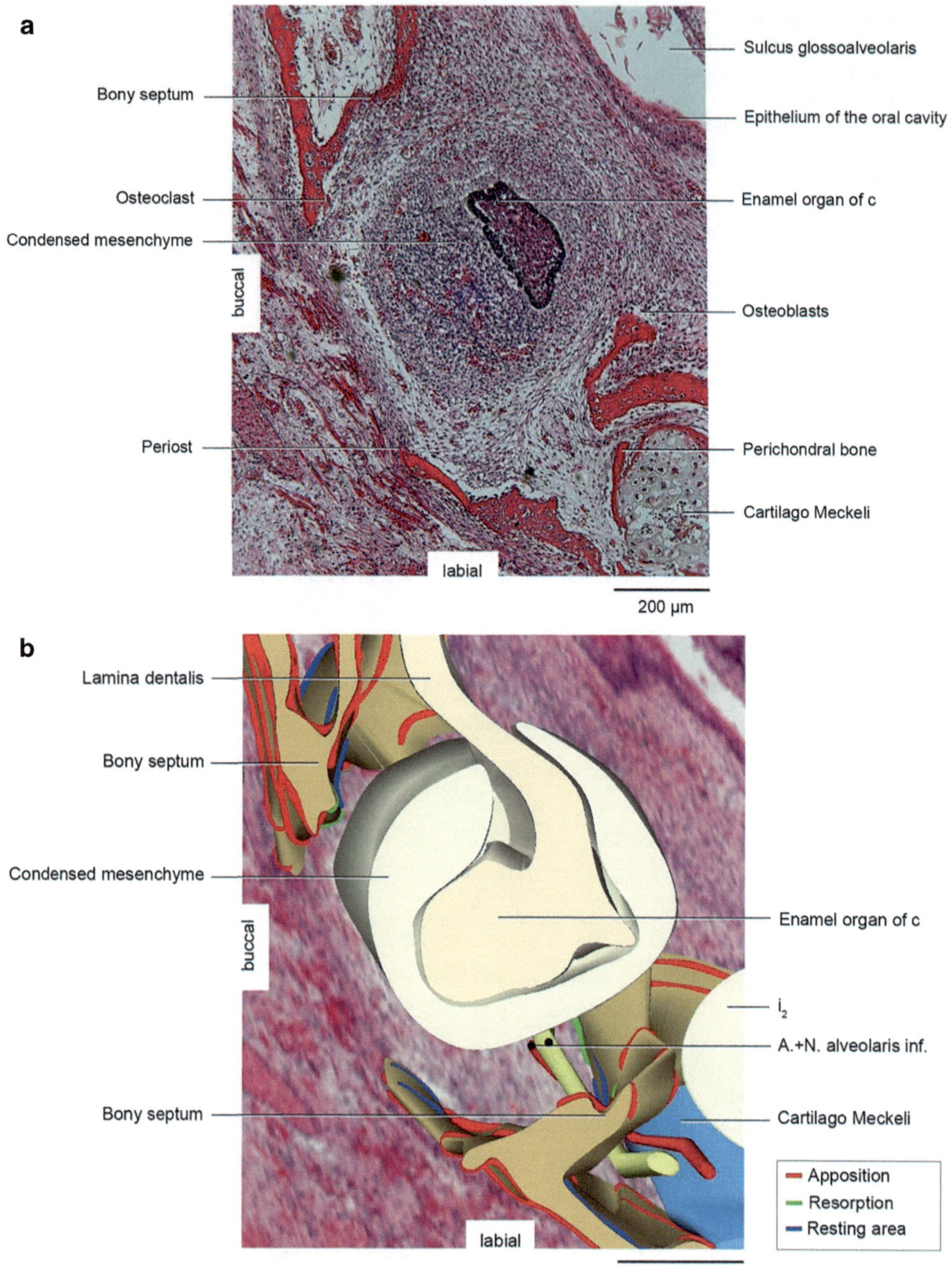

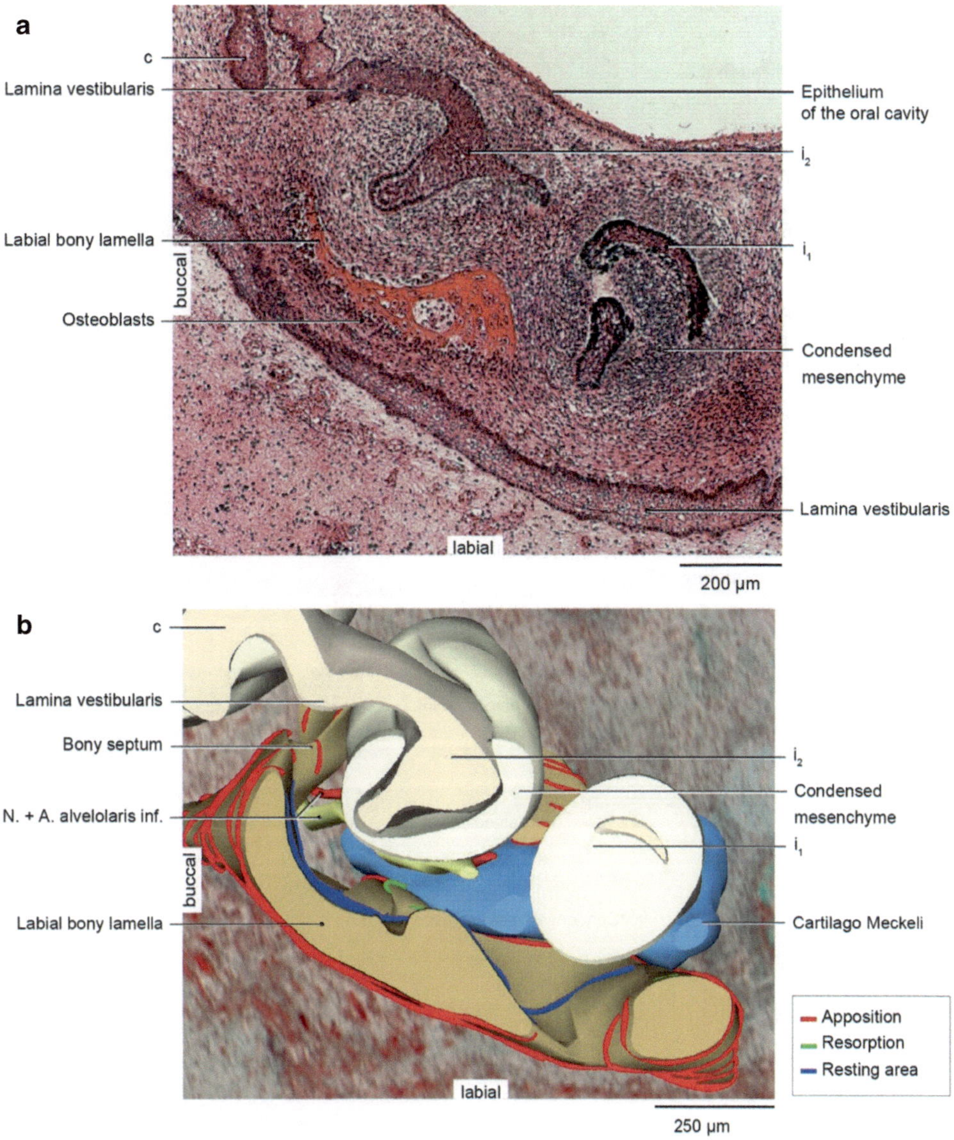

Fig. 24.10 (**a**) Human embryo, 68 mm CRL, horizontal section through the mandibular region of the right deciduous first and second incisor primordia (staining: hematoxylin-eosin). (**b**) 3D partial reconstruction of the right deciduous first and second incisor primordia of the mandible, histological section (staining: trichrome) in the background (appears blurry due to software issues). Cranial view. The dental primordia are sectioned horizontally. Bone remodeling pattern superimposed

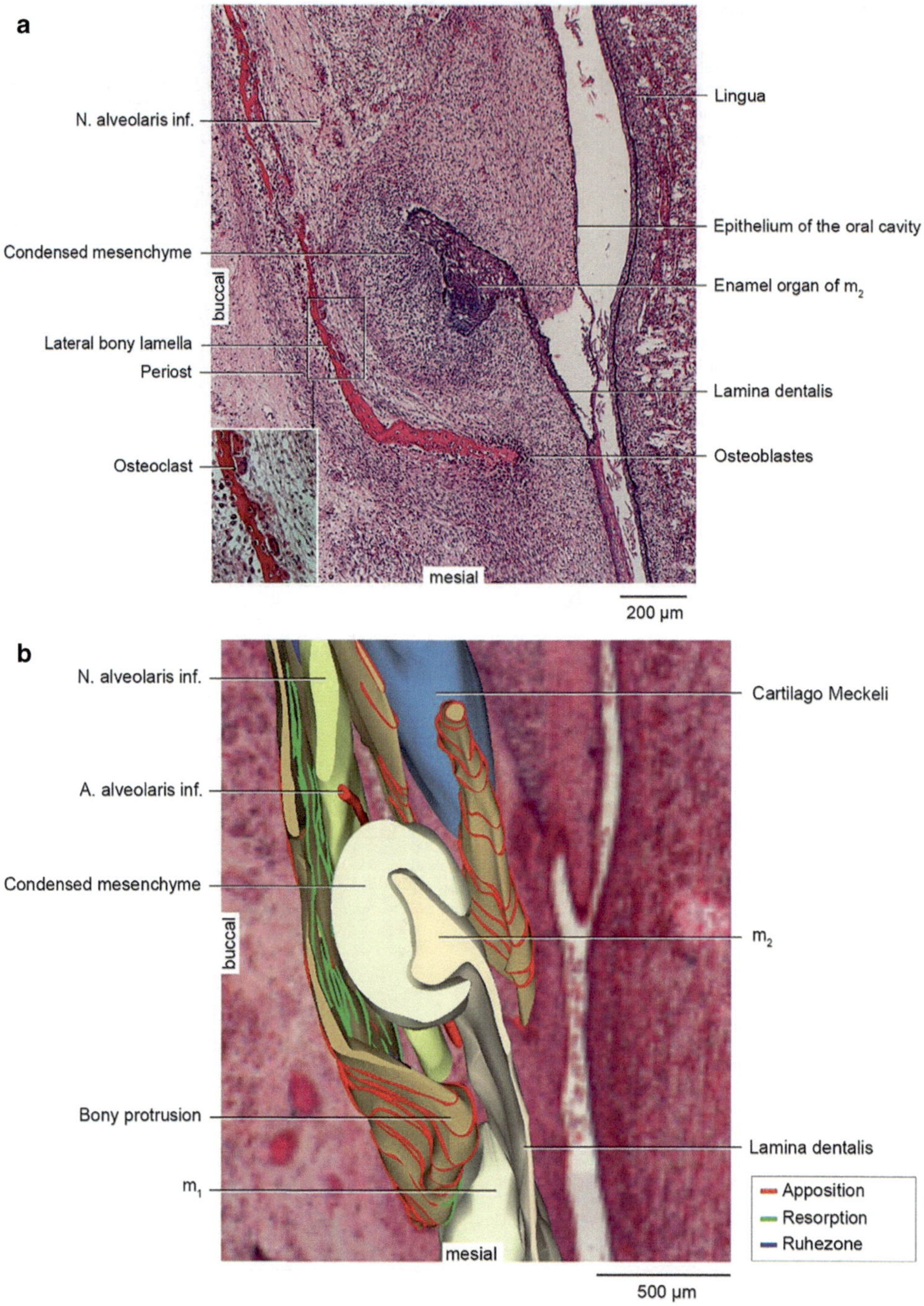

Fig. 24.11 (**a**) Human embryo, 68 mm CRL, horizontal section through the mandibular region of the right deciduous first and second molar primordia (staining: hematoxylin-eosin). The enlarged view (insert) shows the resorption of the buccal bone lamella by osteoclasts. (**b**) 3D partial reconstruction of the right deciduous first and second molar primordia of the mandible and their surrounding structures, histological section (staining: hematoxylin-eosin) in the background (appears blurry due to software issues). Cranial view. The dental primordium is sectioned horizontally. Bone remodeling pattern superimposed

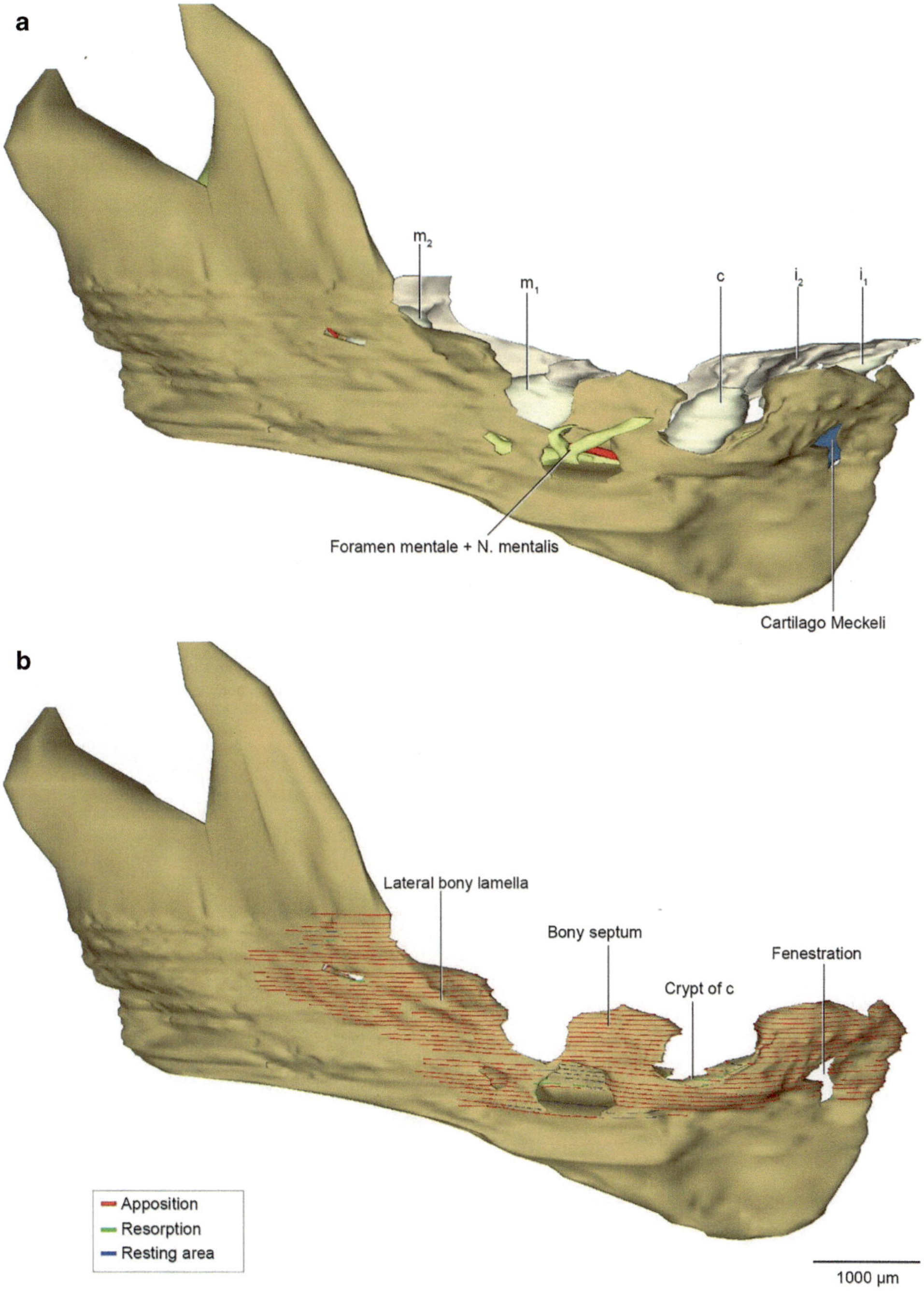

Fig. 24.12 (**a**) Human embryo, 68 mm CRL, right half of the mandible, lateral view. (**b**) The same reconstruction of the half of the mandible with bone remodeling pattern superimposed in the region of dental primordia

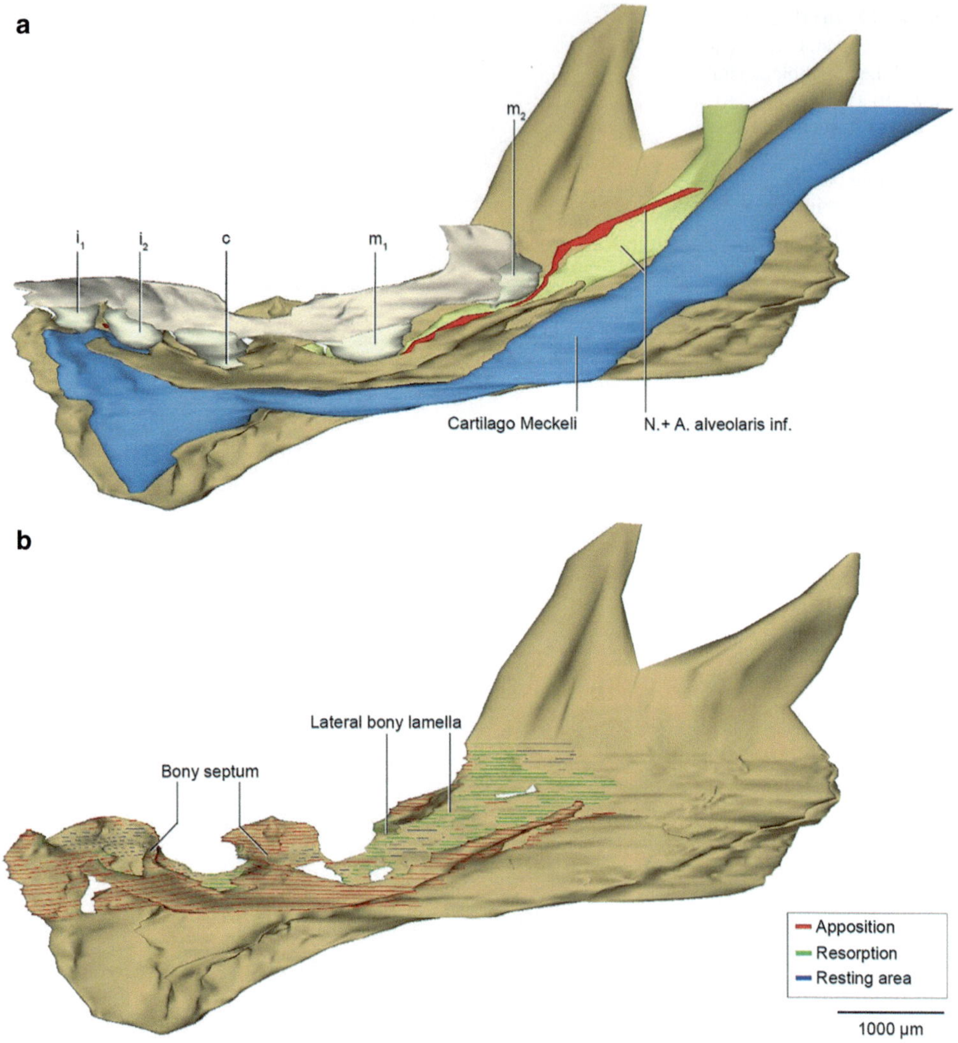

Fig. 24.13 (**a**) Human embryo, 68 mm CRL, right half of the mandible, medial view. (**b**) The same reconstruction of the half of the mandible with bone remodeling pattern superimposed in the region of dental primordia

Fig. 24.14 Human embryo, 68 mm CRL, maxilla (brown) with dental primordia (pink) and infraorbital n. (yellow, only on right side), frontal view

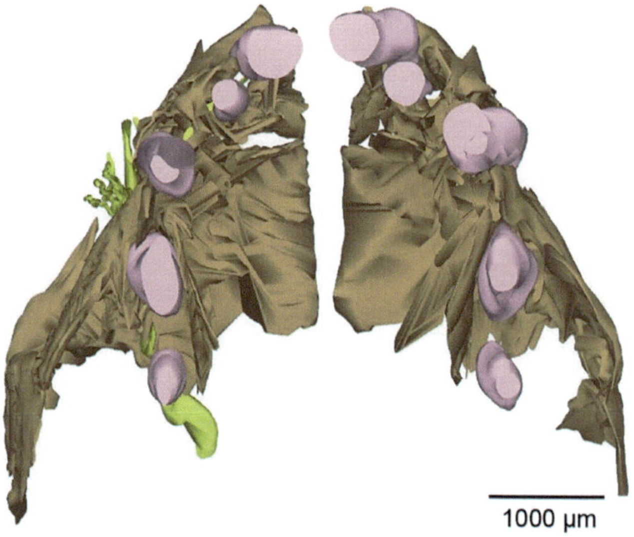

Fig. 24.15 Human embryo, 68 mm CRL, maxilla (brown) with dental primordia (pink) and infraorbital n. (yellow, only on right side), caudal view

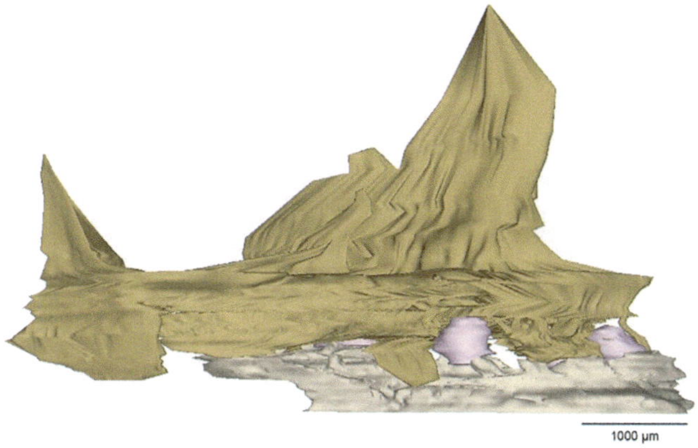

Fig. 24.16 Human embryo, 68 mm CRL, maxilla (brown) with dental primordia (pink), and dental lamina (gray), right lateral view

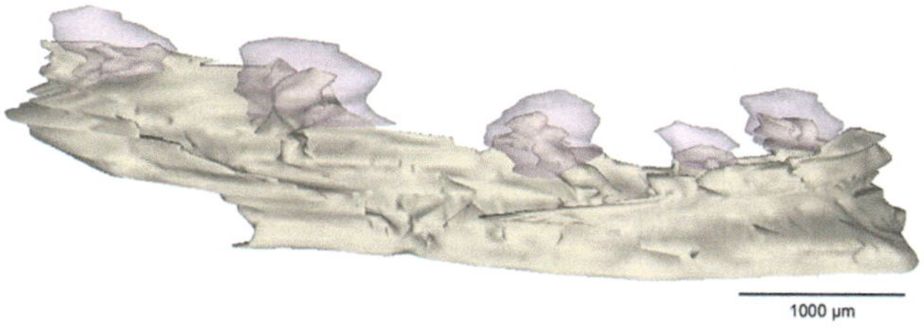

Fig. 24.17 Human embryo, 68 mm CRL, dental primordia (pink) and dental lamina (gray), right lateral view

Fig. 24.18 Human embryo, 68 mm CRL, right half of the maxilla with dental primordia, and bone remodeling pattern superimposed, caudal view, and corresponding histological section (staining: hematoxylin-eosin)

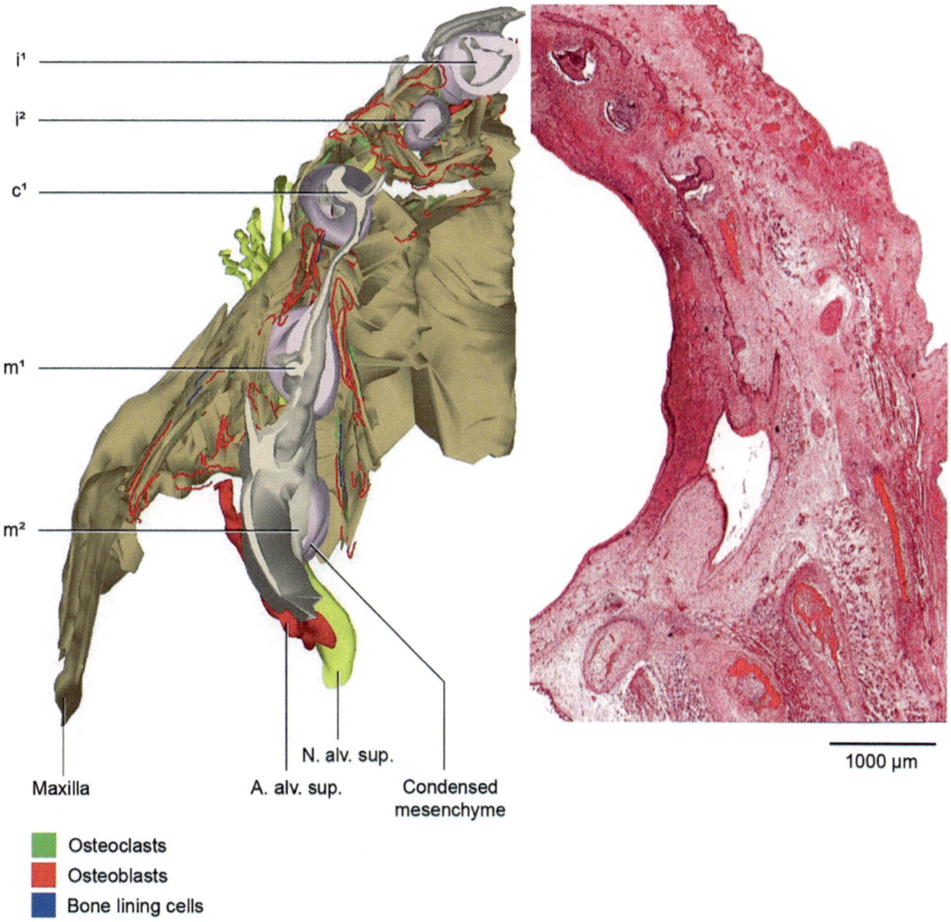

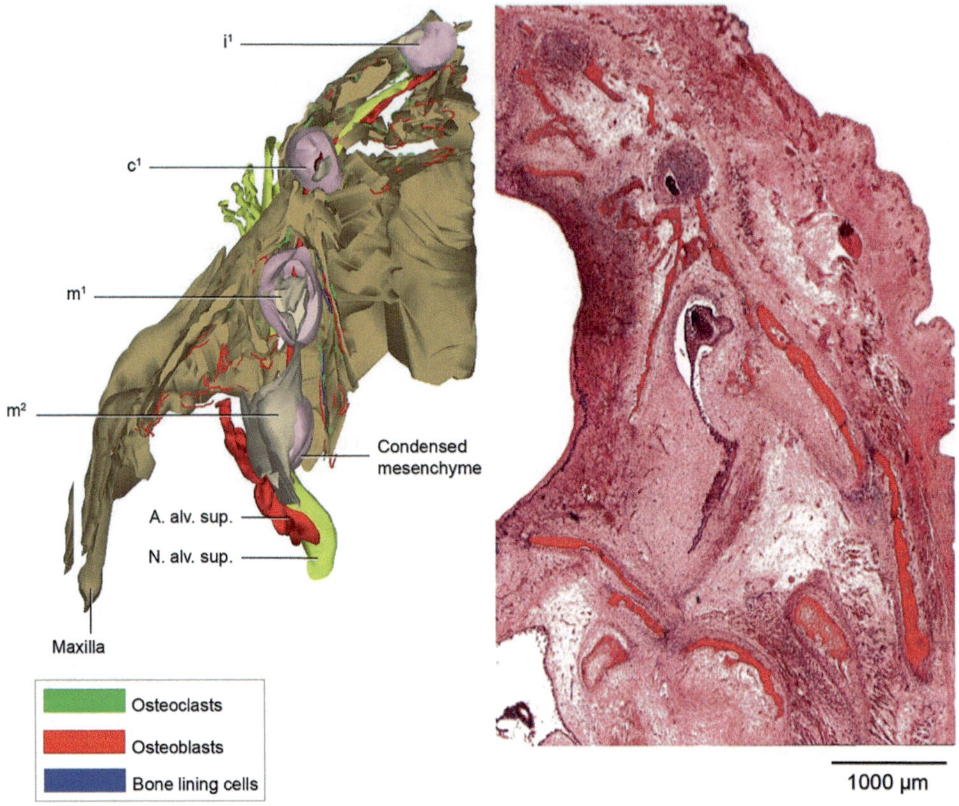

Fig. 24.19 Human embryo, 68 mm CRL, right half of the maxilla with dental primordia, caudal view, bone remodeling pattern superimposed, and corresponding histological section (staining: hematoxylin-eosin)

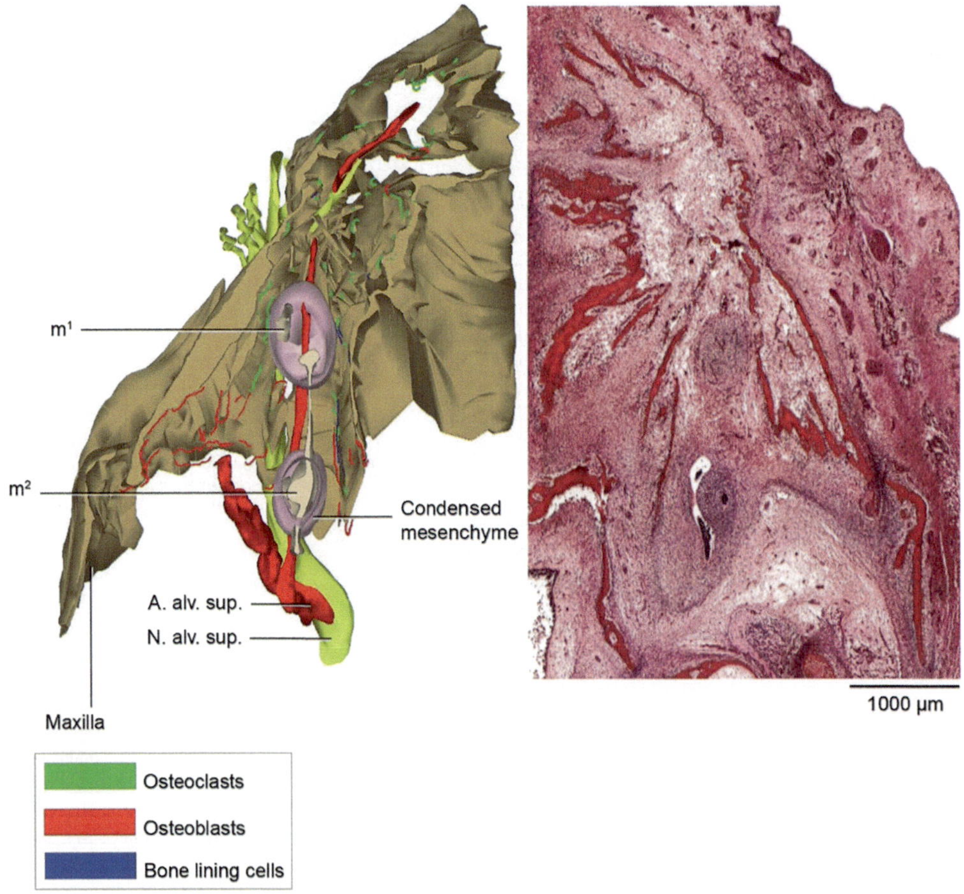

Fig. 24.20 Human embryo, 68 mm CRL, right half of the maxilla with dental primordia, more focused on m^2, caudal view, bone remodeling pattern superimposed, and corresponding histological section (staining: hematoxylin-eosin)

Further Reading

Kalinke U. Embryologische Untersuchungen zur Entwicklung des Alveolarknochens im Oberkiefer beim Menschen zur Zeit der vorgeburtlichen Entwicklungsstadien von 19–68 mm SSL. Med. Diss. Charité-Universitätsmedizin Berlin; 2014.

Konietzny M. Zur Entwicklung des Kieferknochens im Bereich der Zahnanlagen—Histologische und morphologische Untersuchungen an der fetalen Maxilla des Menschen während des Glockenstadiums der Zahnanlagen (160–270 mm SSL). Med. Diss. Charité-Universitätsmedizin Berlin; 2014.

Radlanski RJ. Oral structure and biology. Batavia: Quintessence Publishing; 2018.

Radlanski RJ, Renz H, Klarkowski MC. Prenatal development of the human mandible. 3D reconstructions, morphometry and bone remodelling pattern, sizes 12–117 mm CRL. Anat Embryol (Berl). 2003;207:221–32.

Schuster FP. Zur Entwicklung des Kieferknochens im Bereich der Zahnanlagen. Histologische und morphologische Untersuchungen an der fetalen Mandibula des Menschen während des Glockenstadiums der Zahnanlagen (68–160 mm SSL). Med. Diss. Charité-Universitätsmedizin Berlin; 2013.

Tsengelsaikhan N. Zur Entwicklung des Kieferknochens im Bereich der Zahnanlagen—Histologische und morphologische Untersuchung an der embryonalen Mandibula des Menschen während der frühen Entwicklungsstadien der Zahnanlagen (19–68 mm SSL). Med. Diss. Charité-Universitätsmedizin Berlin; 2014.

Zimmermann C. Zur Entwicklung des Kieferknochens im Bereich der Zahnanlagen. Histologische und morphologische Untersuchungen an der fetalen Mandibula des Menschen während des späten Glockenstadiums der Zahnanlagen (160–270 mm SSL). Med. Diss. Charité-Universitätsmedizin Berlin; 2013.

Embryo 76 mm CRL (ca. 11th wk): FUL140489

25

Overall Remarks

In this largely complete head, we can see how the cranial skeleton has continued to develop. Parts of the frontal bone are present, the frontal processes of the maxilla reach far up to the cranial side; there is even a paired nasal bone already present. The mandible also shows both, the condylar and the coronoid process, but a functioning secondary temporomandibular joint is not yet formed. Meckel's cartilage with its malleus still articulates with the incus during the movements of the mandible. Of course, all the nerves are already very well developed. This reconstruction shows the superior alveolar nerves particularly well, which branch widely into the region of the maxillary teeth (Figs. 25.1, 25.2, 25.3, 25.4, 25.5, 25.6, 25.7, 25.8, 25.9, 25.10, 25.11, 25.12, 25.13, 25.14, 25.15, 25.16, 25.17, 25.18, 25.19, and 25.20).

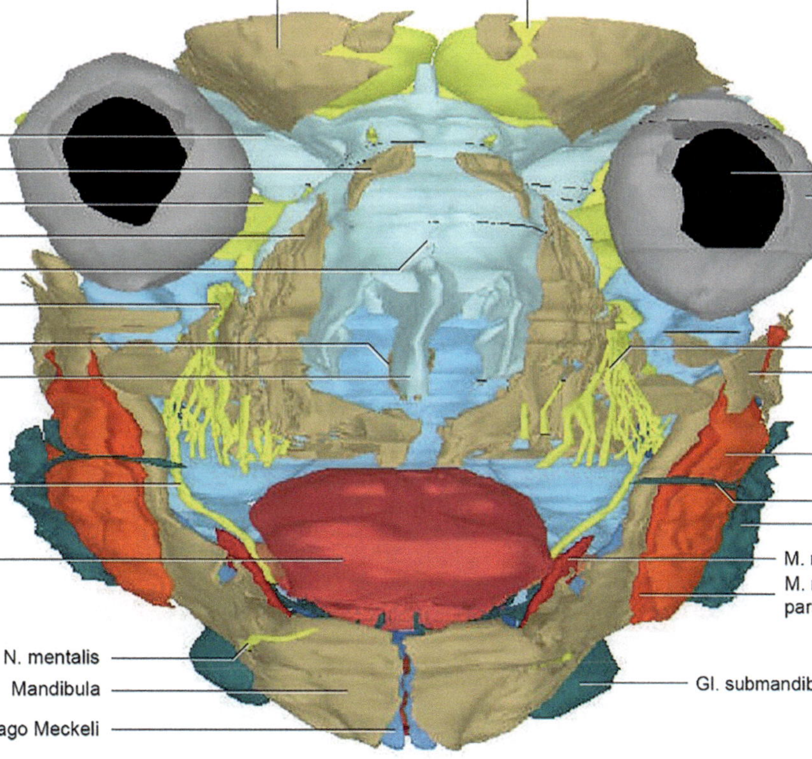

Fig. 25.1 Human embryo, 76 mm CRL, head, frontal view, skin removed, layer 1 of preparation. (*From* Radlanski and Renz [1]; *with permission*)

R. J. Radlanski, *Atlas of Human Prenatal Craniofacial Development*, https://doi.org/10.1007/978-3-031-88812-0_25

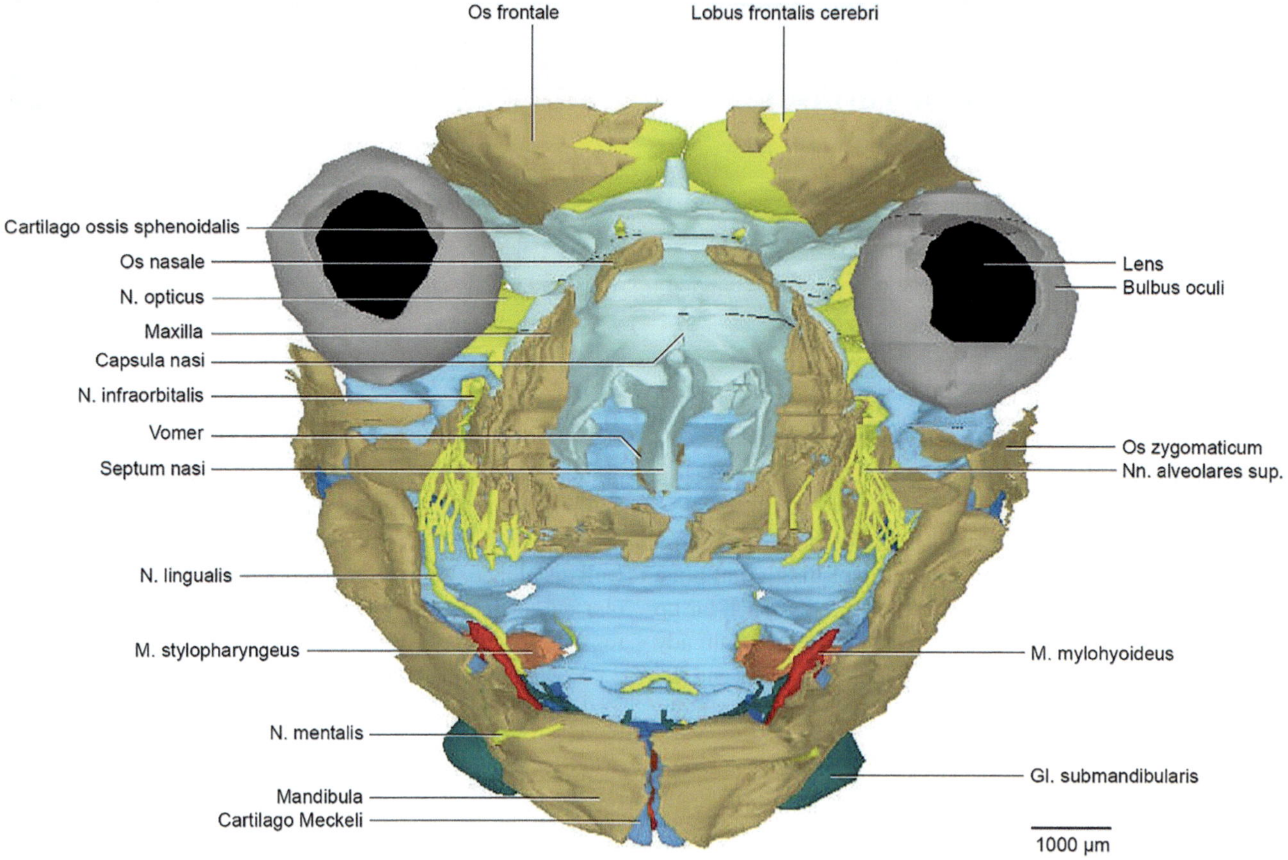

Fig. 25.2 Human embryo, 76 mm CRL, head, frontal view, layer 2 of preparation. (*From* Radlanski and Renz [1]; *with permission*)

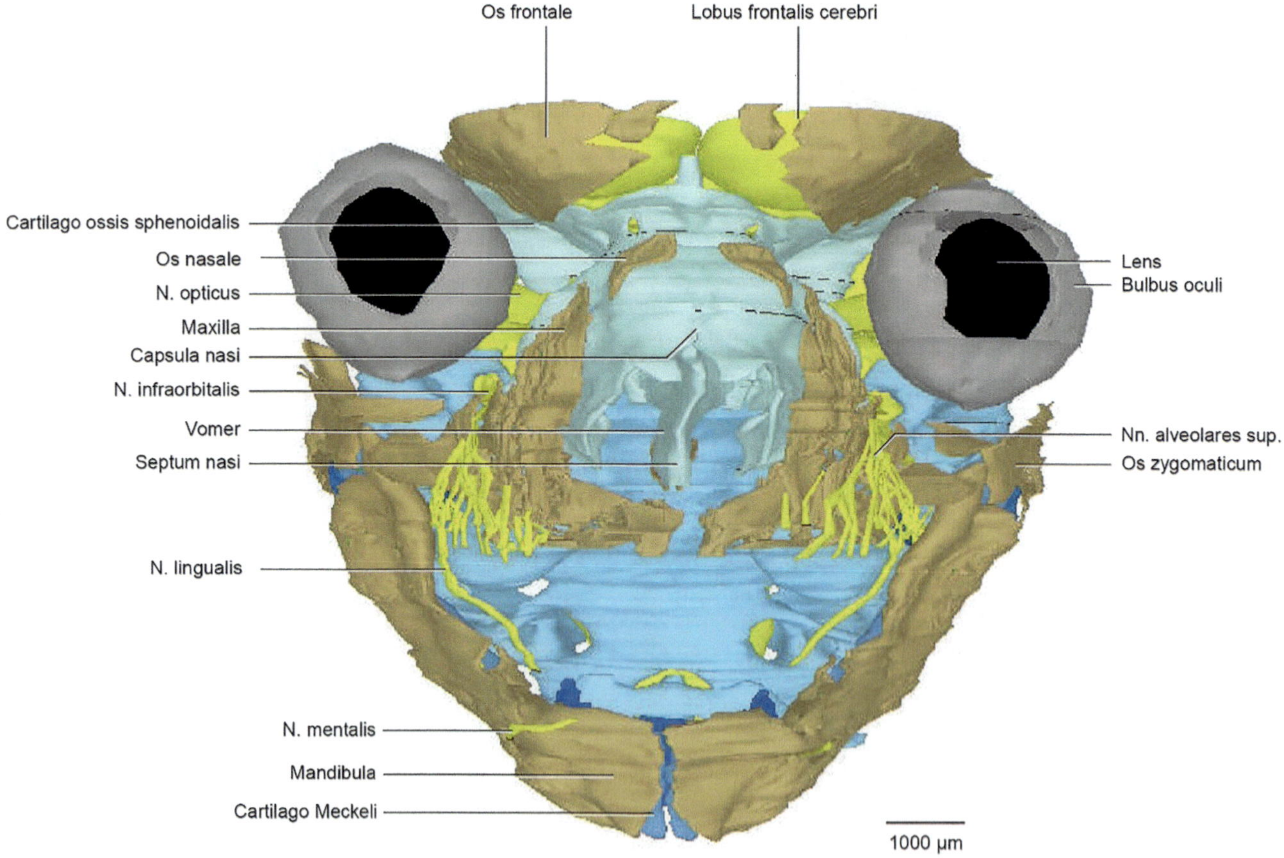

Fig. 25.3 Human embryo, 76 mm CRL, head, frontal view, layer 3 of preparation. (*From* Radlanski and Renz [1]; *with permission*)

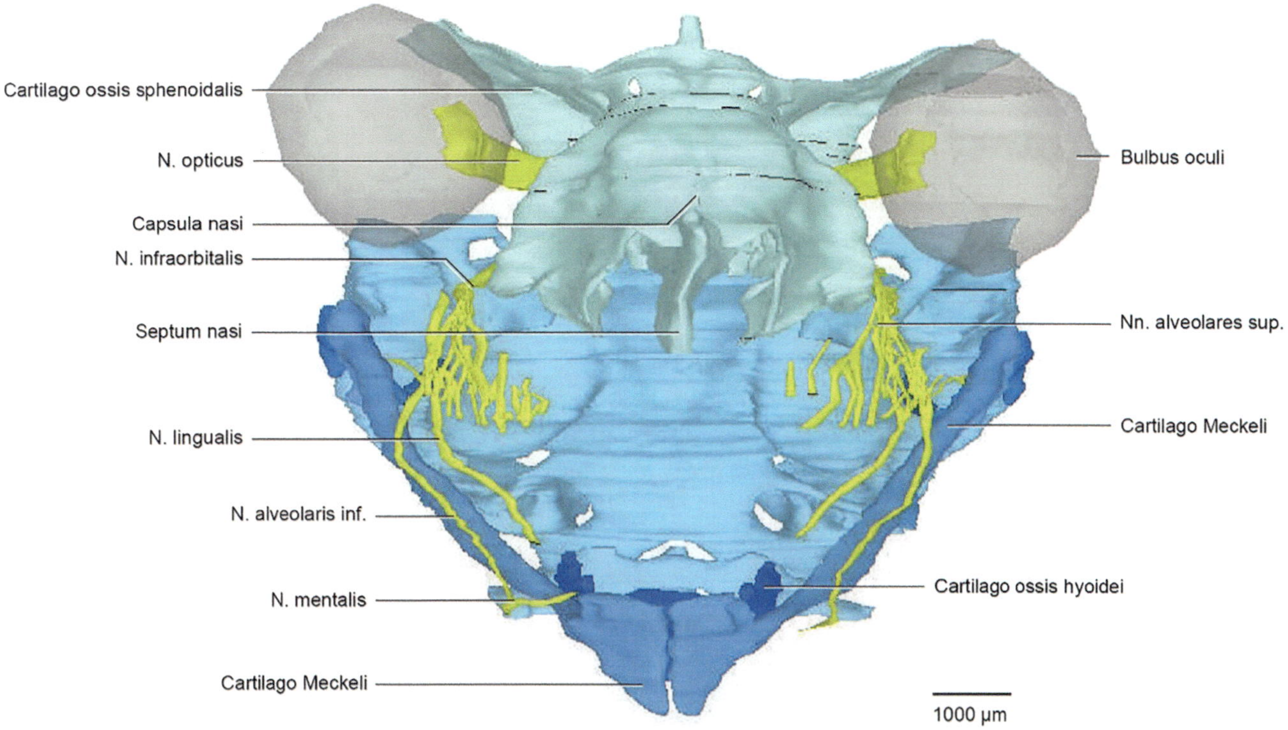

Fig. 25.4 Human embryo, 76 mm CRL, head, frontal view, cartilage, nerves, and eyeballs. (*From* Radlanski and Renz [1]; *with permission*)

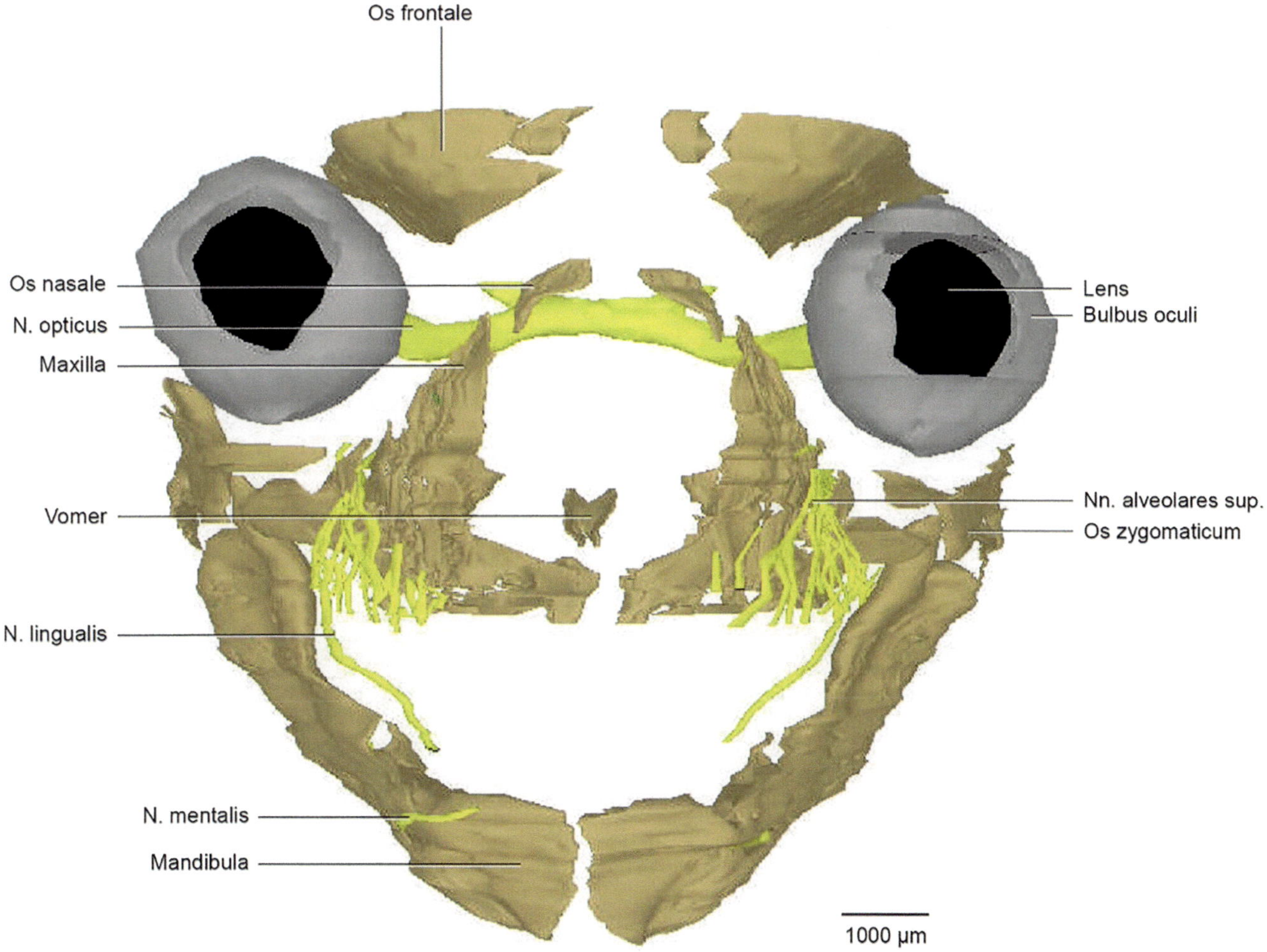

Fig. 25.5 Human embryo, 76 mm CRL, head, frontal view, bone, nerves, and eyeballs. (*From* Radlanski and Renz [1]; *with permission*)

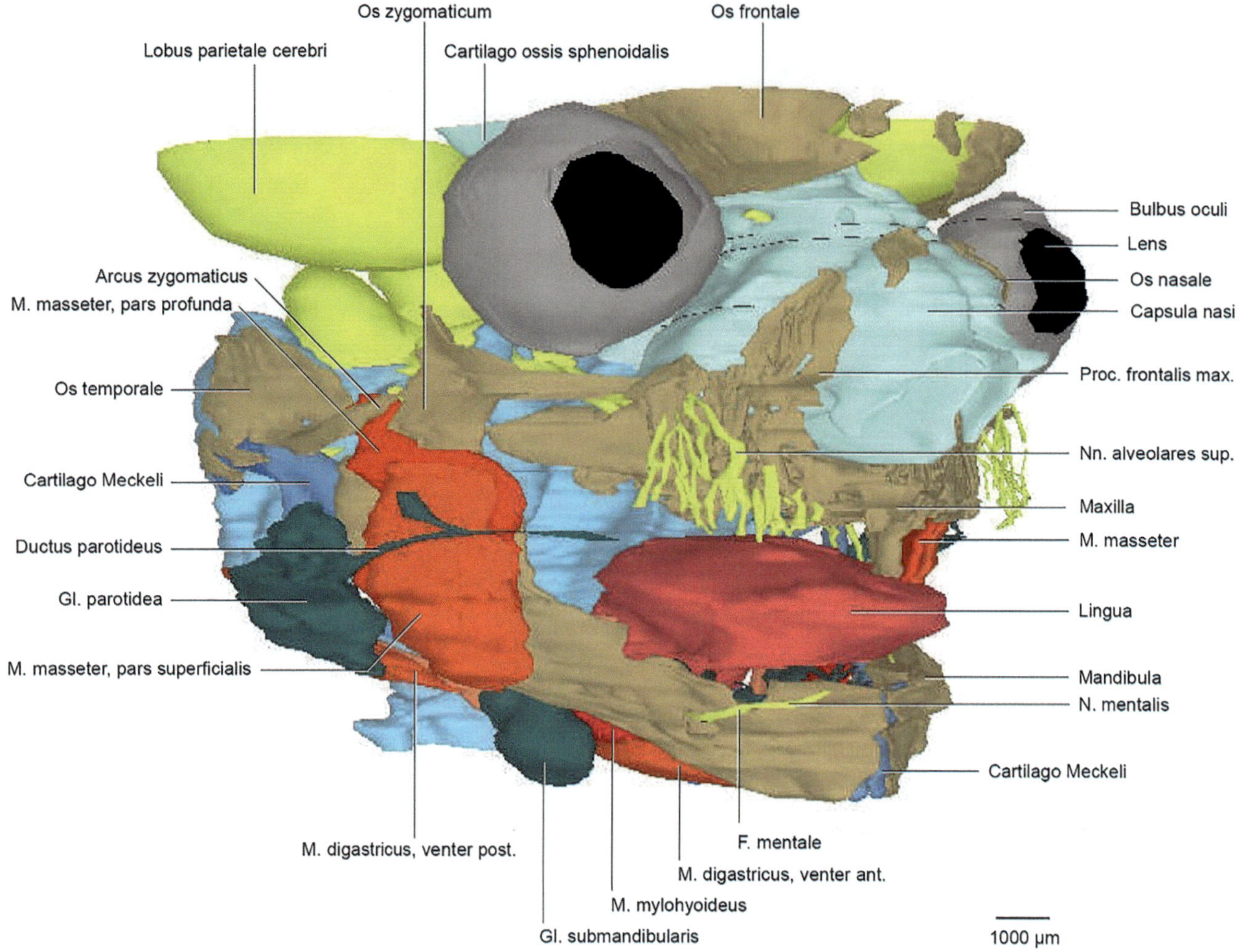

Fig. 25.6 Human embryo, 76 mm CRL, head, 45° lateral view, skin removed, layer 1 of preparation. (*From* Fernández Rubio and Radlanski [2]; *with permission*)

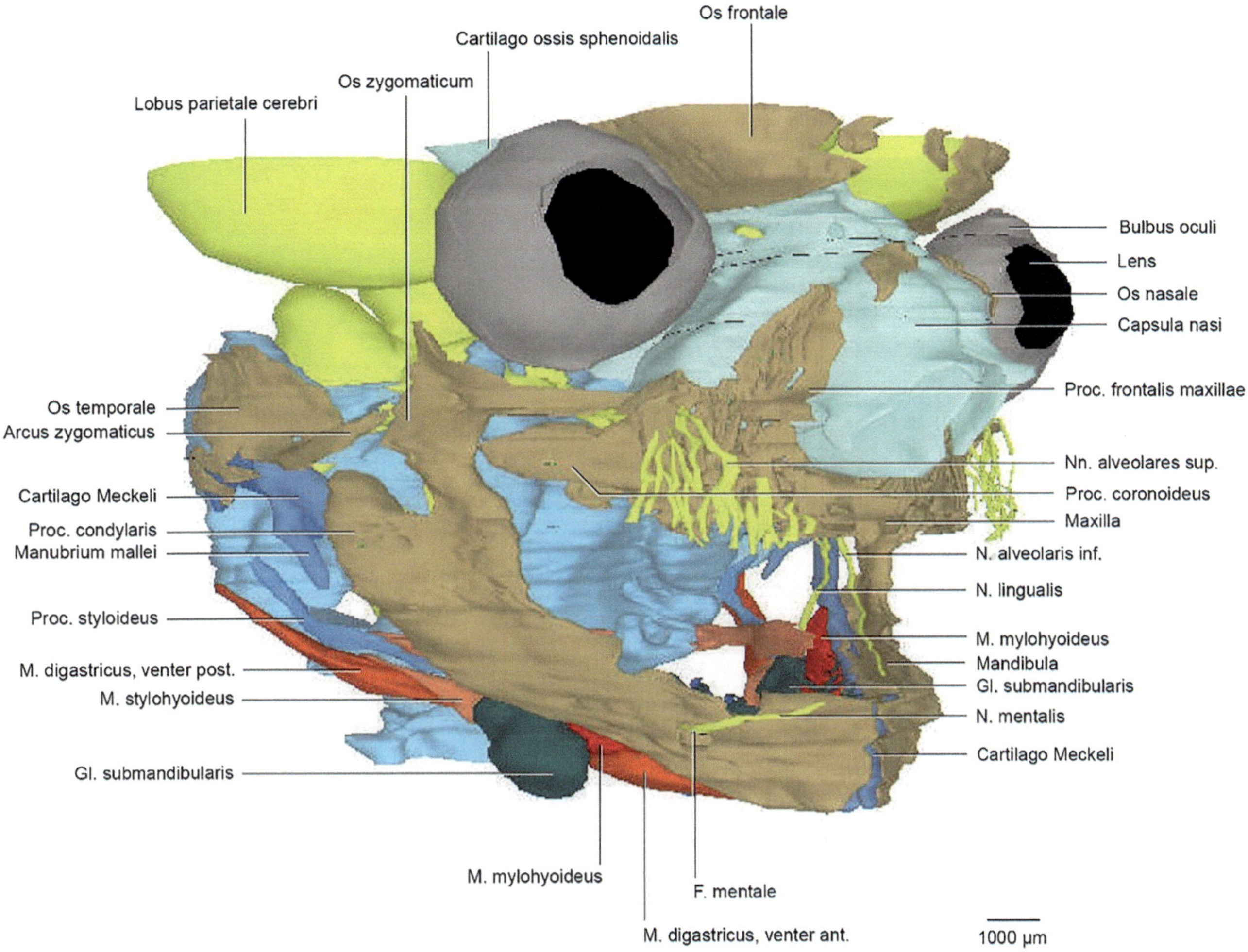

Fig. 25.7 Human embryo, 76 mm CRL, head, 45° lateral view, layer 2 of preparation

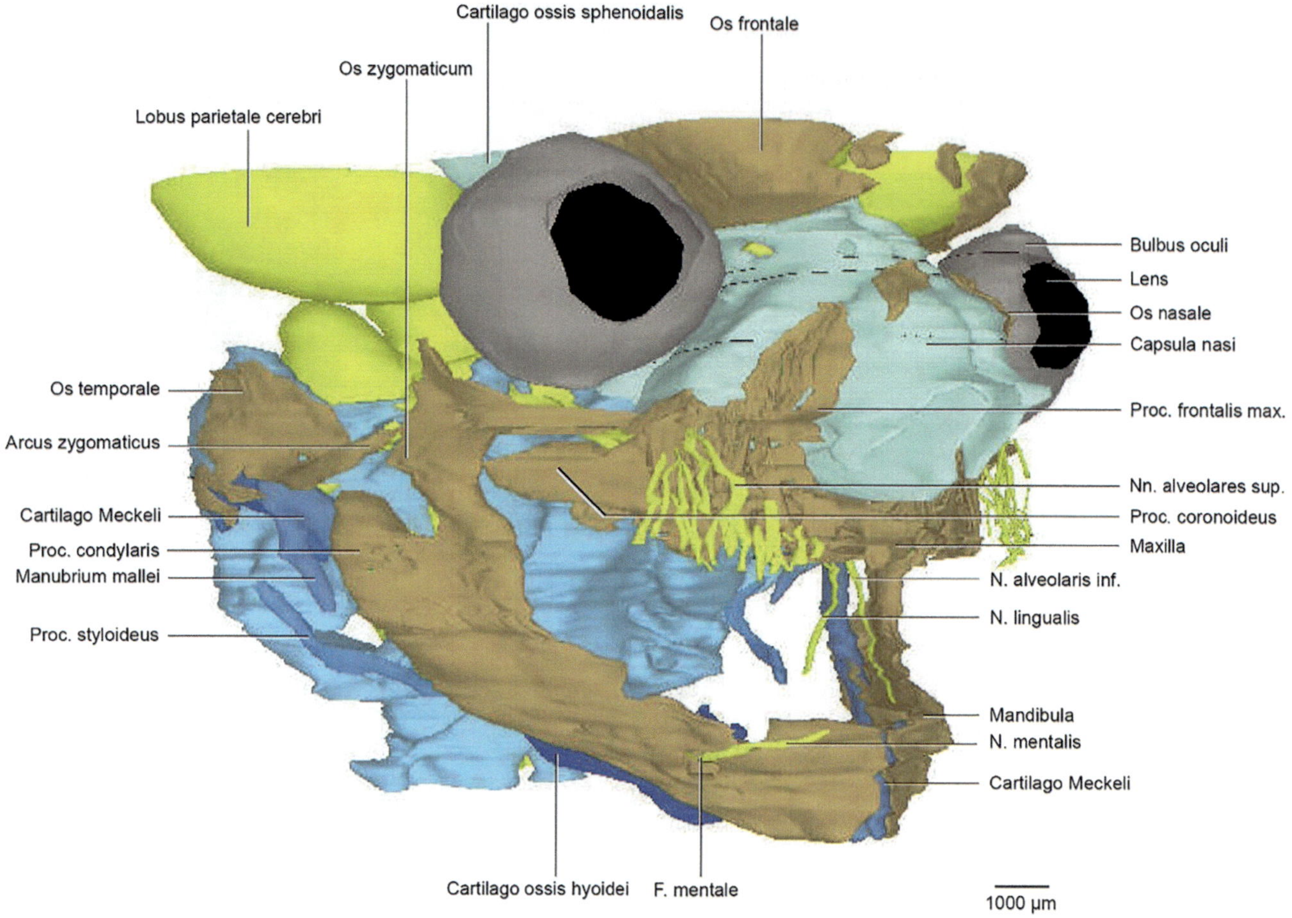

Fig. 25.8 Human embryo, 76 mm CRL, head, 45° lateral view, layer 3 of preparation

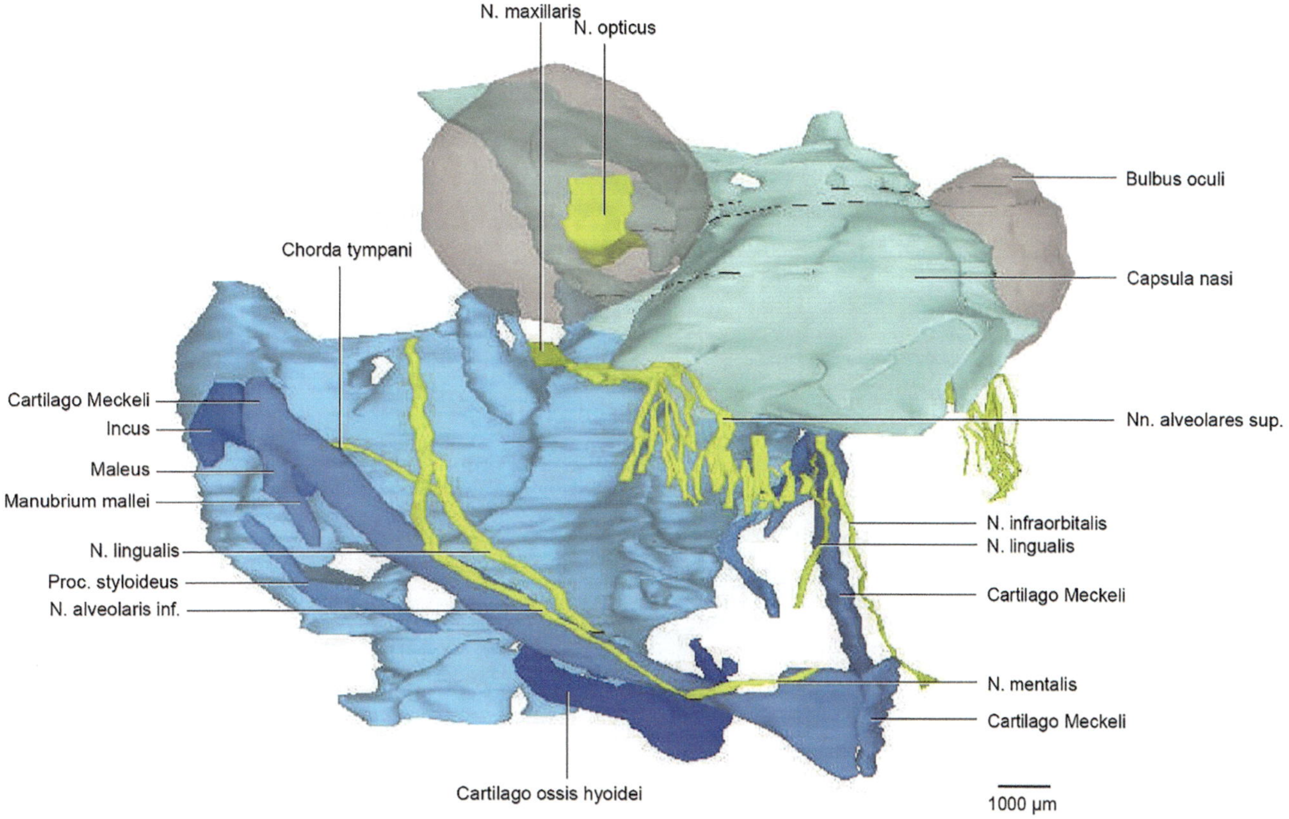

Fig. 25.9 Human embryo, 76 mm CRL, head, 45° lateral view, cartilage, nerves, and eyeballs

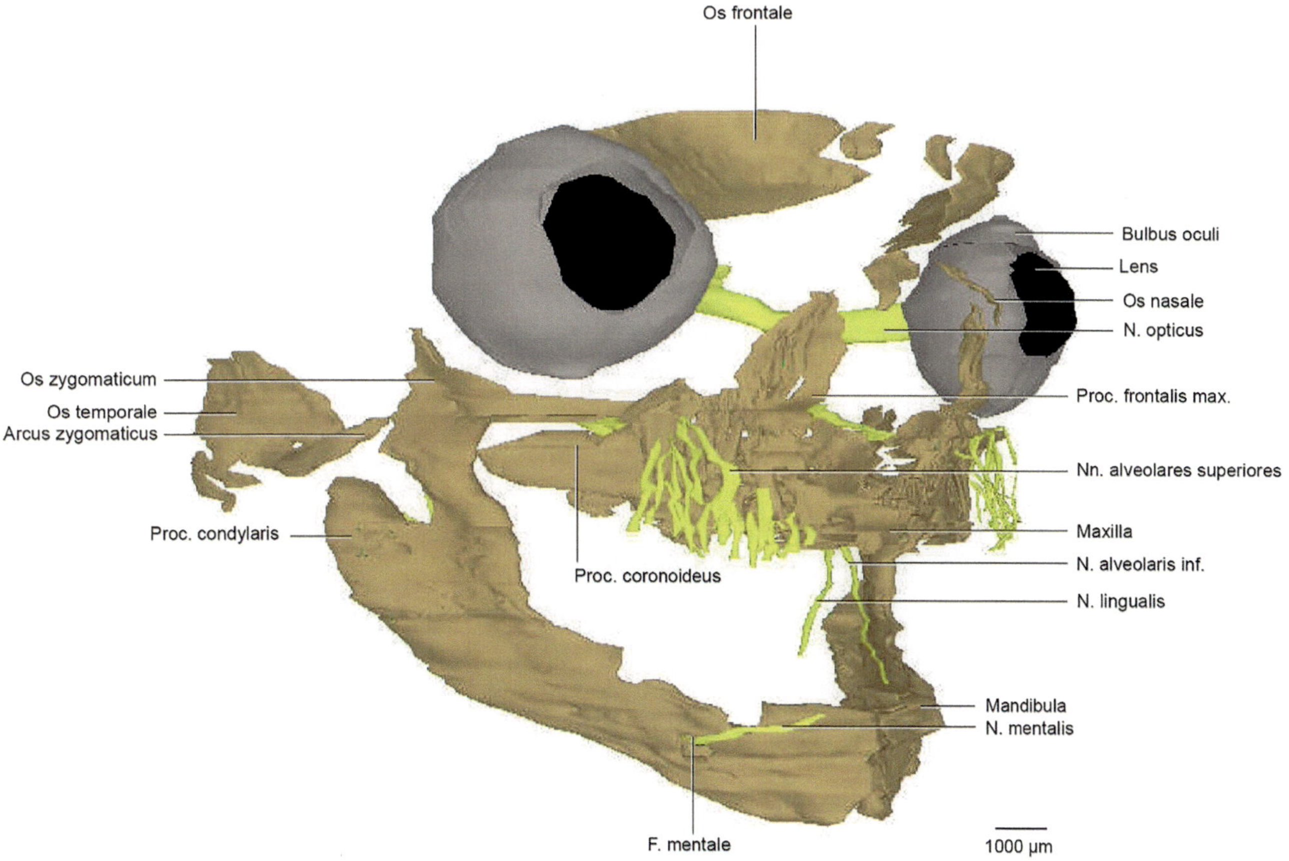

Fig. 25.10 Human embryo, 76 mm CRL, head, 45° lateral view, bone, nerves, and eyeballs

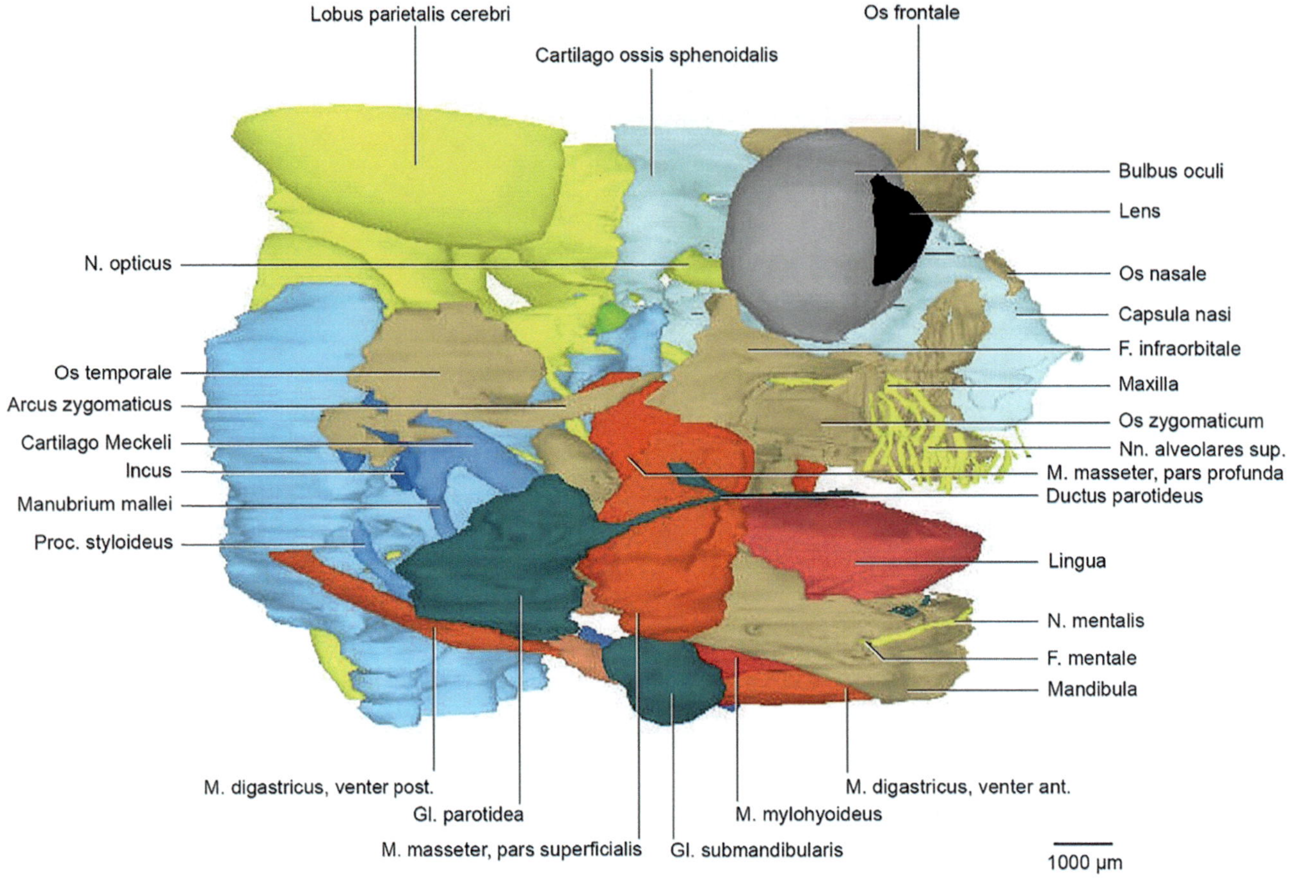

Fig. 25.11 Human embryo, 76 mm CRL, head, lateral view, skin removed, layer 1 of preparation

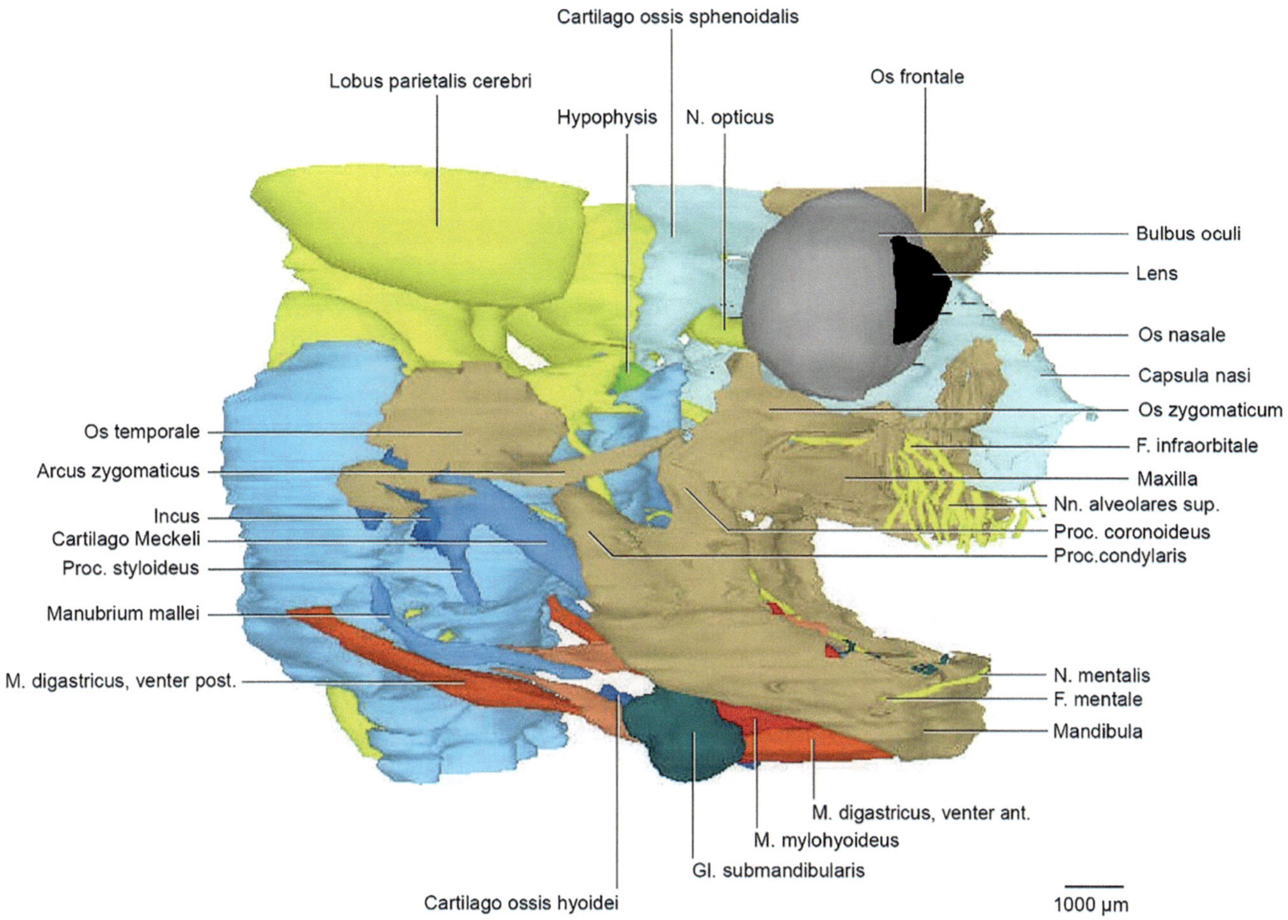

Fig. 25.12 Human embryo, 76 mm CRL, head, lateral view, layer 2 of preparation

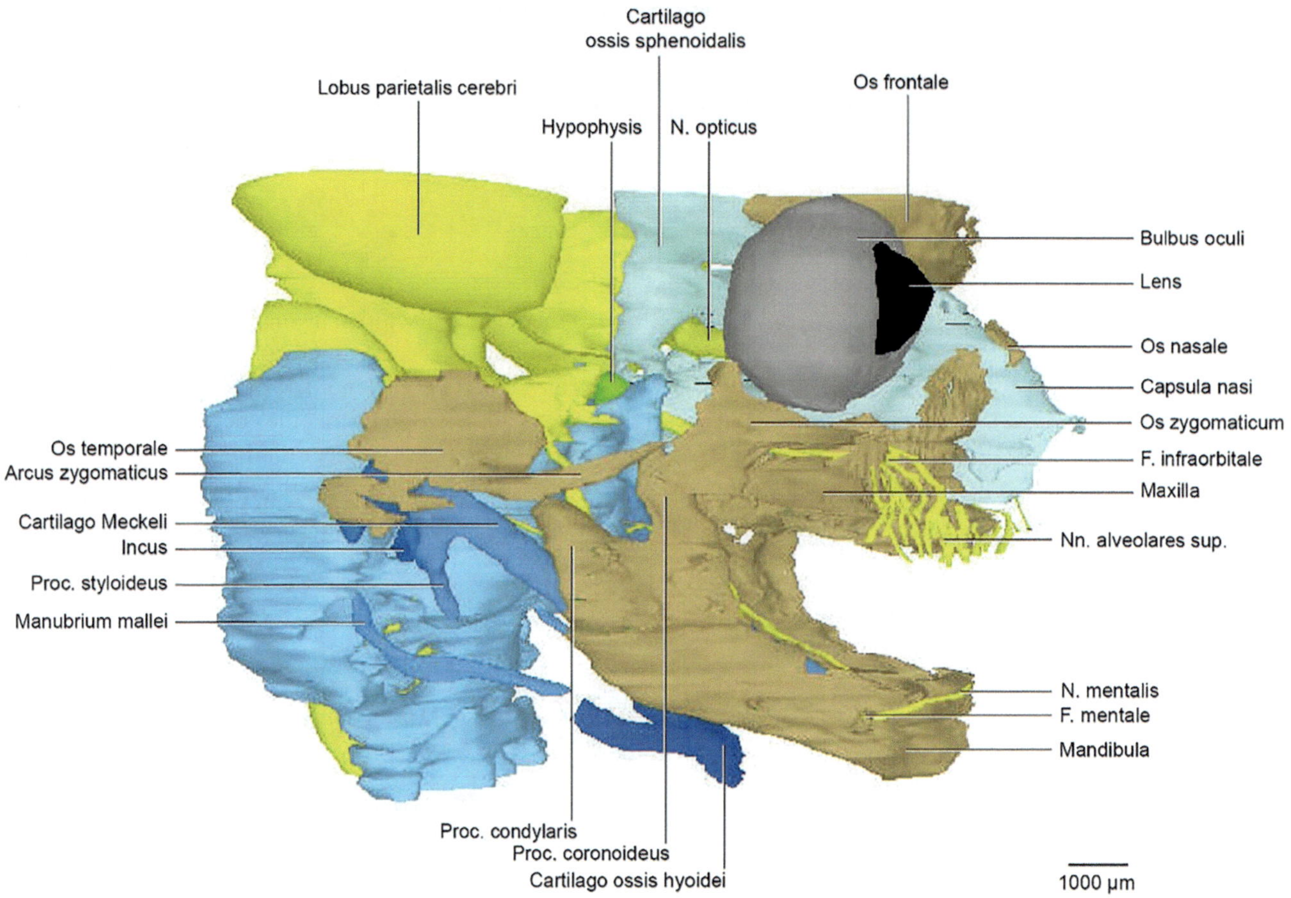

Fig. 25.13 Human embryo, 76 mm CRL, head, lateral view, layer 3 of preparation

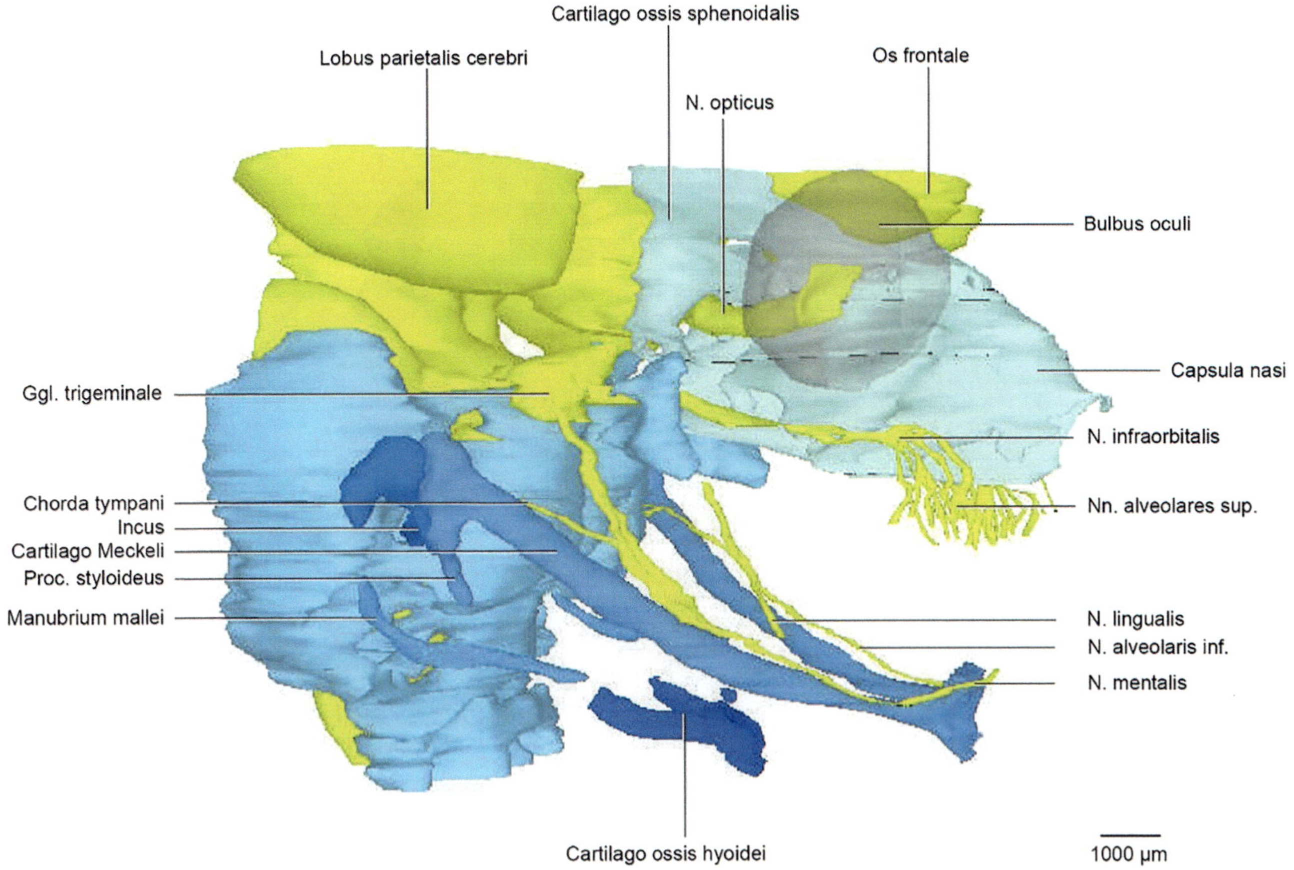

Fig. 25.14 Human embryo, 76 mm CRL, head, lateral view, layer 4 of preparation

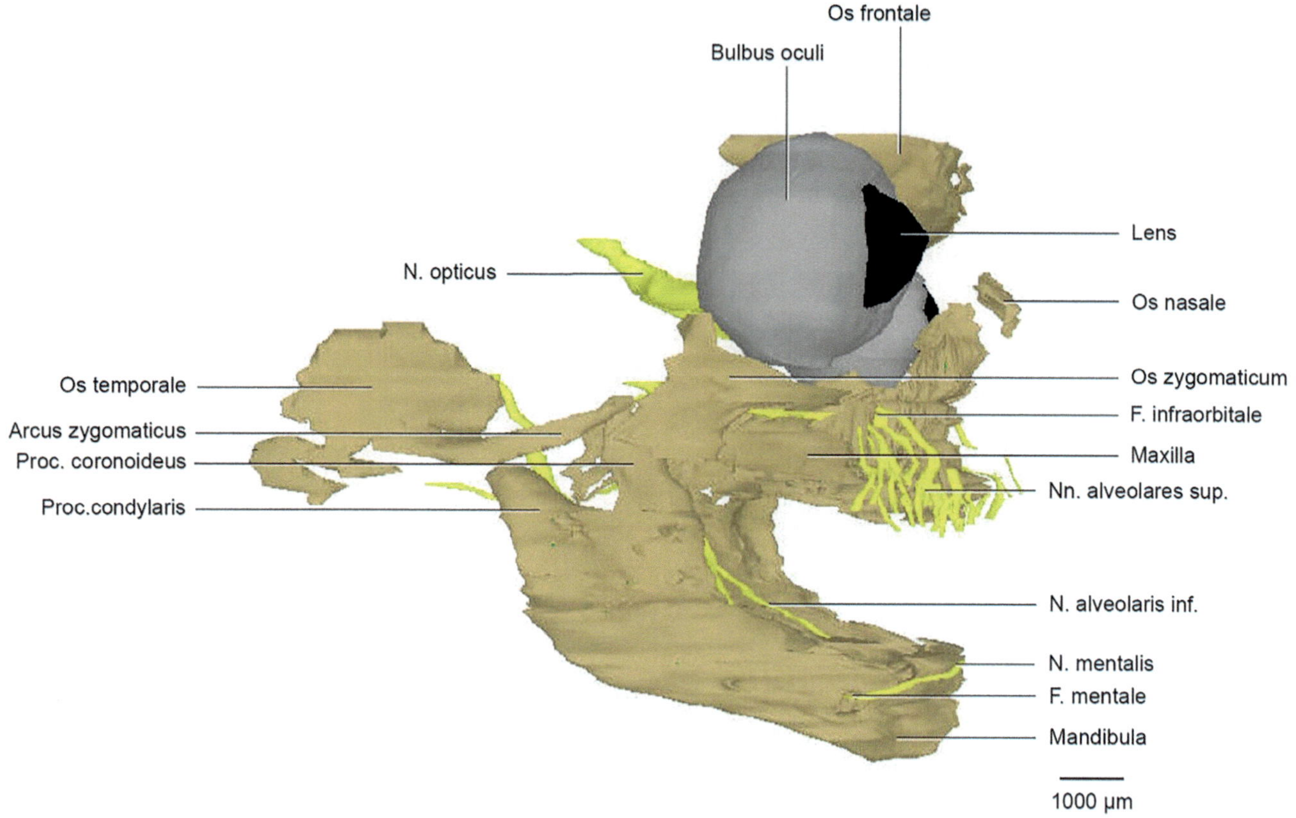

Fig. 25.15 Human embryo, 76 mm CRL, head, lateral view, bone, nerves, and eyeballs

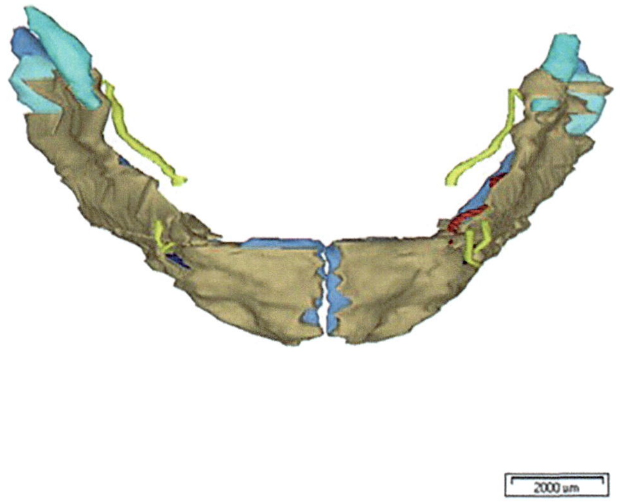

Fig. 25.16 Human embryo, 76 mm CRL, mandible, with Meckel's cartilage, cartilaginous cores of condylar and coronoid processes, and inferior alveolar artery and nerve (and mentalis n.), anterior view

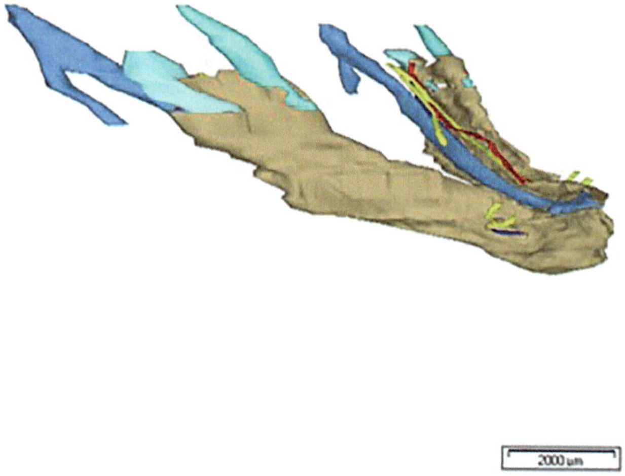

Fig. 25.17 Human embryo, 76 mm CRL, mandible, with Meckel's cartilage, cartilaginous cores of condylar and coronoid processes, and inferior alveolar artery and nerve (and mentalis n.), lateral view

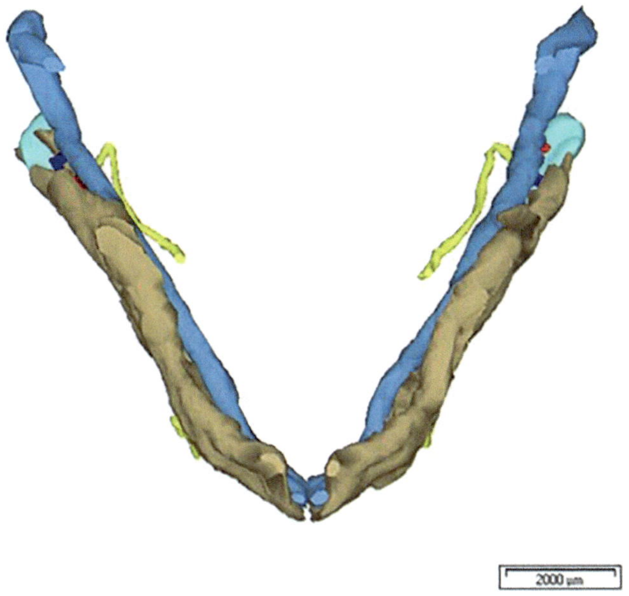

Fig. 25.18 Human embryo, 76 mm CRL, mandible, with Meckel's cartilage, cartilaginous cores of condylar and coronoid processes, and inferior alveolar artery and nerve (and mentalis n.), caudal view

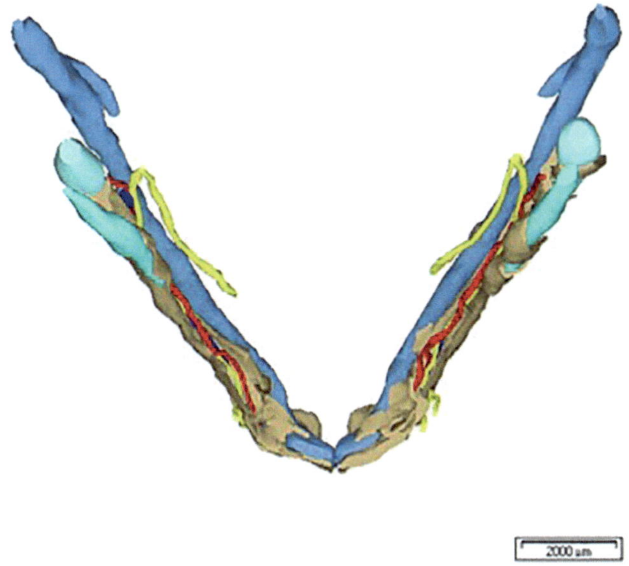

Fig. 25.19 Human embryo, 76 mm CRL, mandible, with Meckel's cartilage, cartilaginous cores of condylar and coronoid processes, and inferior alveolar artery and nerve (and mentalis n.), cranial view

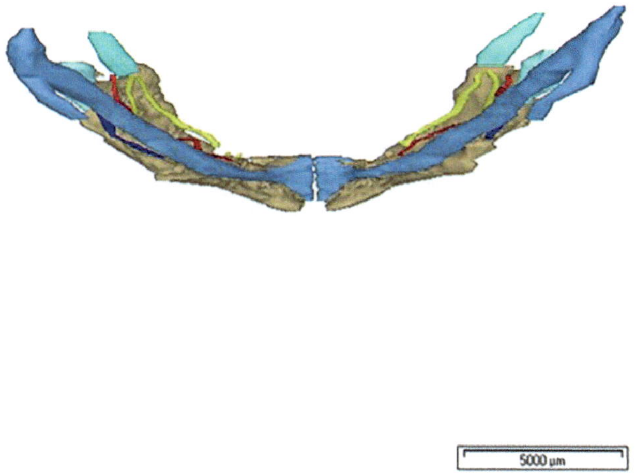

Fig. 25.20 Human embryo, 76 mm CRL, mandible with Meckel's cartilage, cartilaginous cores of condylar and coronoid processes, and inferior alveolar artery and nerve (and mentalis n.), dorsal view

Cartilaginous Cores

At this stage of development, we found such cartilaginous nuclei in all embryos of this age in the condylar process, but also in the coronoid process. They are also addressed as "secondary cartilages." Particularly in the coronoid process, this nucleus leads to rapid growth of the condyle. This is not only important for prenatal development, but also represents a growth reservoir for the adaptive growth of the mandible during the pubertal growth spurt. Even in adulthood, adaptive remodeling can be triggered by orthodontic measures—they also originate from this early cartilage tissue (Figs. 25.21 and 25.22).

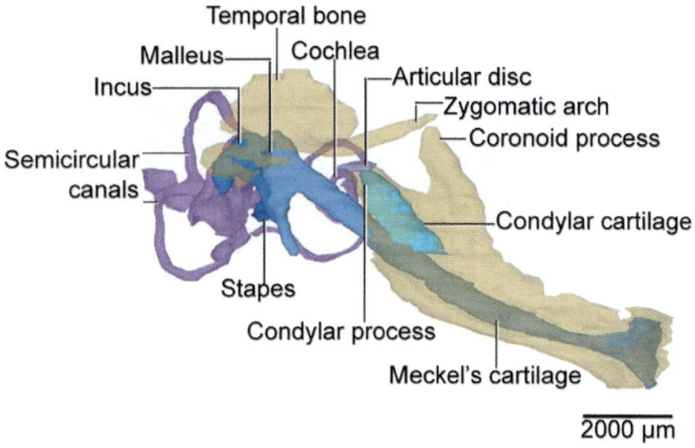

Fig. 25.21 3D reconstruction of selected parts of the temporomandibular joint, middle, and inner ear. Mandible and temporal bone made transparent. (*From* Fernández Rubio and Radlanski [2]; *with permission*)

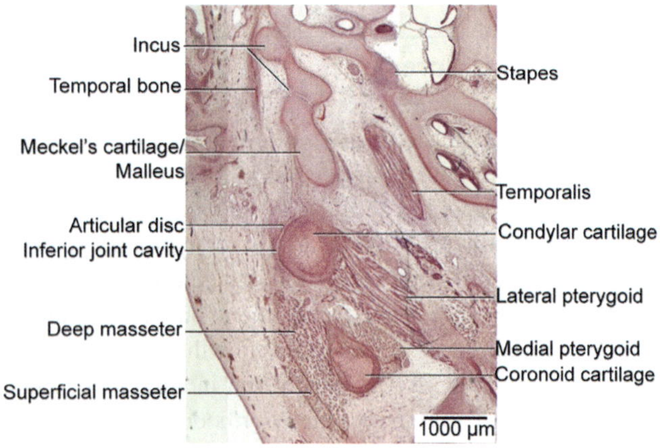

Fig. 25.22 Histological section of the region of the right middle ear and selected parts of the temporomandibular joint. Horizontal plane. Staining: Hematoxylin-eosin

Temporomandibular Joint

Although the condylar (and the coronoid) processes are very prominent, a functioning proper temporomandibular joint is not yet established. Instead, the primary jaw joint is still well in function.

References

1. Radlanski RJ, Renz H. An atlas of the human prenatal craniofacial development. Eur J Oral Sci. 2010;118:321–4.
2. Fernández Rubio EM, Radlanski RJ. Development of the human primary and secondary jaw joint. Ann Anat. 2024;251:1–8.

Further Reading

Radlanski RJ, Renz H, Klarkowski MC. Prenatal development of the human mandible. 3D reconstructions, morphometry and bone remodelling pattern, sizes 12–117 mm CRL. Anat Embryol (Berl). 2003;207:221–32.

Embryo 87 mm CRL (ca. 12th wk): DES200597

26

Overall Remarks

In this preparation, we can first of all see very clearly how the interaction between the dental primordia and the peri-dental bone is configured: It is indeed the case that the interdental bone forms as an active formation. This is shown by the areas marked in red, which have been identified as appositional zones in the corresponding histological sections. This concerns the lower jaw and also the upper jaw (Fig. 26.1).

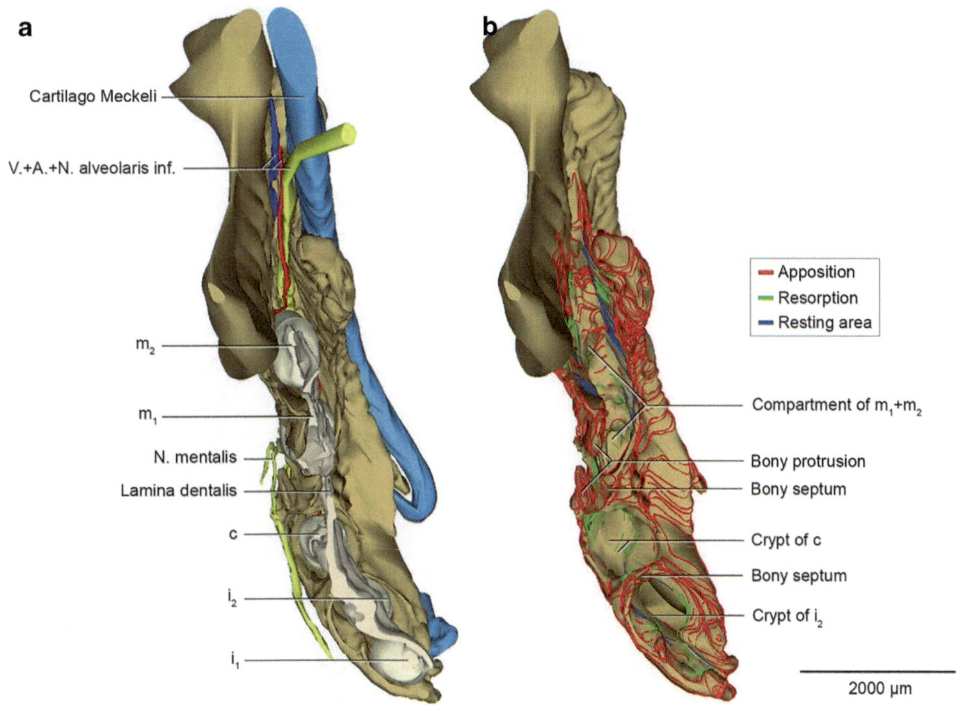

Fig. 26.1 (**a**) Human embryo, 87 mm CRL, partial 3D-reconstruction of the right half of the mandible and surrounding structures, cranial view. (**b**) The same partial 3D-reconstruction dental primordia removed and bone remodeling pattern superimposed next to crypts of dental primordia. Red: apposition, green: resorption, blue: lining cells (resting area)

R. J. Radlanski, *Atlas of Human Prenatal Craniofacial Development*, https://doi.org/10.1007/978-3-031-88812-0_26

Bone Remodeling

At this stage of development, the mandible is dominated by bone growth (marked in red). This affects the large surfaces and also the interdental areas, which develop as bone septa between the tooth layers. These are the first signs of the formation of alveolar crests. Only in the areas inside the crypts, which accommodate the tooth primordium, have bone resorption zones been registered (marked in green) (Fig. 26.2).

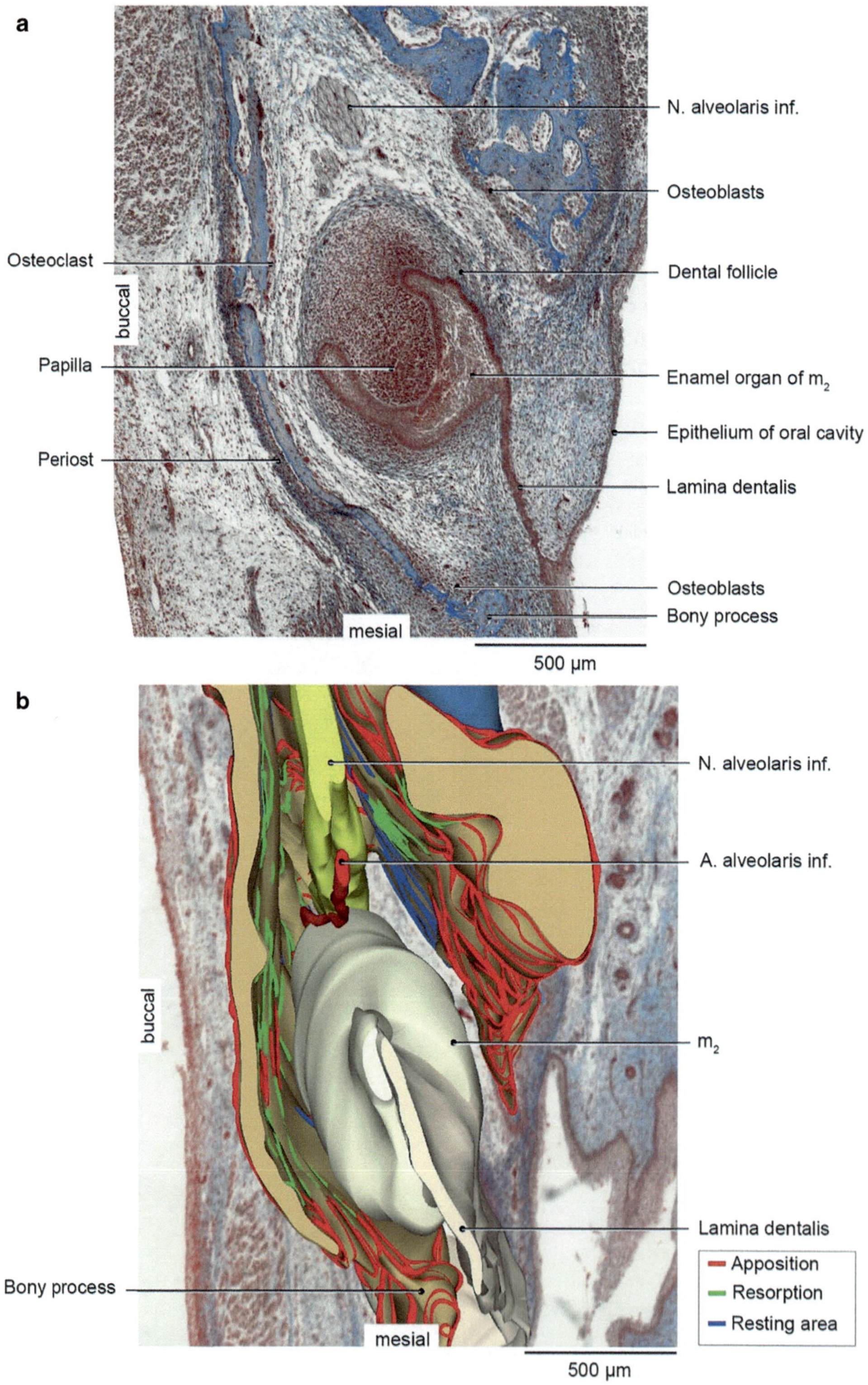

Fig. 26.2 (**a**) Human embryo, 87 mm CRL, horizontal section through the mandibular region of the right deciduous second molar primordium (staining: trichrome). (**b**) 3D partial reconstruction of the right deciduous second molar primordium of the mandible and their surrounding structures, histological section (staining: trichrome) in the background (appears blurry due to software issues). Bone remodeling pattern superimposed. Cranial view

Meckel's Cartilage

The dorsal end (malleus) of Meckel's cartilage is still in articulation with the incus and serves as the primary jaw joint. However, its anterior third is already clearly curved in an S-shape and shows signs of degeneration.

Alveolar Bone

At this stage of development, the two incisor primordia share an extended crypt, while the canine occupies a crypt of its own. The molars also share a longish crypt (Fig. 26.1). The voluminous bony protrusion, active formation of bone between the canine and the molar crypts (Fig. 26.3), is impressive.

Maxilla

At this stage of development, we can see how far the maxilla itself has developed with its frontal process, also mainly with apposition of bone (Fig. 26.4). And it becomes clear how the nasopalatine duct is developing. It is embedded in the primary palatal process from the very beginning and is not at all a result of the fusion point of all three palatal processes.

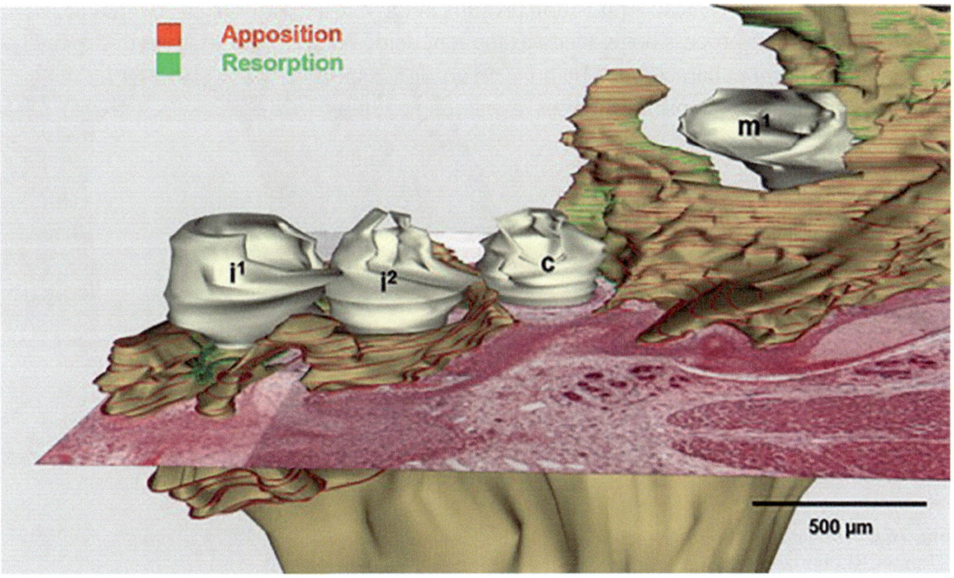

Fig. 26.3 Human embryo, 87 mm CRL, reconstruction of the right half of the mandible, medial view, with superimposition of a histological section at level of the canine dental primordium (staining: hematoxylin-eosin)

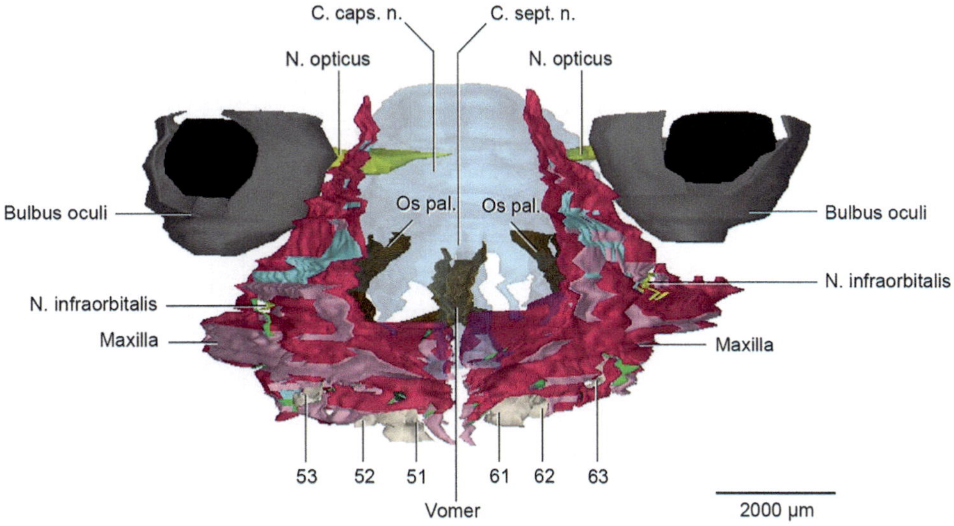

Fig. 26.4 Human embryo, 87 mm CRL, maxilla, palatal bone, nasal capsule, and dental primordia, frontal view. Bone remodeling pattern superimposed (color code: dark red: apposition, pink: minor apposition, turquoise: resorption). C. caps. n.: Cartilago capsulae nasi, cartilaginous primordium of nasal capsule. C. sept. n.: Cartilago septi nasi, cartilaginous primordium of nasal septum

Incisive Canal

The bone formation of the primary palate is so far advanced medially that the nasopalatine duct running lateral to the nasopalatine nerve is no longer visible from the front. The anterior wall of the incisive canal is not completely closed; there is an average distance of 200 μm between the bony parts of the left and right primary palate. The nasopalatine nerve has a very steep, slightly dorsal course through the maxillary bone toward the nasal cavity. The nasopalatine nerves of both sides surround a centrally located epithelial pearl (Figs. 26.5, 26.6, 26.7, 26.8, 26.9, 26.10, 26.11, and 26.12).

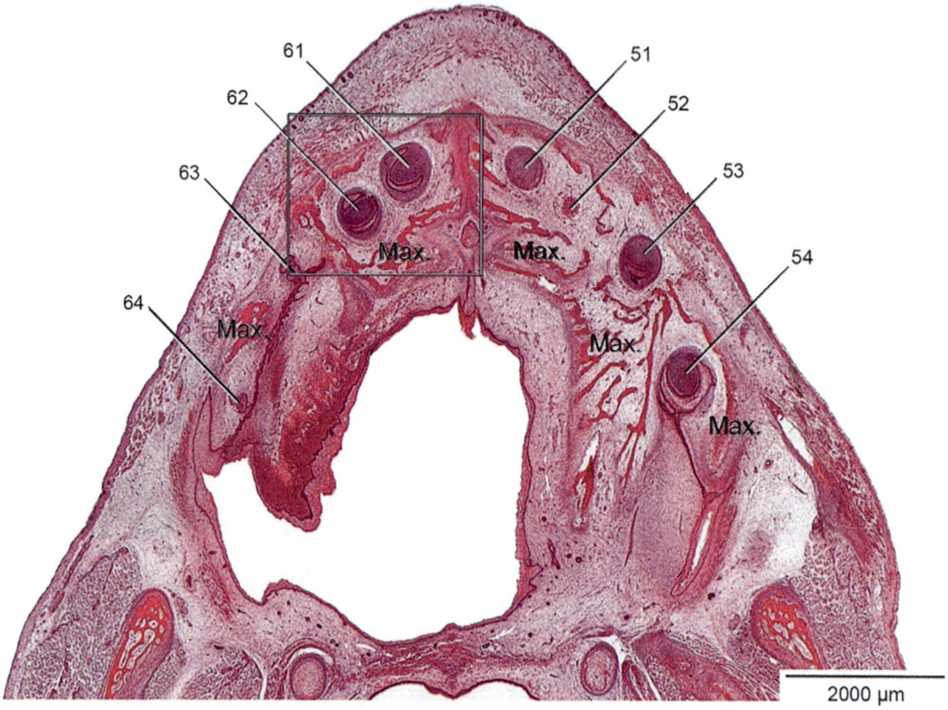

Fig. 26.5 Human embryo, 87 mm CRL, histological section (staining: hematoxylin-eosin), maxillary region through dental primordia

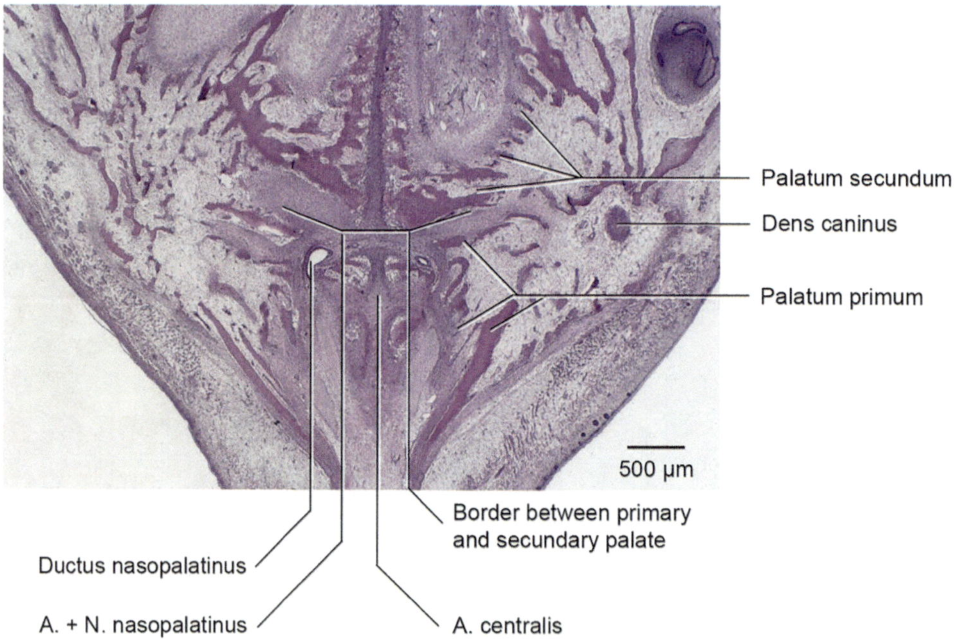

Fig. 26.6 Human embryo, 87 mm CRL, horizontal section at level of the floor of the nasal cavity (staining: hematoxylin-eosin)

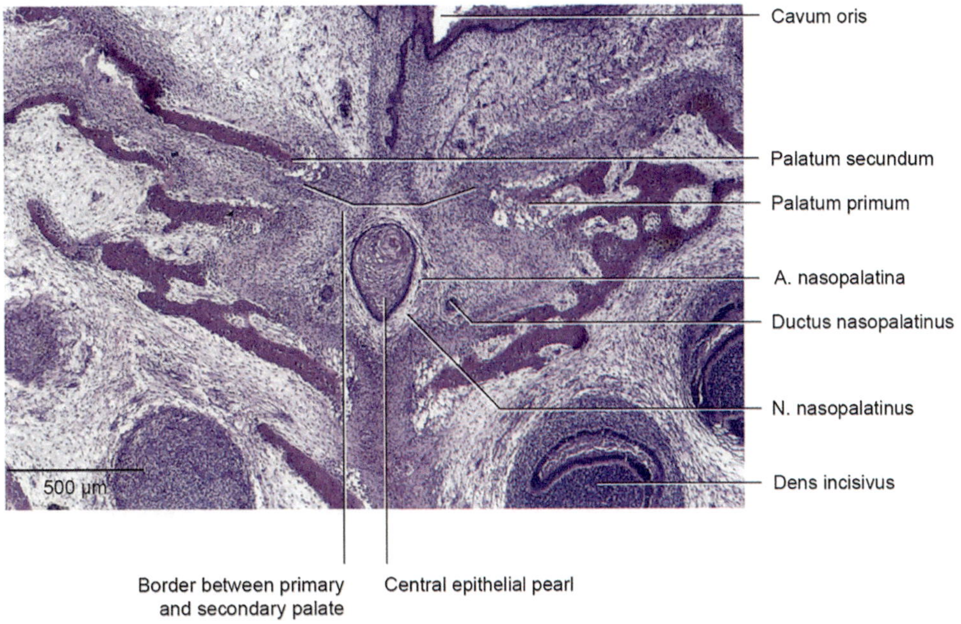

Fig. 26.7 Human embryo, 87 mm CRL, horizontal section at level of palate (staining: hematoxylin-eosin)

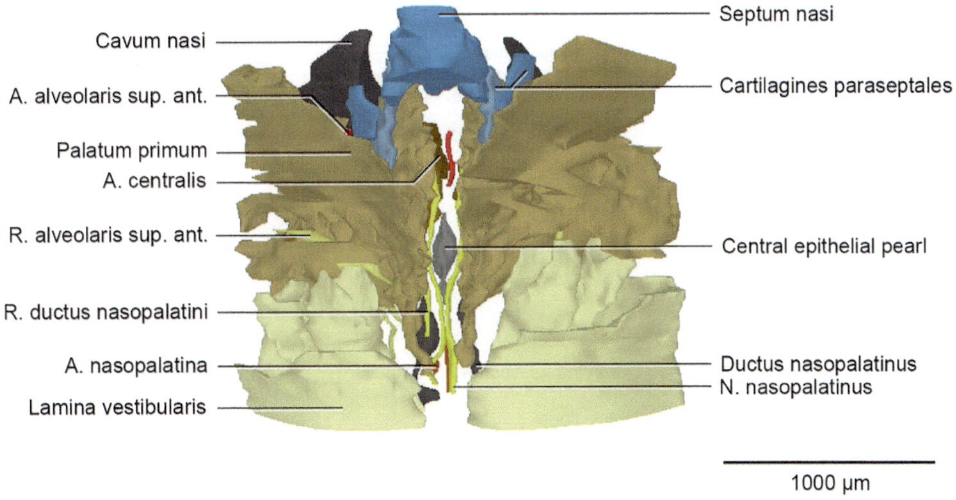

Fig. 26.8 Human embryo, 87 mm CRL, 3D-reconstruction of the maxilla and dental primordia, frontal view

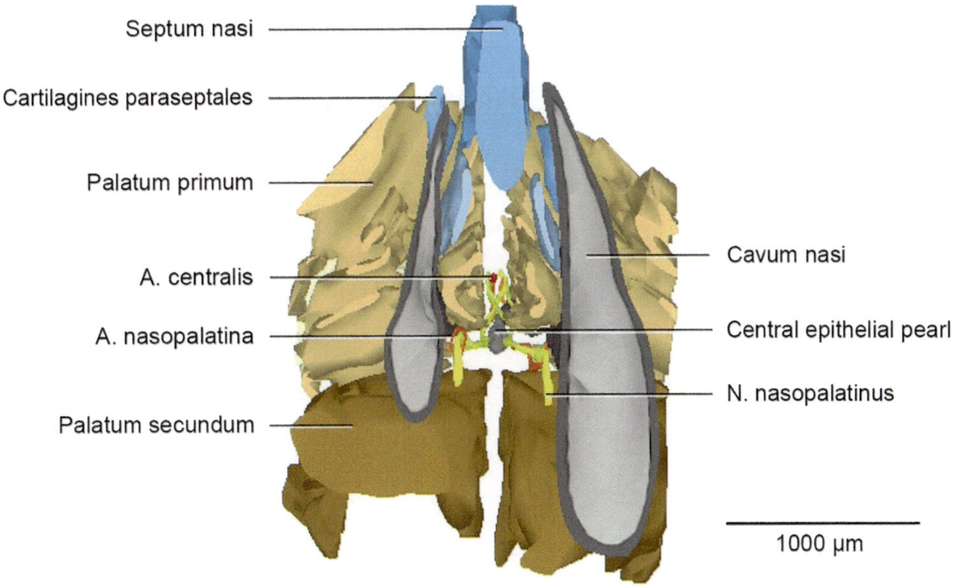

Fig. 26.9 Human embryo, 87 mm CRL, 3D-reconstruction of the maxilla, cranial view

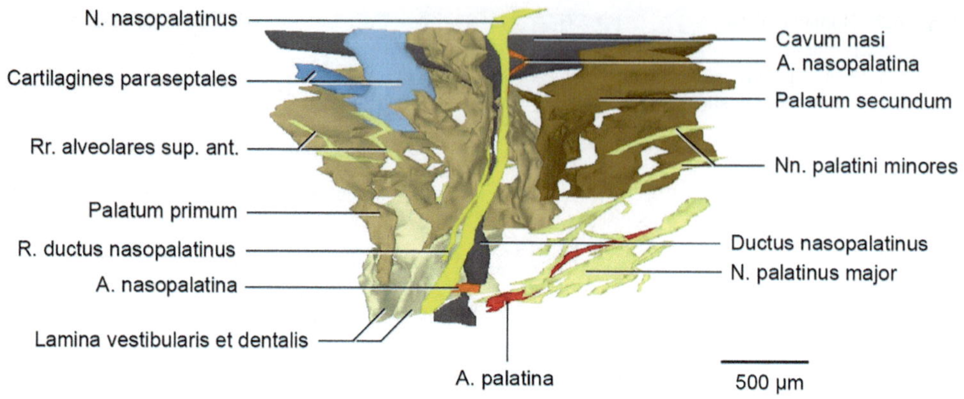

Fig. 26.10 Human embryo, 87 mm CRL, 3D-reconstruction of the maxilla, right half, lateral view

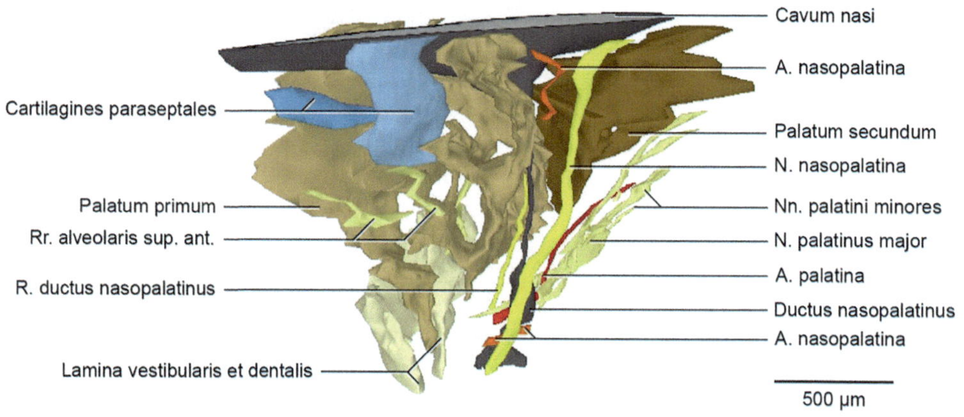

Fig. 26.11 Human embryo, 87 mm CRL, 3D-reconstruction of the maxilla, right half, 20° cranial, 30° frontal view

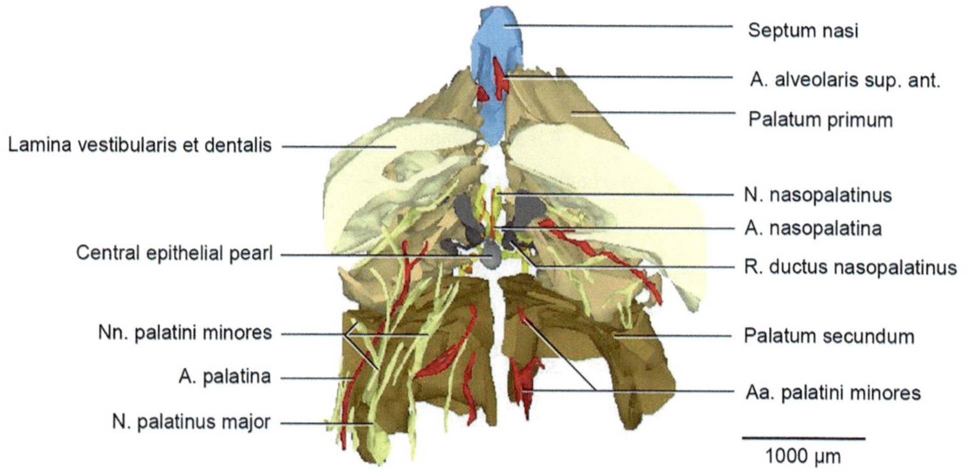

Fig. 26.12 Human embryo, 87 mm CRL, 3D-reconstruction of the maxilla, caudal view

Middle Ear

The ossification of the tympanic cavity has progressed. The lateral, squamous bony border has grown ventrally toward the acoustic duct. The tympanic ring is characterized by a growth in thickness, with the medio-ventral part being quite prominent. Its latero-cranial area can now also be easily distinguished from the squamous part of the temporal bone. Furthermore, the emerging canal for the facial nerve can be seen as a fibrous structure (Fig. 26.13).

At this stage of development, the tympanic membrane does not yet consist of the typical three-layered structure, but initially only of the ectoderm of the ear canal plate. A stapedic artery is no longer present. The ossification of the primordia of the auditory ossicles has not yet begun (Figs. 26.14, 26.15, 26.16, and 26.17).

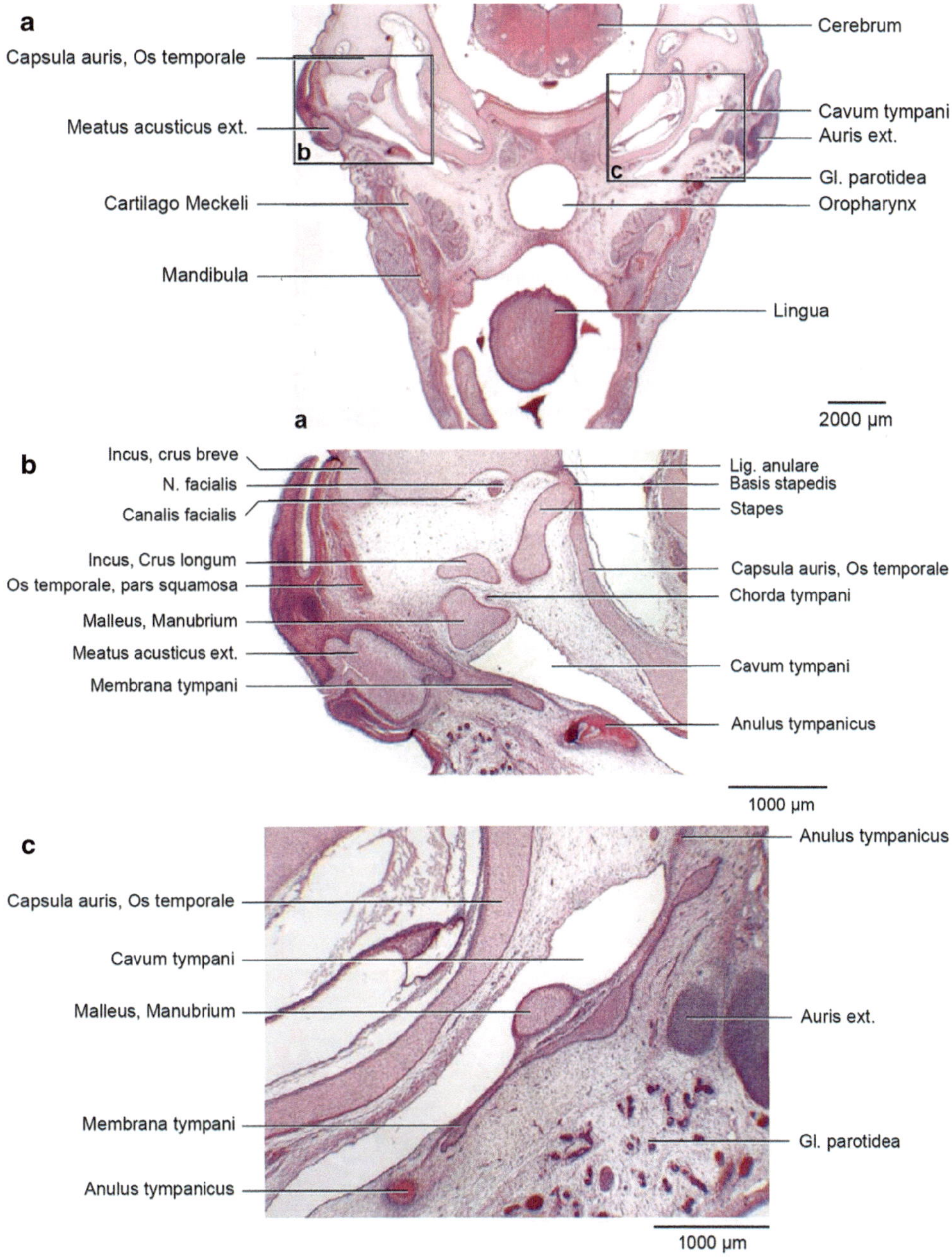

Fig. 26.13 (**a**) Human embryo, 87 mm CRL, viscerocranium horizontally sectioned in the plane of the right stapes (staining: hematoxylin-eosin). Left box: b; right box: c. (**b**) Detailed view of the marked region in (a). (**c**) Detailed view of the marked region in (a). Human embryo, 87 mm CRL, histological section, horizontally sectioned at the level of the tympanic membrane, detailed view of the marked region (on the right) in (a)

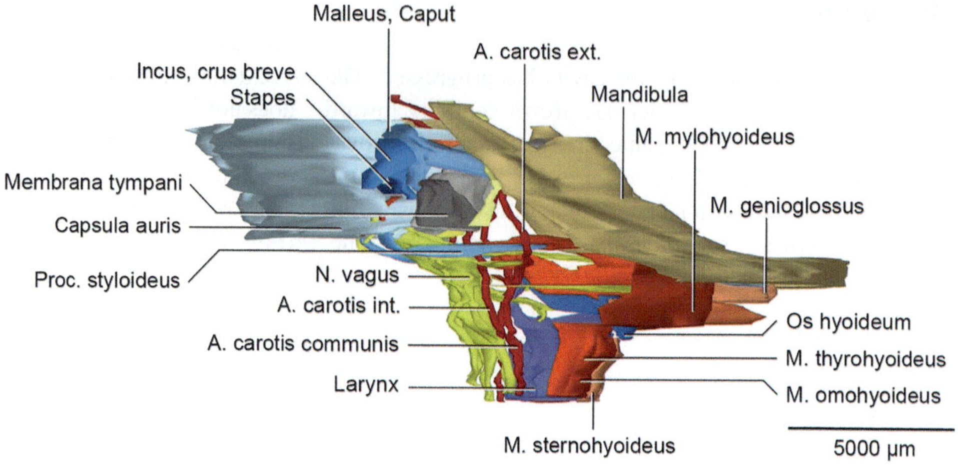

Fig. 26.14 Human embryo, 87 mm CRL mandible, primary jaw joint, and otic capsule, lateral view

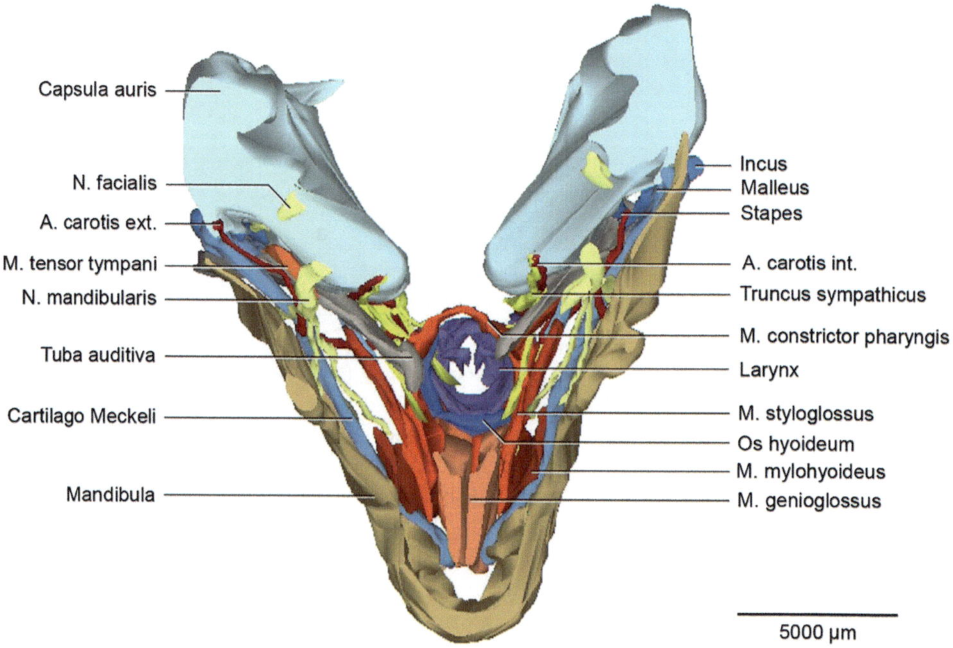

Fig. 26.15 Human embryo, 87 mm CRL, mandible, surrounding structures, and otic capsules, caudal view

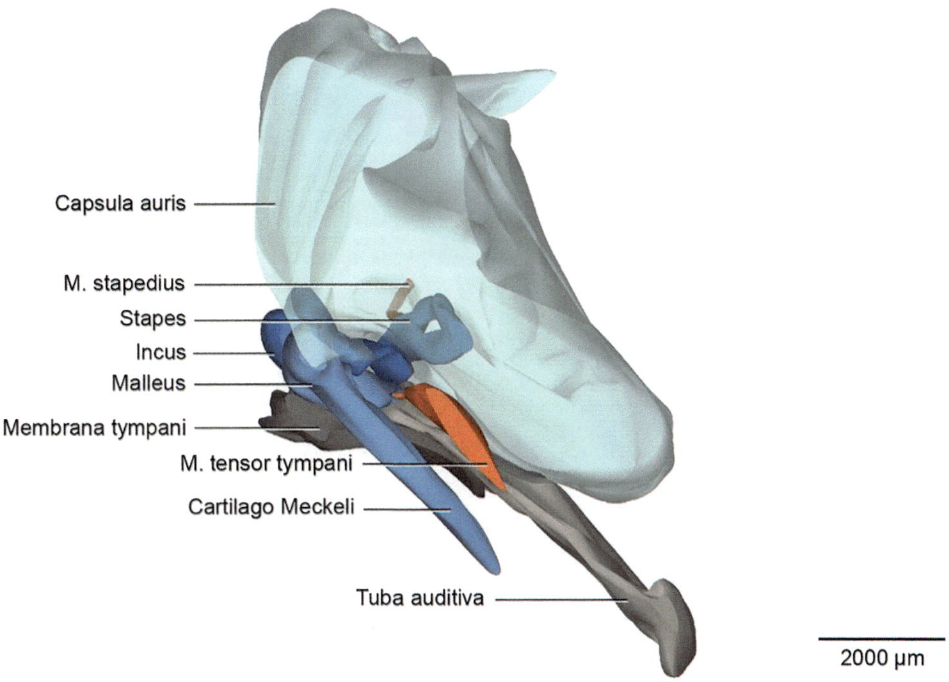

Fig. 26.16 Human embryo, 87 mm CRL, right auditory system, cranial view, cartilaginous otic capsule transparent

Fig. 26.17 Human embryo, 87 mm CRL, right ossicular chain (cartilaginous primordia), lateral, medial and cranial view (from left)

Further Reading

Beckert K. Beiträge zur pränatalen Morphogenese der Region des Mittelohres des Menschen—Computergestützte 3D-Rekonstruktionen und histologische Untersuchungen menschlicher Embryonen/Feten von 19 bis 150 mm CRL (7. bis 18. Woche). Med. Diss. Charité-Universitätsmedizin Berlin; 2010.

Benedikter B. Beiträge zur pränatalen Morphogenese des Foramen infraorbitale des Menschen. Computergestützte 3D-Rekonstruktionen und histologische Untersuchungen menschlicher Embryonen und Feten von 19–250 mm Scheitel-Steiß-Länge. Med. Diss. Charité Universitätsmedizin Berlin; 2005.

Emmerich S. Die pränatale Morphogenese des humanen Canalis incisivus. Med. Diss. FU Berlin; 2004.

Radlanski RJ. Oral structure and biology. Batavia: Quintessence Publishing; 2018.

Radlanski RJ, Emmerich S, Renz H. Prenatal morphogenesis of the human incisive canal. Anat Embryol. 2004;208:265–71.

Schuster FP. Zur Entwicklung des Kieferknochens im Bereich der Zahnanlagen. Histologische und morphologische Untersuchungen an der fetalen Mandibula des Menschen während des Glockenstadiums der Zahnanlagen (68–160 mm SSL). Med. Diss. Charité-Universitätsmedizin Berlin; 2013.

Zimmermann C. Zur Entwicklung des Kieferknochens im Bereich der Zahnanlagen. Histologische und morphologische Untersuchungen an der fetalen Mandibula des Menschen während des späten Glockenstadiums der Zahnanlagen (160–270 mm SSL). Med. Diss. Charité-Universitätsmedizin Berlin; 2013.

Fetus 95 mm CRL (ca. 12–13th wk): FLO010689

27

Overall Remarks

Many textbooks classically hold the view that the mandible arises from purely desmal ossification. This is certainly also due to the idea that wants to derive the formation of the mandible in humans also phylogenetically—and it is, therefore, described as a purely desmal ossifying bone. However, if we look at the biological facts, then at this stage of development we can clearly see many small cartilaginous centers in the ossifying mandible, which are certainly not any part of Meckel's cartilage, which is of course also gradually replaced by bone (Radlanski et al. 2016). There are isolated cartilage centers in the anterior region and others are preferably located in the spandrel between the dental units, on the buccal side. More details hereto in Chap. 30. And then there are the cartilaginous cores in the condylar and coronoid processes (Fig. 27.1).

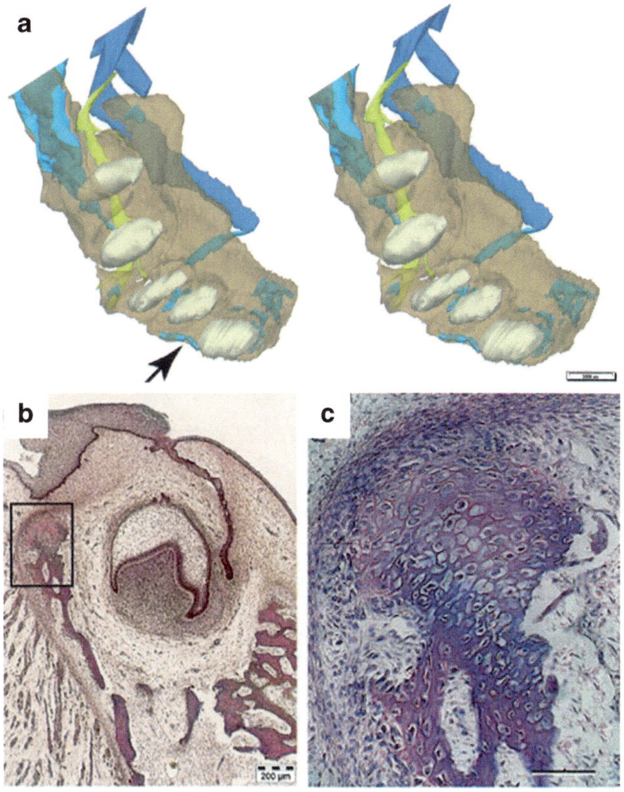

Fig. 27.1 (**a**) Stereoscopical pair of images. Human fetus 95 mm CRL (ca. 12–13th week), mandible (right half) with dental primordia, Meckel's cartilage. Structures made transparent in order to visualize the single cartilaginous structures, located at the brim of the crestal bone buccally at all dental primordia (arrow) and lingually at the central incisor primordium, cranial view. Bar: 2000 µm. (**b**) Human fetus, 95 mm CRL, horizontal section (staining: hematoxylin-eosin) at level of dental primordium i^1 with cartilaginous tissue in the buccal brim of the surrounding bone. Scale bar: 200 µm. Box: Location of figure (c). (**c**) Region of box in figure (b). Cartilage cells in ossifying matrix, covered by a thin layer of bone towards the dental primordium. Scale bar: 50 µm

R. J. Radlanski, *Atlas of Human Prenatal Craniofacial Development*, https://doi.org/10.1007/978-3-031-88812-0_27

Cartilaginous Centers in the Mandible

The frontal histological sections show the location and the typical chondrocytes at the brim of the bone around the dental primordium i_2. The 3-D reconstructions made it possible to clearly localize the islands of cartilaginous tissue and to distinguish their position from any other cartilaginous structures such as Meckel's cartilage or its insular decay products.

Further Reading

Radlanski RJ, Renz H, Zimmermann CA, Schuster FP, Voigt A, Heikinheimo K. Chondral ossification centers next to dental primordia in the human mandible. A study of the prenatal development ranging between 68 to 270 mm CRL. Ann Anat. 2016;208:49–57.

Schuster FP. Zur Entwicklung des Kieferknochens im Bereich der Zahnanlagen. Histologische und morphologische Untersuchungen an der fetalen Mandibula des Menschen während des Glockenstadiums der Zahnanlagen (68–160 mm SSL). Med. Diss. Charité-Universitätsmedizin Berlin; 2013.

Zimmermann C. Zur Entwicklung des Kieferknochens im Bereich der Zahnanlagen. Histologische und morphologische Untersuchungen an der fetalen Mandibula des Menschen während des späten Glockenstadiums der Zahnanlagen (160–270 mm SSL). Med. Diss. Charité-Universitätsmedizin Berlin; 2013.

Fetus 100 mm CRL (ca. 13th wk): BEL170294

28

Overall Remarks

In this developmental stage, bone has encapsulated the nasopalatine nerve in such a way that the term "canal" is appropriate from now on. The outline of the canal is ovoid in a transversal histological section. The sagittal diameter is about 1 mm and the transversal diameter is about 750 µm. This nasopalatine duct is clearly located in the bony extension of the primary palate. The nasopalatine vessels can be found to run within the incisive canal, lateral of the nasopalatine nerves (Fig. 28.1).

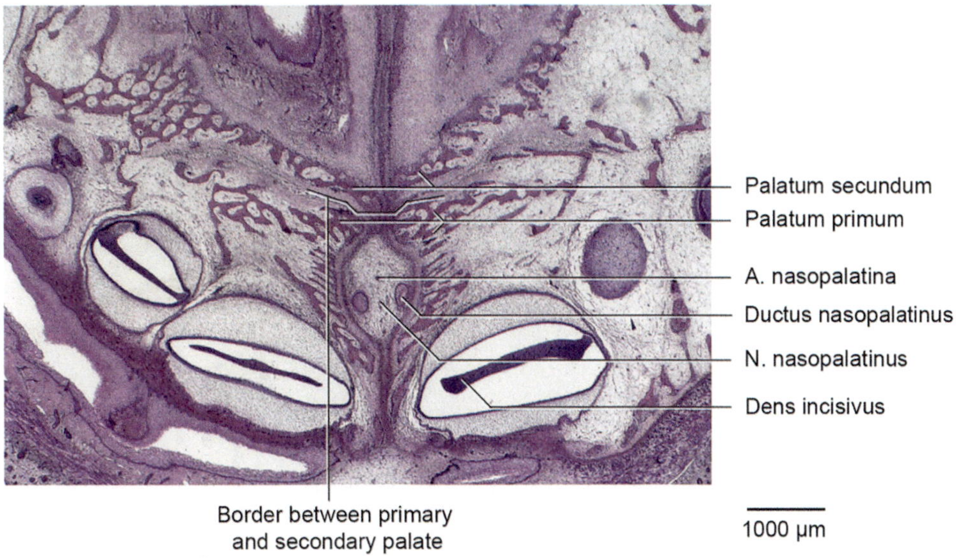

Fig. 28.1 Human fetus, 100 mm CRL, histological section, horizontal section at level of incisive canal (staining: hematoxylin-eosin)

R. J. Radlanski, *Atlas of Human Prenatal Craniofacial Development*, https://doi.org/10.1007/978-3-031-88812-0_28

Further Reading

Emmerich S. Die pränatale Morphogenese des humanen Canalis incisivus. Med. Diss. FU Berlin; 2004.

Radlanski RJ, Emmerich S, Renz H. Prenatal morphogenesis of the human incisive canal. Anat Embryol. 2004;208:265–71.

Fetus 117 mm CRL (ca. 14th wk): HUL110589

29

Overall Remarks

The maxilla completely surrounds the cartilaginous nasal capsule; there is even a well-defined nasal bone. Laterally, however, there is still a distance to the zygomatic arch. Each individual tooth has its own bony crypt. This also applies to the mandible, which also has a very prominent coronoid process and an equally prominent condylar process. Inside the condylar process, an extensive cartilaginous core is imposing. The resorption of Meckel's cartilage has now started at its anterior end. The remaining cartilage appeared slightly thinner than in the previous fetuses (Figs. 29.1, 29.2, 29.3, 29.4, 29.5, 29.6, 29.7, 29.8, 29.9, 29.10, 29.11, 29.12, 29.13, 29.14, 29.15, 29.16, 29.17, 29.18, 29.19, 29.20, 29.21, 29.22, 29.23, and 29.24).

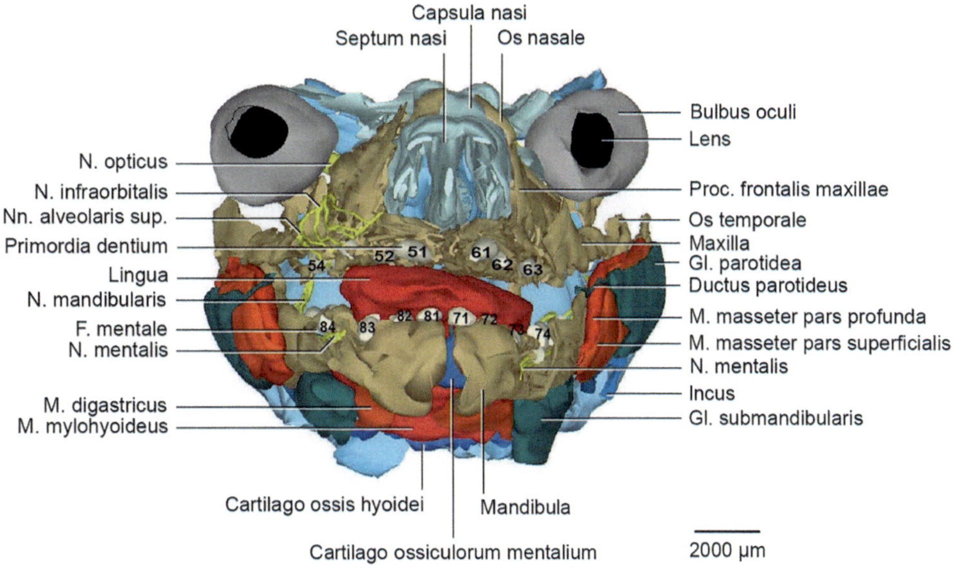

Fig. 29.1 Human fetus, 117 mm CRL, head, frontal view, skin removed, layer 1 of preparation. (*From* Radlanski and Renz [1]; *with permission*)

R. J. Radlanski, *Atlas of Human Prenatal Craniofacial Development*, https://doi.org/10.1007/978-3-031-88812-0_29

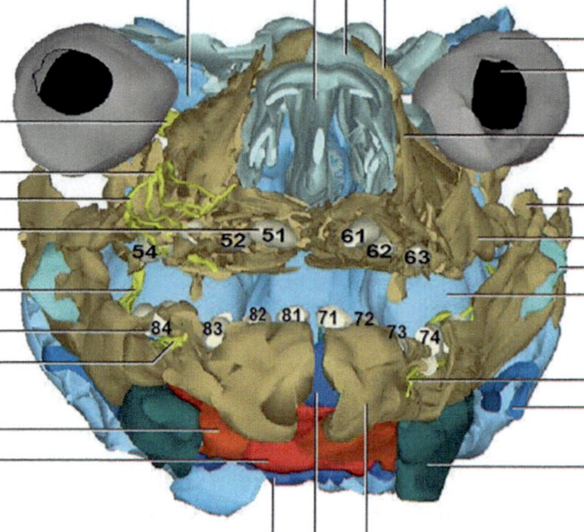

Fig. 29.2 Human fetus, 117 mm CRL, head, frontal view, layer 2 of preparation. (*From* Radlanski and Renz [1]; *with permission*)

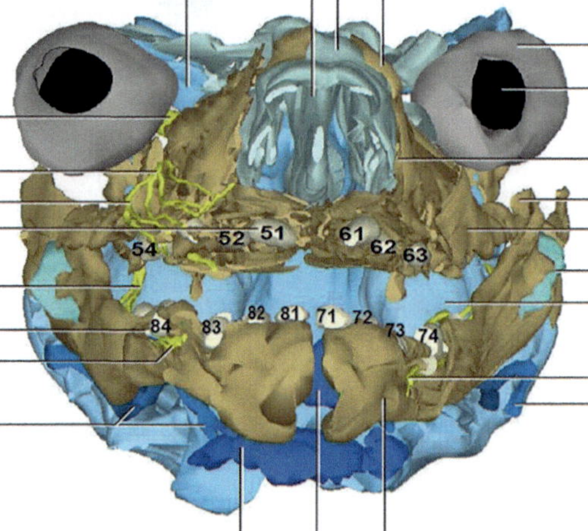

Fig. 29.3 Human fetus, 117 mm CRL, head, frontal view, layer 3 of preparation. (*From* Radlanski and Renz [1]; *with permission*)

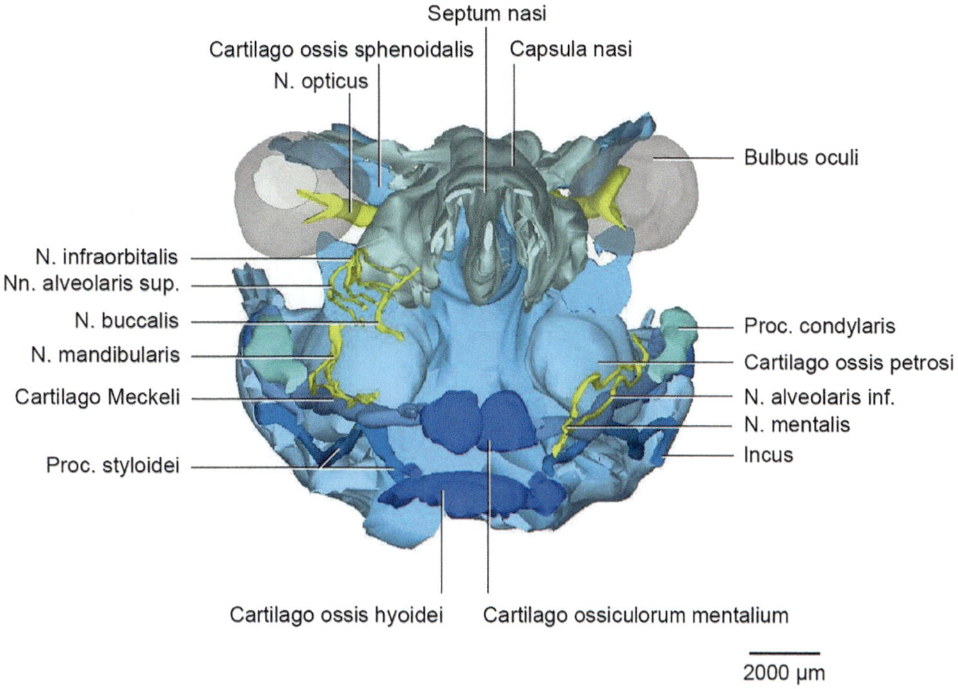

Fig. 29.4 Human fetus, 117 mm CRL, head, frontal view, cartilage, nerves, and eyeballs. (*From* Radlanski and Renz [1]; *with permission*)

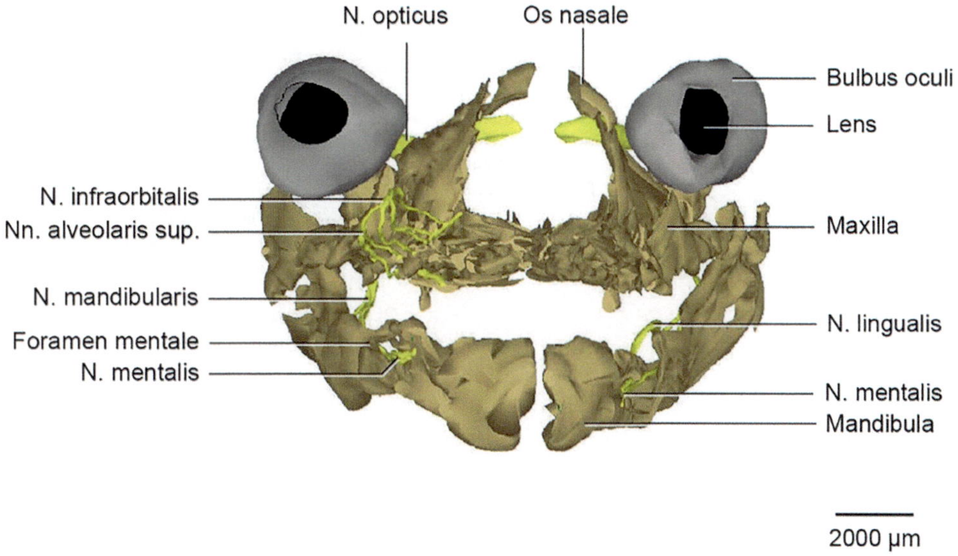

Fig. 29.5 Human fetus, 117 mm CRL, head, frontal view, bone, nerves, and eyeballs. (*From* Radlanski and Renz [1]; *with permission*)

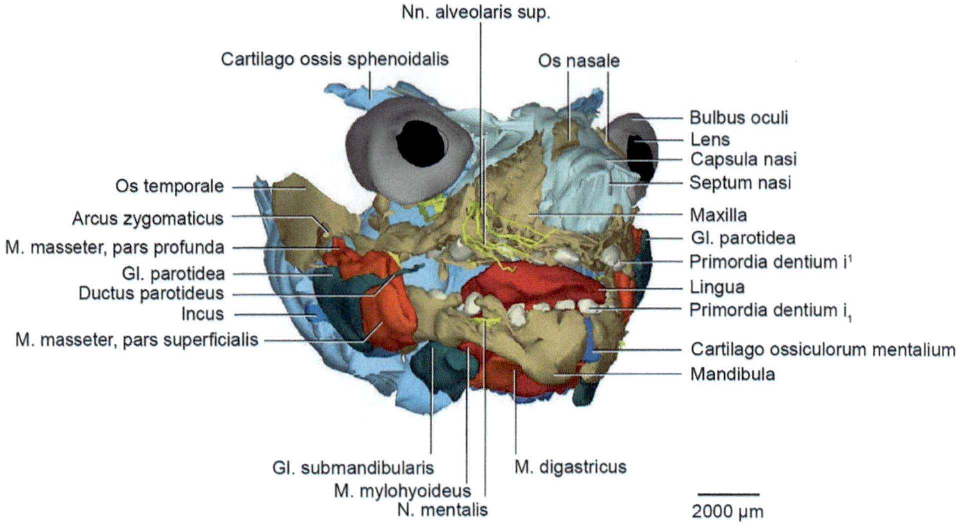

Fig. 29.6 Human fetus, 117 mm CRL, head, 45° lateral view, skin removed, layer 1 of preparation. (*From* Fernández Rubio and Radlanski [2]; *with permission*)

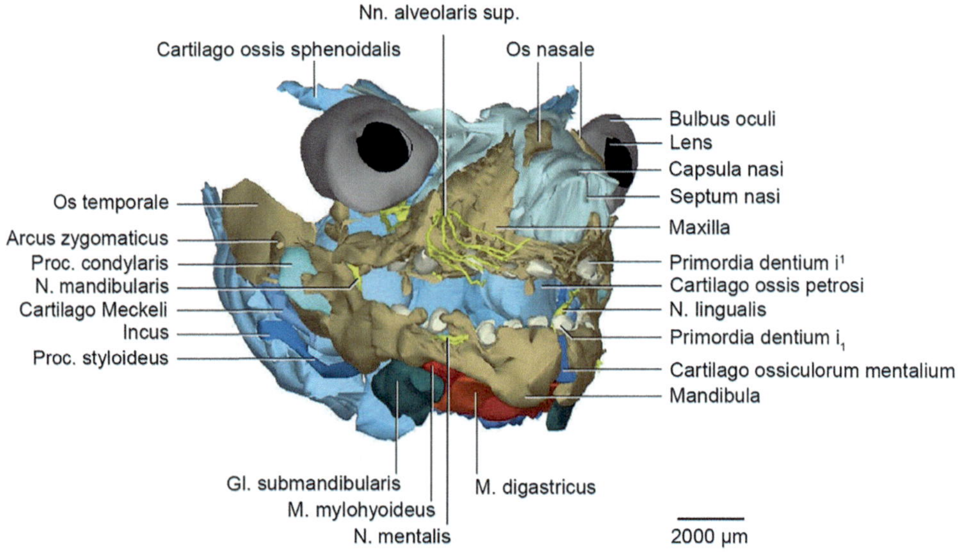

Fig. 29.7 Human fetus, 117 mm CRL, head, 45° lateral view, layer 2 of preparation

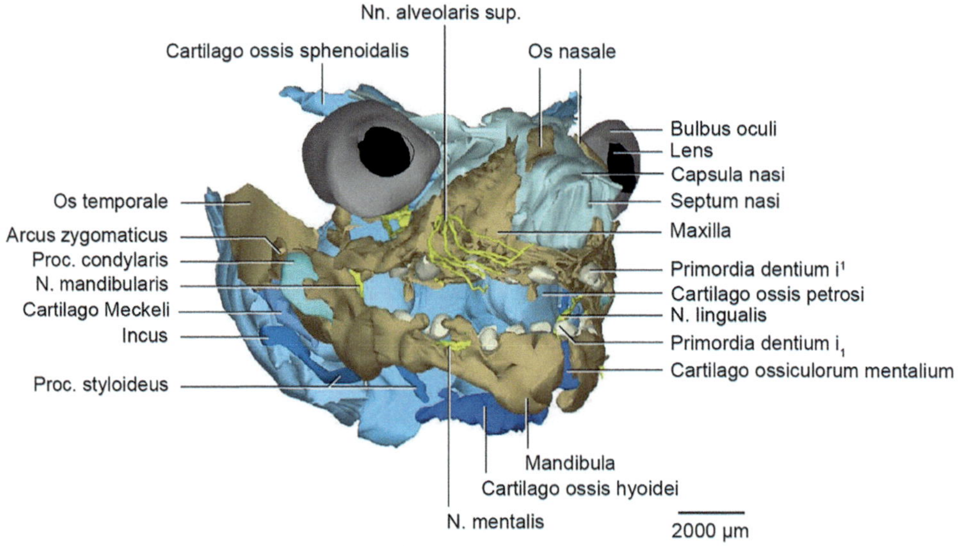

Fig. 29.8 Human fetus, 117 mm CRL, head, 45° lateral view, layer 3 of preparation

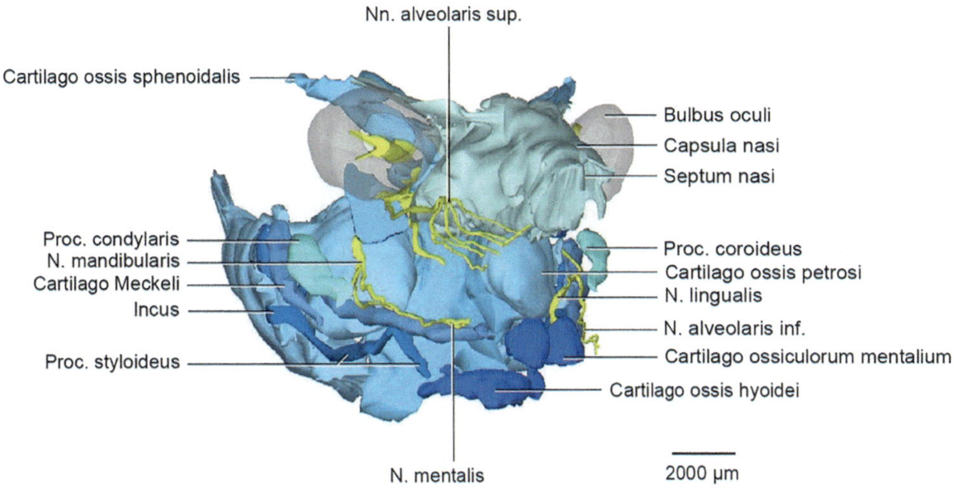

Fig. 29.9 Human fetus, 117 mm CRL, head, 45° lateral view, cartilage, nerves, and eyeballs

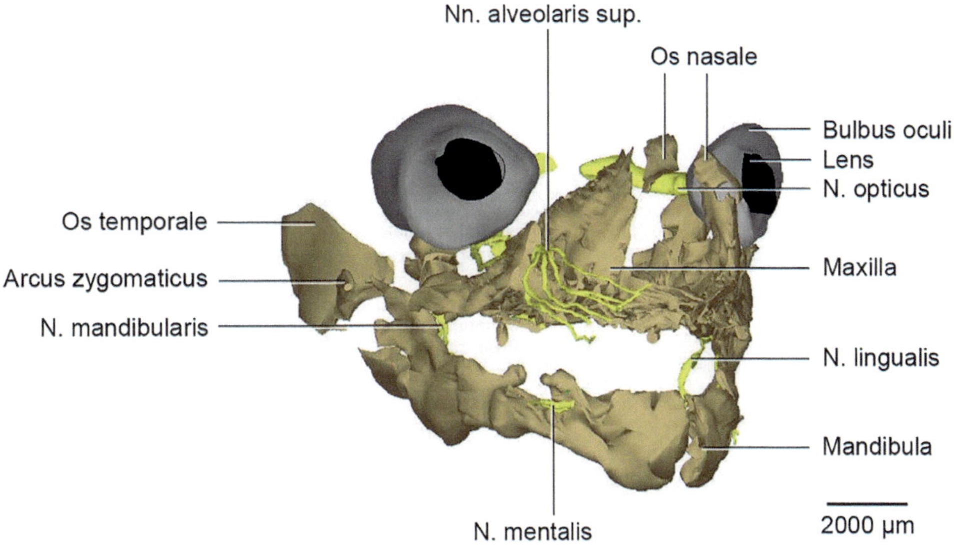

Fig. 29.10 Human fetus, 117 mm CRL, head, 45° lateral view, bone, nerves, and eyeballs

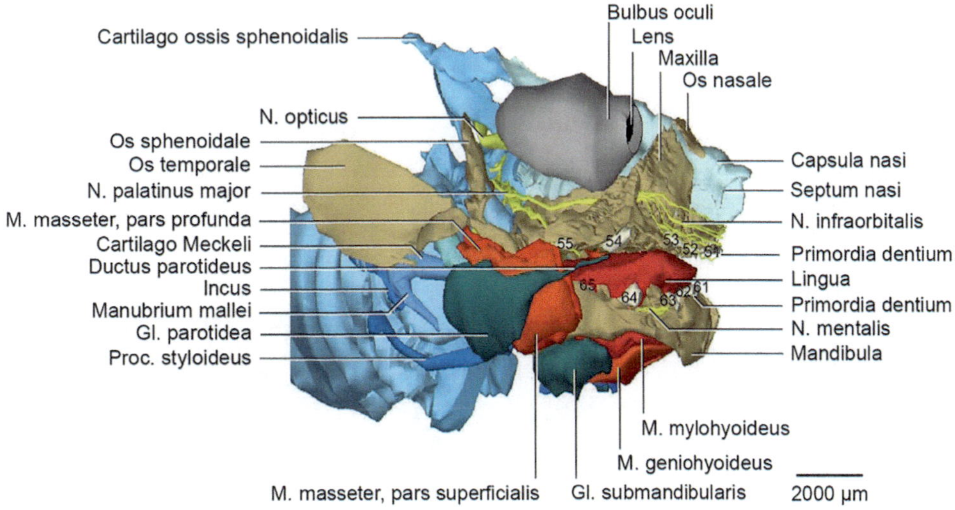

Fig. 29.11 Human fetus, 117 mm CRL, head, lateral view, skin removed, layer 1 of preparation

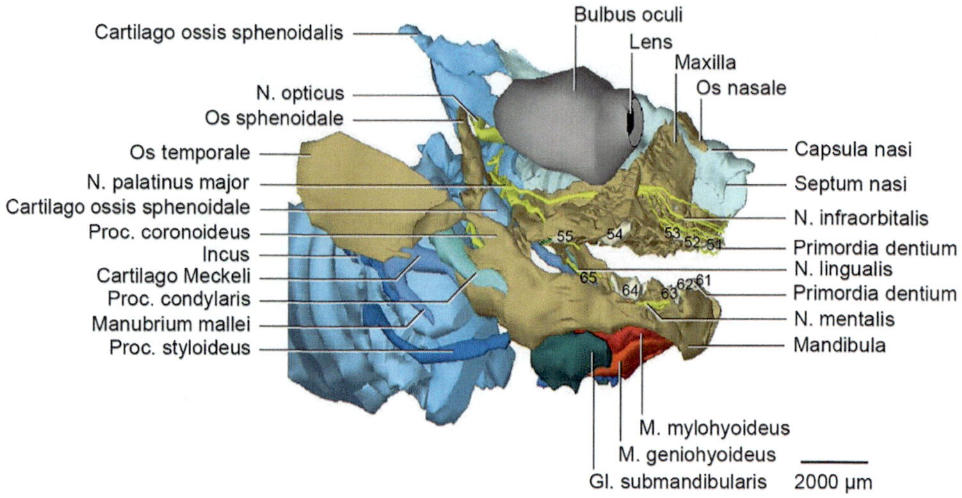

Fig. 29.12 Human fetus, 117 mm CRL, head, lateral view, layer 2 of preparation

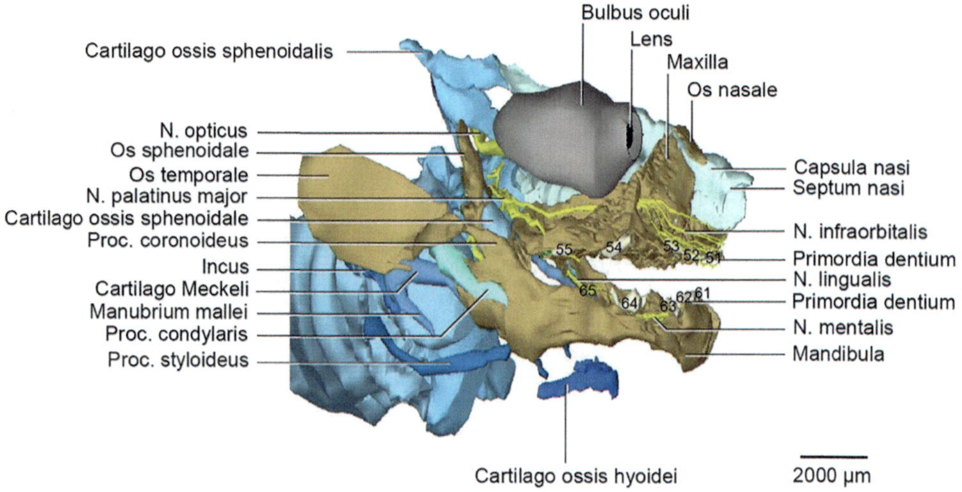

Fig. 29.13 Human fetus, 117 mm CRL, head, lateral view, layer 3 of preparation

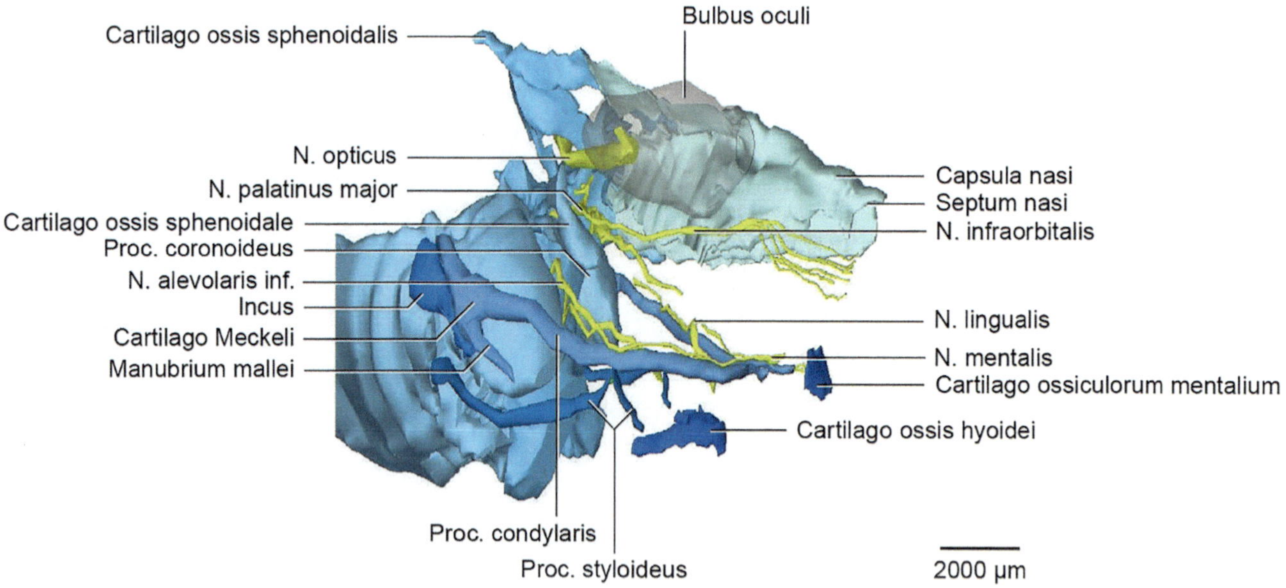

Fig. 29.14 Human fetus, 117 mm CRL, head, lateral view, cartilage, nerves, and eyeballs

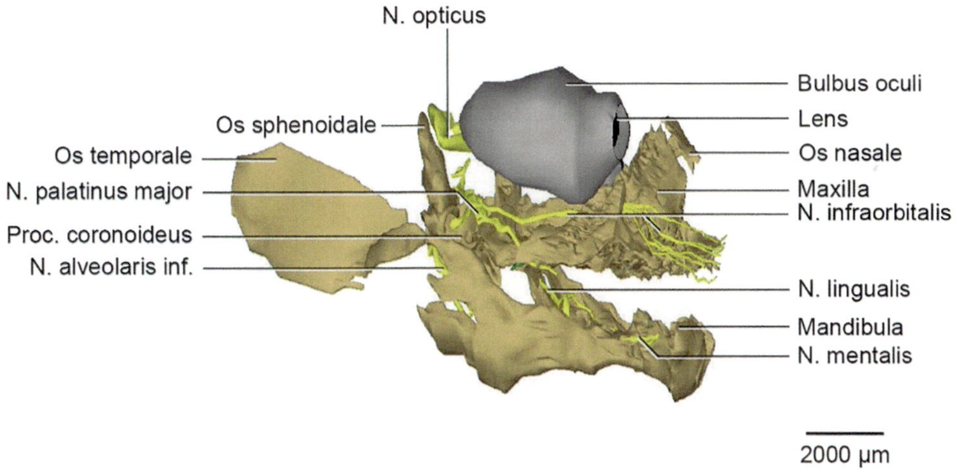

Fig. 29.15 Human fetus, 117 mm CRL, head, lateral view, bone, nerves, and eyeballs

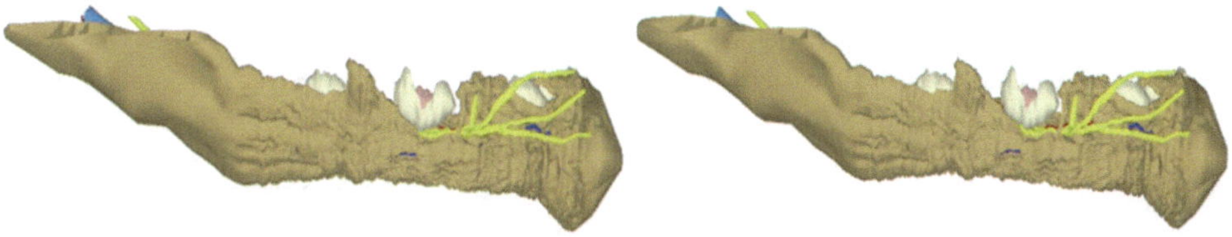

Fig. 29.16 Stereoscopical pair of images. Human fetus 117 mm CRL (ca. 14th week), mandible with dental primordia and mental nerve, lateral view

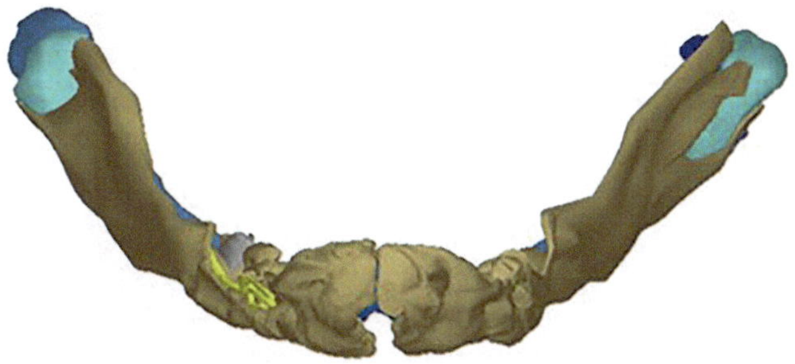

Fig. 29.17 Human fetus, 117 mm CRL, mandible, with mental nerve and cartilaginous core in condylar process, frontal view

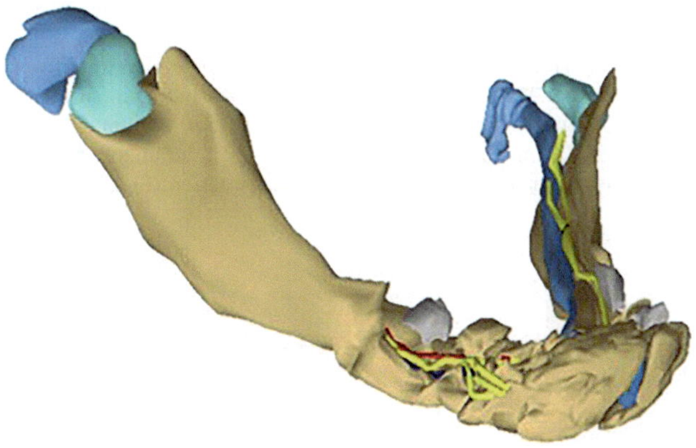

Fig. 29.18 Human fetus, 117 mm CRL, mandible with dental primordia, inferior alveolar nerves and artery, and cartilaginous core in condylar process, 45° lateral view

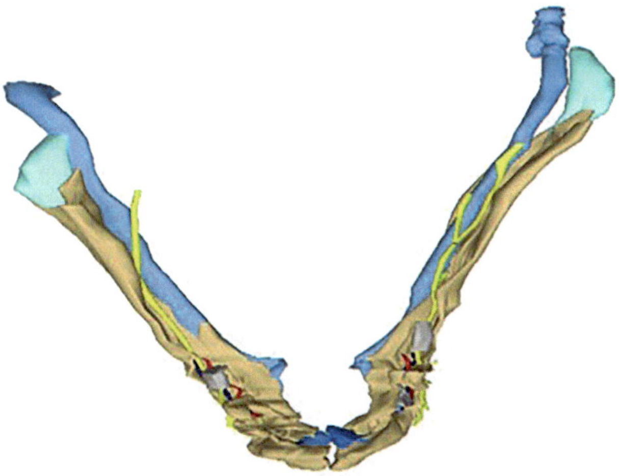

Fig. 29.19 Human fetus, 117 mm CRL, mandible with dental primordia, inferior alveolar nerves and artery, and cartilaginous core in condylar process, cranial view

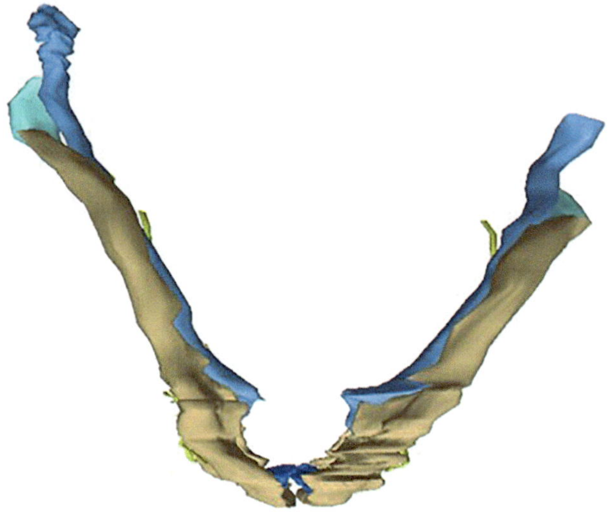

Fig. 29.20 Human fetus, 117 mm CRL, mandible with inferior alveolar nerves and cartilaginous core in condylar process, caudal view

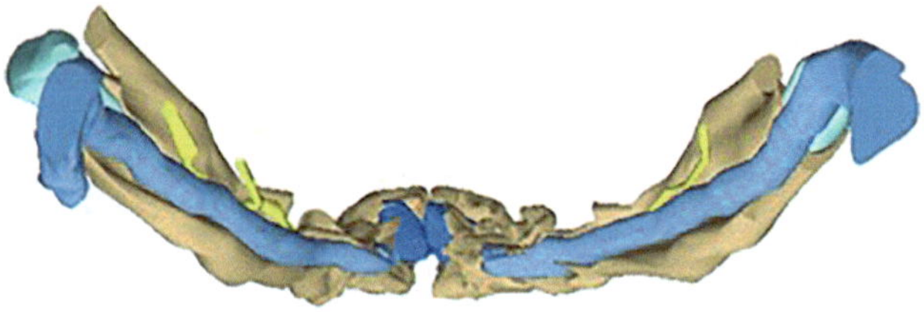

Fig. 29.21 Human fetus, 117 mm CRL, mandible with inferior alveolar nerves and cartilaginous core in condylar process, dorsal view

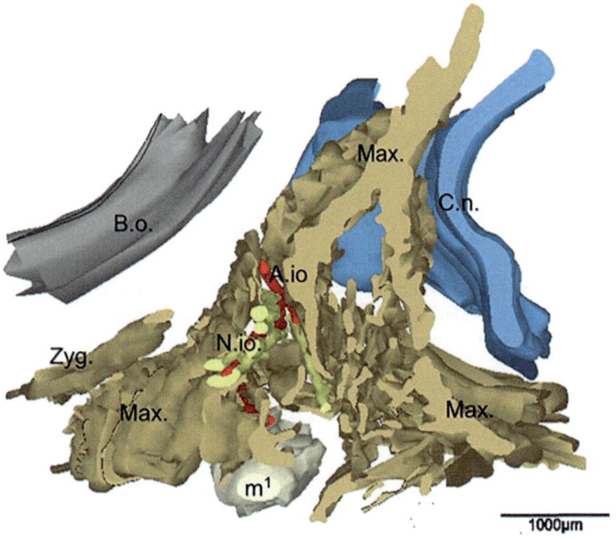

Fig. 29.22 Human fetus, 117 mm CRL, partial reconstruction of the right half of the maxillary region, frontal view. *B.o.* bulbus oculi, *Max* maxilla, *Zyg.* zygomatic bone, *C.n.* cartilago nasi, *A.io* infraorbital a., *N.io* infraorbital n., m^1 primordium of upper first deciduous molar

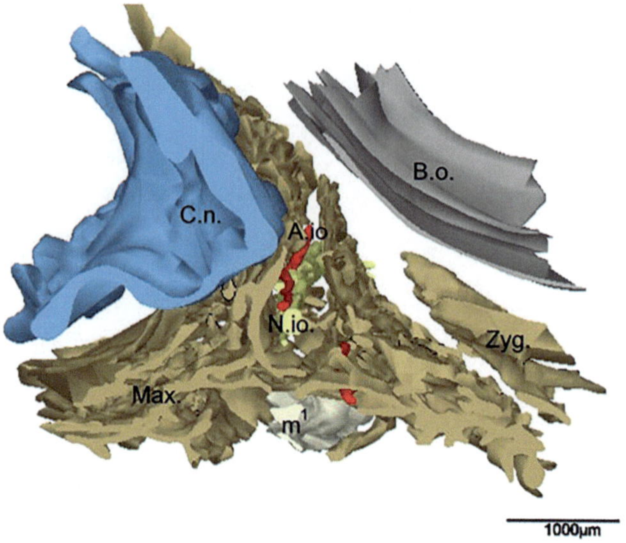

Fig. 29.23 Human fetus, 117 mm CRL, partial reconstruction of the right half of the maxillary region, dorsal view. *B.o.* bulbus oculi, *Max* maxilla, *Zyg.* zygomatic bone, *C.n.* cartilago nasi, *A.io* infraorbital a., *N.io* infraorbital n., m^1 primordium of upper first deciduous molar

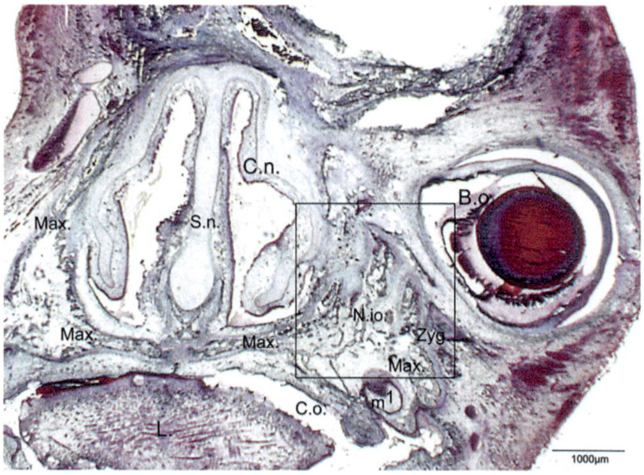

Fig. 29.24 Human fetus, 117 mm CRL, histological section, frontal plane, staining: Masson-Goldner staining. *B.o.* bulbus oculi, *Max* maxilla, *Zyg.* zygomatic bone, *C.n.* cartilago nasi, *S.n.* septum nasi, *C.o.* cavum oris, *L.* lingua, *N.io* infraorbital n., m^1 primordium of upper first deciduous molar. Box: Region of the reconstruction (Figs. 29.22 and 29.23)

Temporomandibular Joint

At this stage of development, both, the primary and the secondary jaw joints, were still concomitant.

The auditory ossicles have not yet started to ossify.

The resorption of Meckel's cartilage has now started at its anterior end. The remaining cartilage appeared slightly thinner than in the previous fetuses.

In the temporomandibular joint, the cartilaginous core in the condylar process imposes very voluminous, and the convexity of the glenoid fossa has become slightly more pronounced than in the developmental stages before. The distance between both structures is smaller here. The articular disc is visibly enlarged (Figs. 29.25, 29.26, and 29.27).

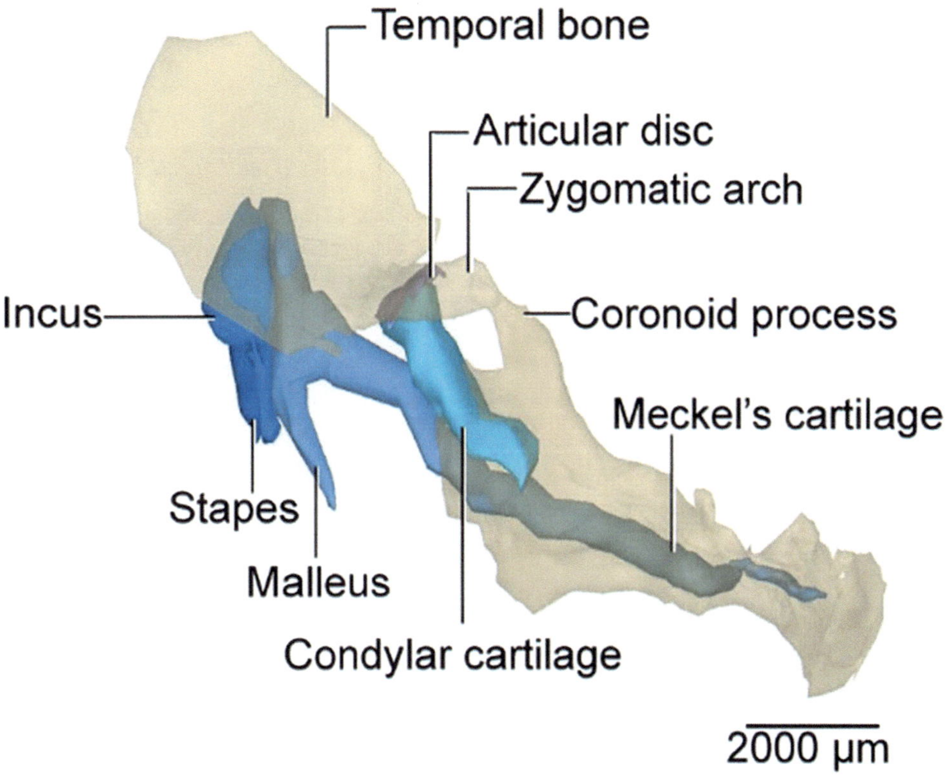

Fig. 29.25 Human 3D reconstruction of selected parts of the temporomandibular joint and middle ear. Mandible and temporal bone made transparent. (*From* Fernández Rubio and Radlanski [2]; *with permission*)

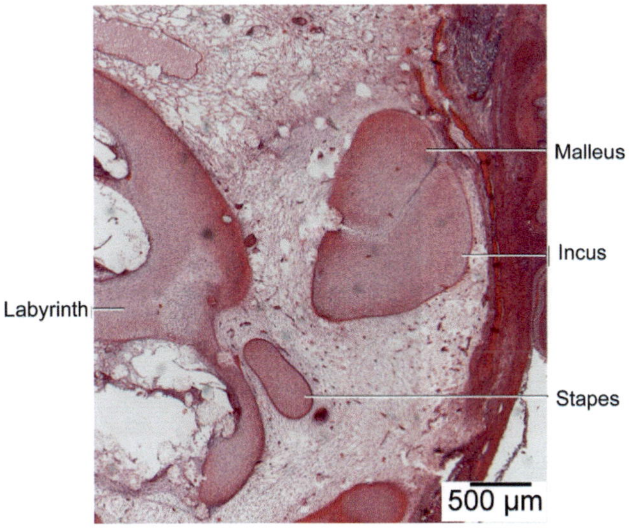

Fig. 29.26 Histological section of the right middle ear. Frontal plane (staining: hematoxylin-eosin)

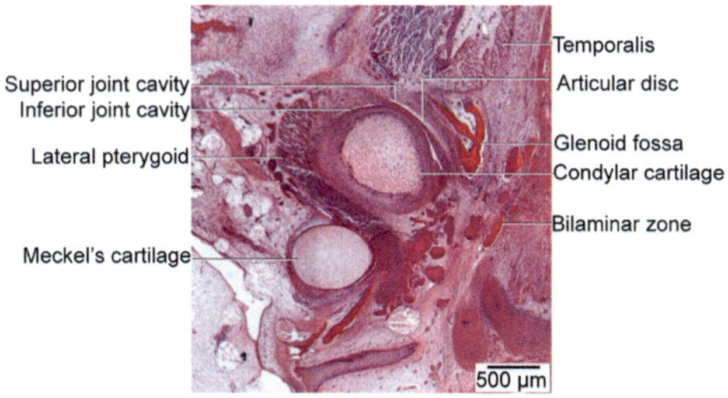

Fig. 29.27 Histological section of the right temporomandibular joint. Frontal plane (staining: hematoxylin-eosin)

References

1. Radlanski RJ, Renz H. An atlas of the human prenatal craniofacial development. Eur J Oral Sci. 2010;118:321–4.
2. Fernández Rubio EM, Radlanski RJ. Development of the human primary and secondary jaw joint. Ann Anat. 2024;251:1–8.

Further Reading

Beckert K. Beiträge zur pränatalen Morphogenese der Region des Mittelohres des Menschen—Computergestützte 3D-Rekonstruktionen und histologische Untersuchungen menschlicher Embryonen/Feten von 19 bis 150 mm CRL (7. bis 18. Woche). Med. Diss. Charité-Universitätsmedizin Berlin; 2010.

Benedikter B. Beiträge zur pränatalen Morphogenese des Foramen infraorbitale des Menschen. Computergestützte 3D-Rekonstruktionen und histologische Untersuchungen menschlicher Embryonen und Feten von 19–250 mm Scheitel-Steiß-Länge. Med. Diss. Charité Universitätsmedizin Berlin; 2005.

Radlanski RJ, Renz H, Klarkowski MC. Prenatal development of the human mandible. 3D reconstructions, morphometry and bone remodelling pattern, sizes 12-117 mm CRL. Anat Embryol (Berl). 2003;207:221–32.

Fetus 118 mm CRL (ca. 14th wk): T55

30

Cartilaginous Structures in the Mandible

As already mentioned in Chap. 27, in many textbooks, it is roughly stated that the mandible arises as a desmal ossification next to Meckel's cartilage. However, there are many further cartilaginous structures, which eventually contribute to the formation of the mandible once it is a completed bone. Meckel's cartilage itself, which first serves as framework for the two lateral bony plates, eventually vanishes almost completely due to resorption and ossification. The only remnants of its existence are parts of the spheno-mandibular ligament and the malleus as its most posterior part. In the anterior region of the mandible, also referred to as the symphyseal region, a number of transient cartilaginous islets have been described. In the posterior part of the mandible, there are also several cartilaginous structures. There are carrot-like cartilaginous cores inside the condylar and the coronoid processes, visible in human fetuses between 50 and 120 mm CRL. In later stages, they are prone to ossification. Also, at the angular process, transient cartilaginous tissue has been described for humans and for several other species. The 3-D reconstructions made it possible to clearly localize the islands of cartilaginous tissue and to distinguish their position from any other cartilaginous structures such as Meckel's cartilage or its insular decay products. Only those cartilaginous islands within bone, which were located in the course of Meckel's cartilage—or where it did run in previous stages—may be remains of it.

This fetus shows clearly stained cartilaginous tissue at the brim of the bone around the primordium i_1 (on the left) and the center of i_2 (on the right) in Fig. 30.1a. Some parts of the medial embrasure and the symphyseal region impresses with intensive turquoise staining, indicating the acid mucins typical of cartilaginous tissue (Fig. 30.1).

R. J. Radlanski, *Atlas of Human Prenatal Craniofacial Development*, https://doi.org/10.1007/978-3-031-88812-0_30

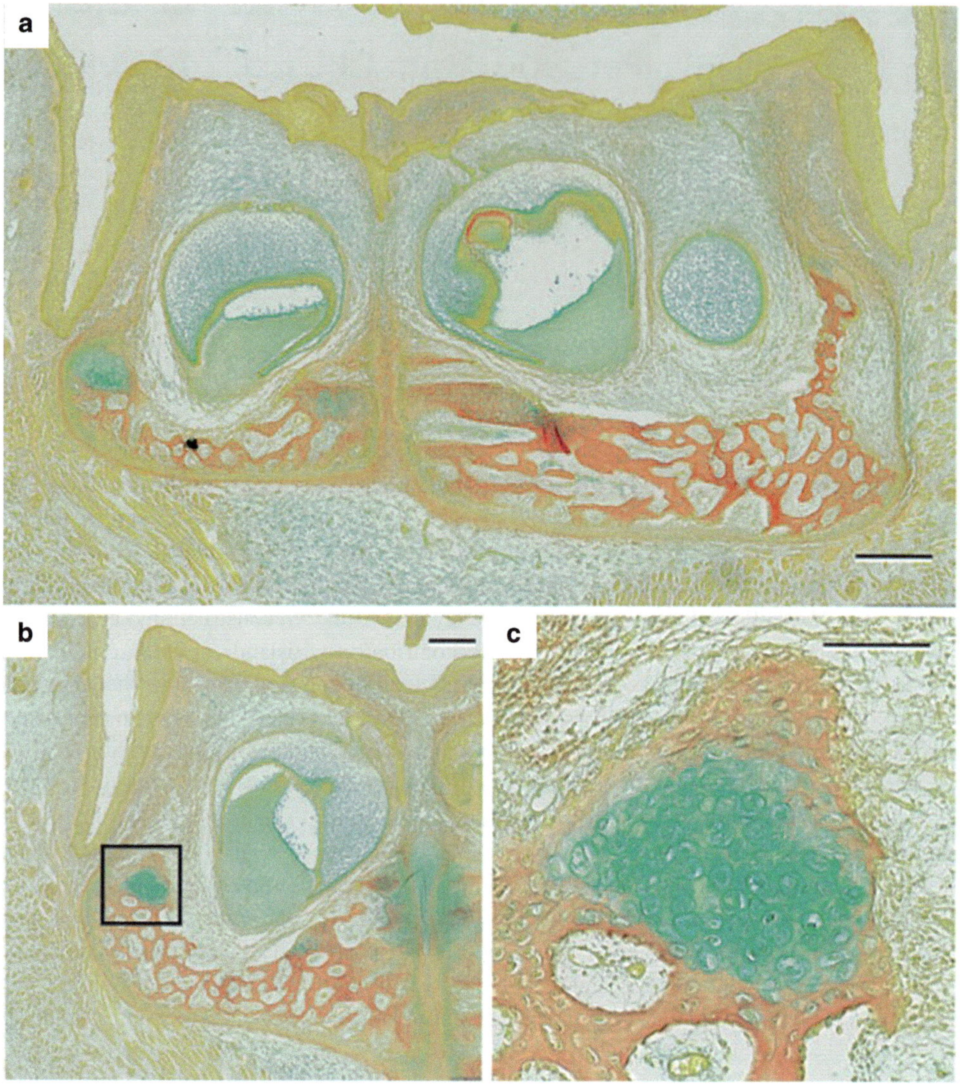

Fig. 30.1 (**a**) Human fetus, 118 mm CRL (ca. 14th week), frontal section (stained alcian blue and van Giesson, modified according to Domagk). Acid mucins and other macromolecules appear turquoise to green. Scale bar: 400 µm. (**b**) Frontal section of the same series, further posteriorly. Scale bar: 400 µm. Box: Indication of region of figure (c). (**c**) Region of figure (b). Numerous cartilage cells were found crowding in a sparse matrix, surrounded by bone. Scale bar: 100 µm. (*From* Radlanski et al. [1]; *with permission*)

Reference

1. Radlanski RJ, Renz H, Zimmermann CA, Schuster FP, Voigt A, Heikinheimo K. Chondral ossification centers next to dental primordia in the human mandible. A study of the prenatal development ranging between 68 to 270 mm CRL. Ann Anat. 2016;208:49–57.

Further Reading

Hinrichsen KV. The early development of morphology and patterns of the face in the human embryo. Berlin: Springer-Verlag; 1985.

Radlanski RJ, Renz H, Klarkowski MC. Prenatal development of the human mandible. 3D reconstructions, morphometry and bone remodelling pattern, sizes 12–117 mm CRL. Anat Embryol (Berl). 2003;207:221–32.

Sperber GH. Craniofacial embryology. Hamilton: B.C. Decker; 2001.

Fetus 150 mm CRL (ca. 18th wk): T26_050265

31

Overall Remarks

With this fetus we can visualize the situation of the middle ear region (Figs. 31.1, 31.2, 31.3, 31.4, 31.5, 31.6, 31.7, 31.8, and 31.9). The dorsal end of Meckel's cartilage still protrudes into the middle ear region as the malleus. However, the incus and stapes already show clear signs of ossification, so does the malleus, but only its caput and collum (see Fig. 31.7). And the tongue and lower tongue region could be depicted (see Figs. 31.8 and 31.9).

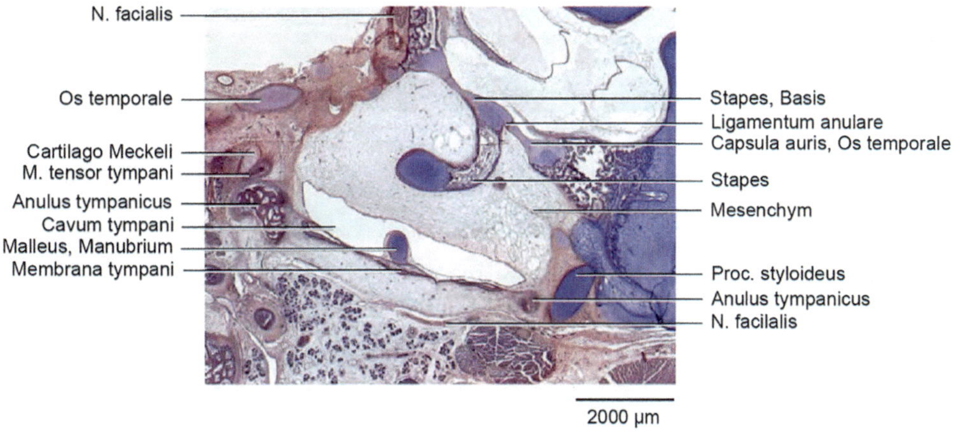

Fig. 31.1 Human fetus, 150 mm CRL, vertical section (staining: Trichrome) at the level of the base of the stapes, left is ventral

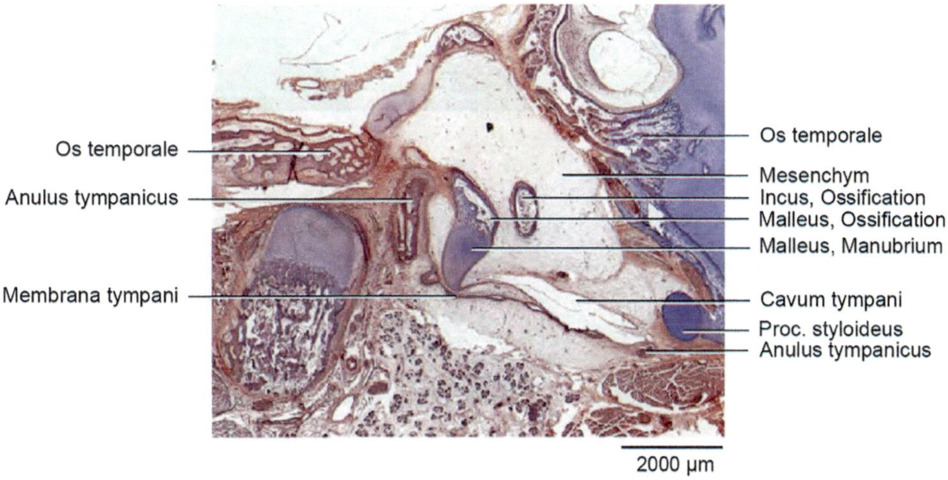

Fig. 31.2 Human fetus, 150 mm CRL, vertical section (staining: Trichrome) at the level of Manubrium mallei, left is ventral. Section plane is laterally to Fig. 31.1

R. J. Radlanski, *Atlas of Human Prenatal Craniofacial Development*, https://doi.org/10.1007/978-3-031-88812-0_31

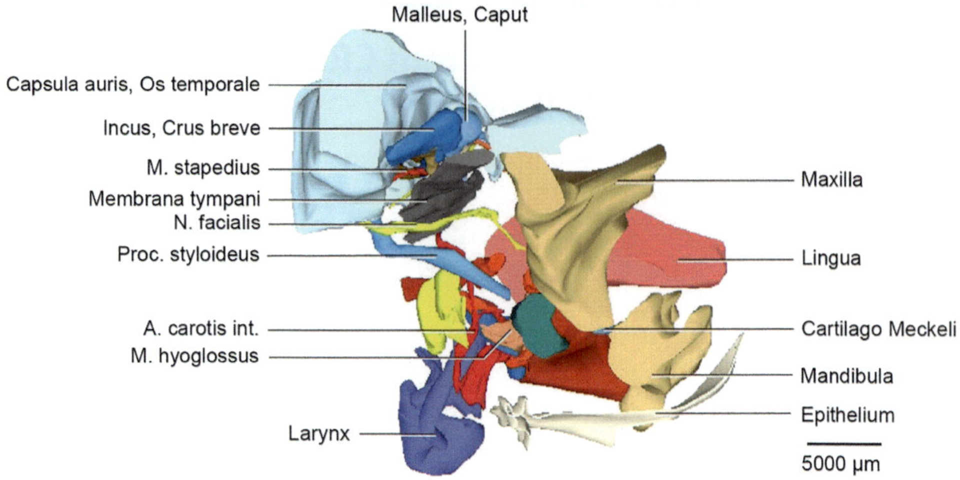

Fig. 31.3 Human fetus, 150 mm CRL, 3D-reconstruction, survey of the oral and otic regions, lateral view

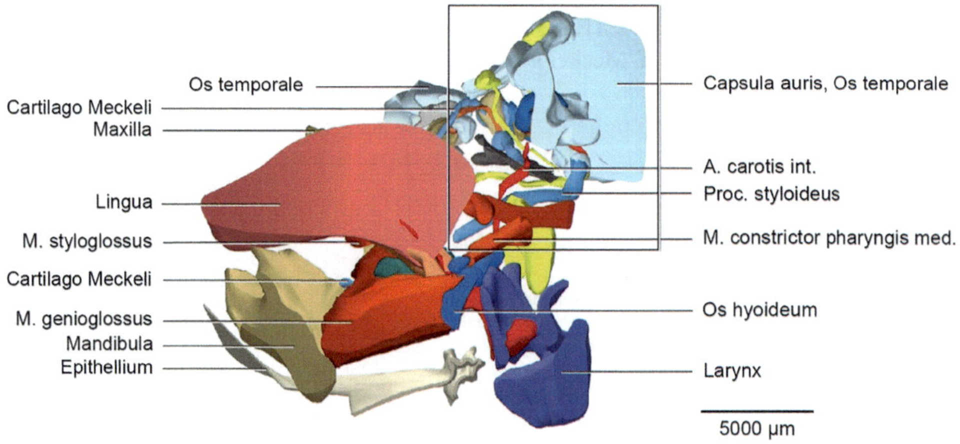

Fig. 31.4 Human fetus, 150 mm CRL, 3D-reconstruction, survey of oral and otic regions, medial view. Box: Fig. 31.5

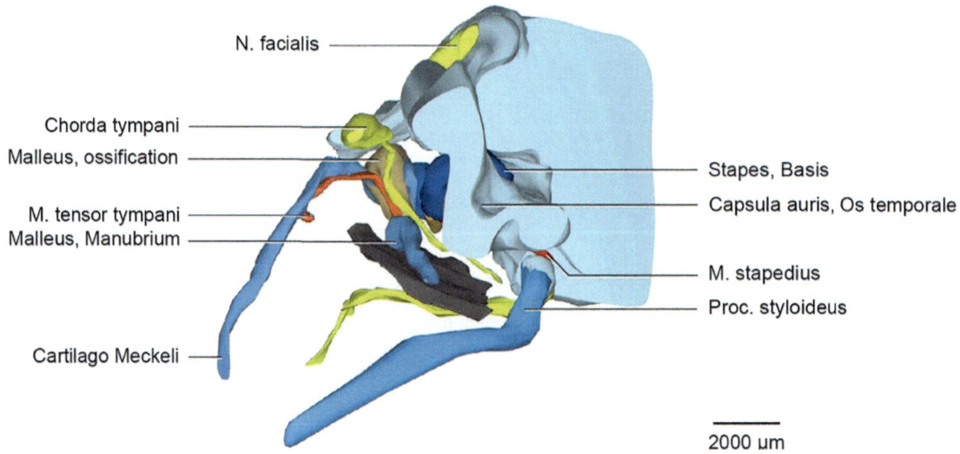

Fig. 31.5 Detailed view of the marked region in 31.4 otic region, medial view

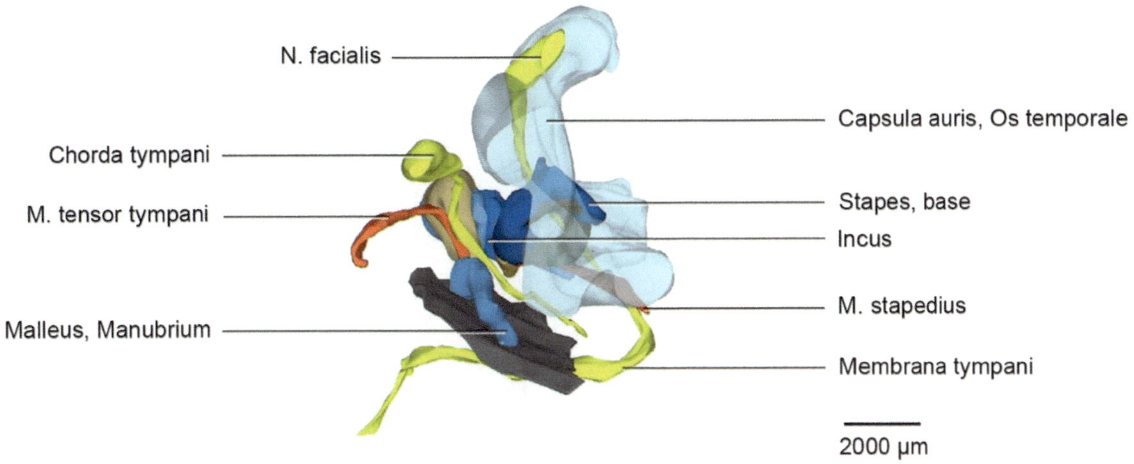

Fig. 31.6 Human fetus, 150 mm CRL, right auditory system, otic capsule transparent, medial view

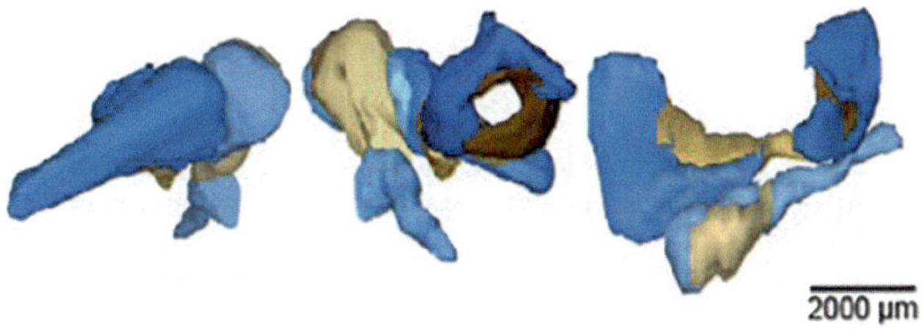

Fig. 31.7 Human fetus, 150 mm CRL, auditory ossicles (right ear), lateral, medial, and cranial view (from left)

Fig. 31.8 Human fetus, 150 mm CRL, vertical section, at the level of the major and lesser horns of the hyoid bone. Staining: Trichrome

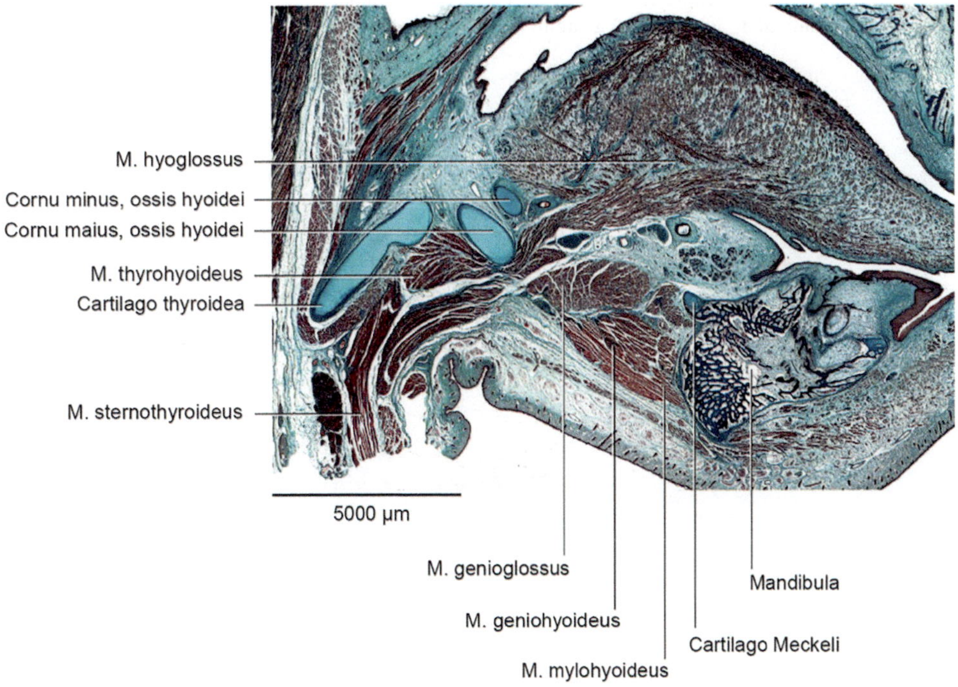

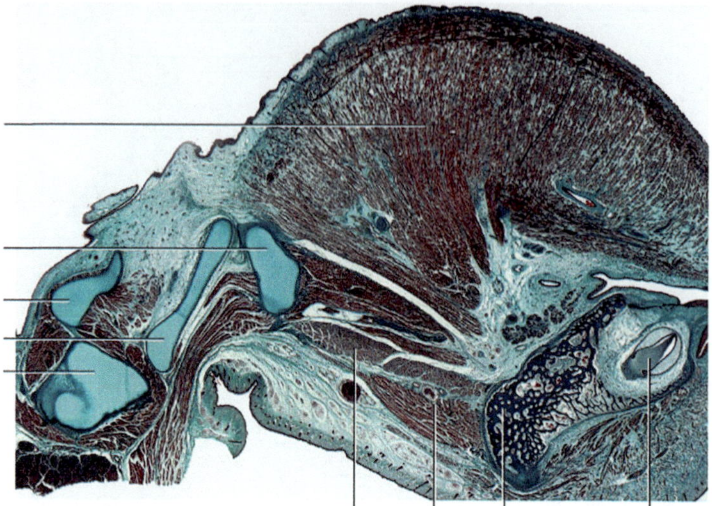

Fig. 31.9 Human fetus, 150 mm CRL, vertical section, at the level of the body of the hyoid bone. Staining: Trichrome

Further Reading

Beckert K. Beiträge zur pränatalen Morphogenese der Region des Mittelohres des Menschen—Computergestützte 3D-Rekonstruktionen und histologische Untersuchungen menschlicher Embryonen/Feten von 19 bis 150 mm CRL (7. bis 18. Woche). Med. Diss. Charité-Universitätsmedizin Berlin. 2010.

Fernández Rubio EM, Radlanski RJ. Development of the Human Primary and Secondary Jaw Joint. Ann Anat. 2024;251:1–8.

Giambartolomei LA, Brunotto M. Gomez de Ferrais ME Human temporomandibular joint disc: anatomy and measurements in prenatal development. Acta odontol. Latino-americana. 2011;24:98–103.

Germelmann AR (née Than) Beitrag zur pränatalen Morphogenese des Os hyoideum des Menschen—Computergestützte 3D-Rekonstruktionen und histologische Untersuchungen menschlicher Embryonen/Feten von 19–150 mm CRL (7. bis 18. Woche) Med. Diss. Charité-Universitätsmedizin Berlin; 2008.

Hinrichsen KV. The early development of morphology and patterns of the face in the human embryo. Berlin Heidelberg New York Tokyo: Springer Verlag; 1985.

32 Fetus 160 mm CRL (ca. 18th wk): PIT2_180294

Overall Remarks

The cartilage islands in the mandible can only be reliably assessed in their spatial relationship if the 3D-reconstructions are displayed transparently. In this fetus, Meckel's cartilage shows a continuity only up to the primordium of m_1. Further anteriorly, singular remnants of Meckel's cartilage can be identified on the lingual aspects of the incisor primordia. Additional small cartilaginous islets, which were not in the course of Meckel's cartilage, are found buccally to all the deciduous primordia and also lingually to the central incisor primordium. They can be seen as single centers within the brim of the alveolar bone. In addition, laterally to the inferior alveolar vein cartilaginous tissue is to be also found within the mandibular bone (Figs. 32.1, 32.2, 32.3, 32.4, and 32.5).

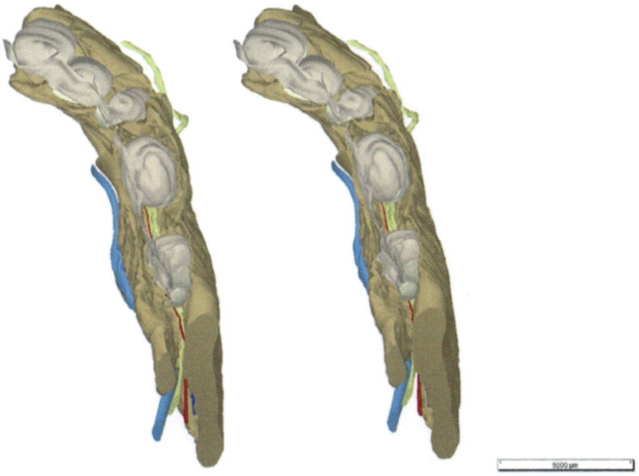

Fig. 32.1 Stereoscopical pair of images. Human fetus 160 mm CRL (ca. 18th week), mandible (right half) with dental primordia, and Meckel's cartilage, cranial view. The condylar process and the coronoid process are not included in the reconstruction. Scale bar: 5000 µm

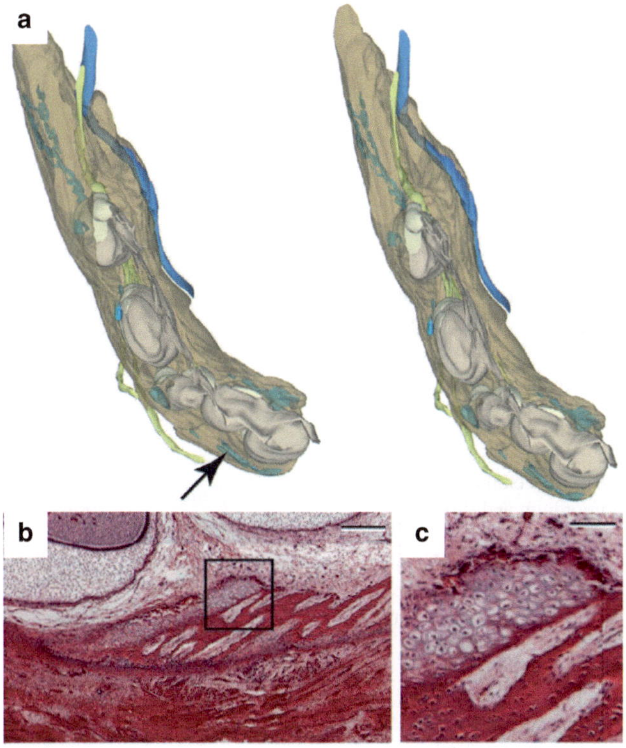

Fig. 32.2 (**a**) Stereoscopical pair of images. Human fetus 160 mm CRL (ca. 18th week), mandible (right half) with dental primordia, Meckel's cartilage. Structures made transparent in order to visualize the single cartilaginous structures, located at the brim of the crestal bone buccally at all dental primordia (arrow) and lingually at the central incisor primordium, cranial view. (**b**) Human fetus, 160 mm CRL, horizontal section (staining: Hematoxylin-eosin) at level of dental primordium i_1 with cartilaginous tissue in the buccal brim of the surrounding bone. Scale bar: 200 µm. Box: Location of (**c**). (**c**) Region of box in (**b**). Cartilage cells in sufficient matrix, covered by a thin layer of bone toward the dental primordium. Scale bar: 50 µm

R. J. Radlanski, *Atlas of Human Prenatal Craniofacial Development*, https://doi.org/10.1007/978-3-031-88812-0_32

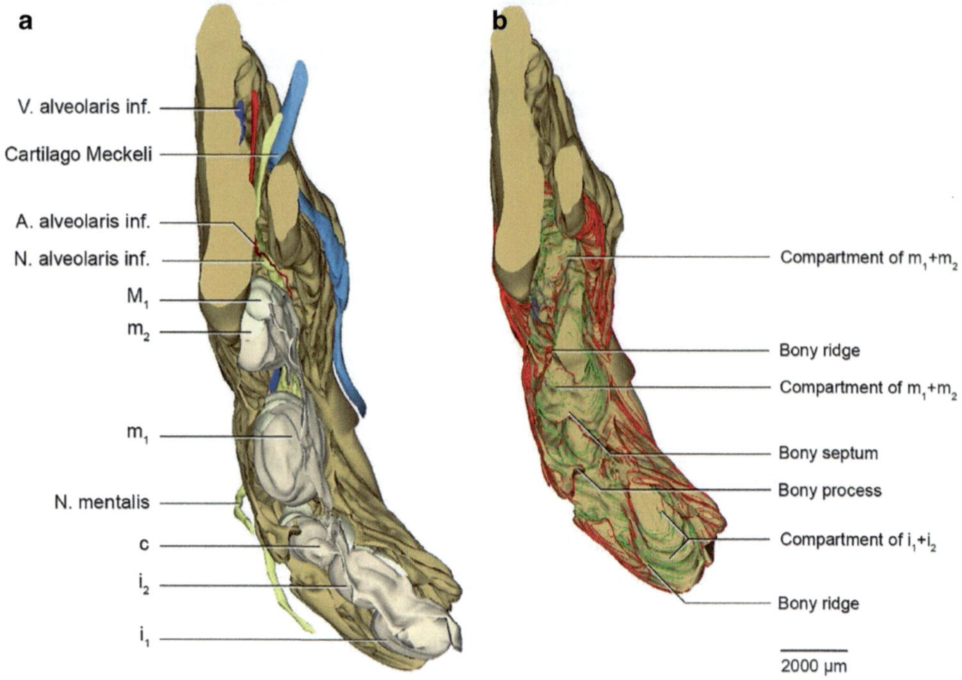

Fig. 32.3 (**a**) Human fetus, 160 mm CRL, 3D-reconstruction of the right half of the mandible with dental primordia. Cranial view. The condylar process and the coronoid process are not included in the reconstruction. (**b**) The same reconstruction of the right half of the mandible with bone remodeling pattern superimposed

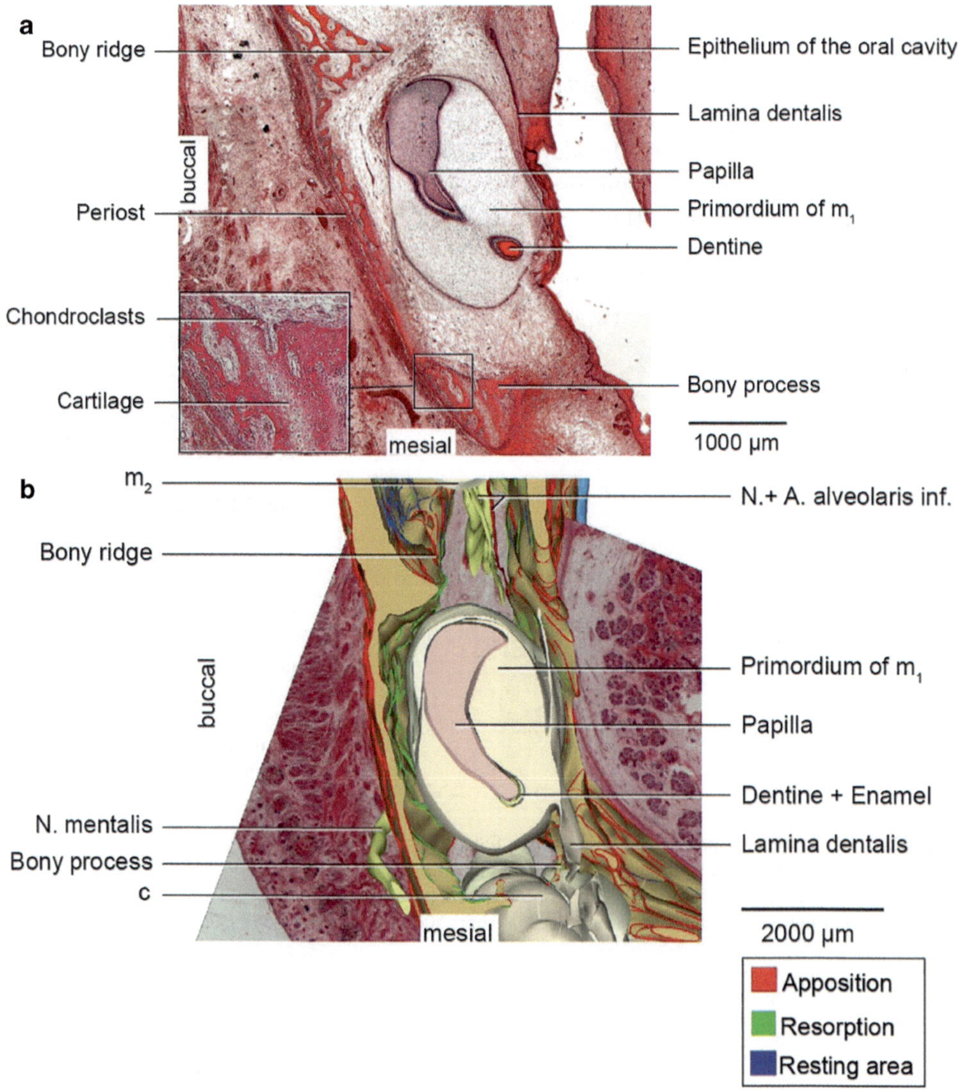

Fig. 32.4 (**a**) Human fetus, 160 mm CRL, horizontal section through the mandibular region of the right first deciduous molar primordium, HE-staining. (**b**) 3D partial reconstruction of the right deciduous first molar primordium and surrounding parts of the mandible, histological section (Staining: Hematoxylin-eosin) in the background (appears blurry due to software issues). Cranial view. The dental primordium is sectioned horizontally. Bone remodeling pattern superimposed

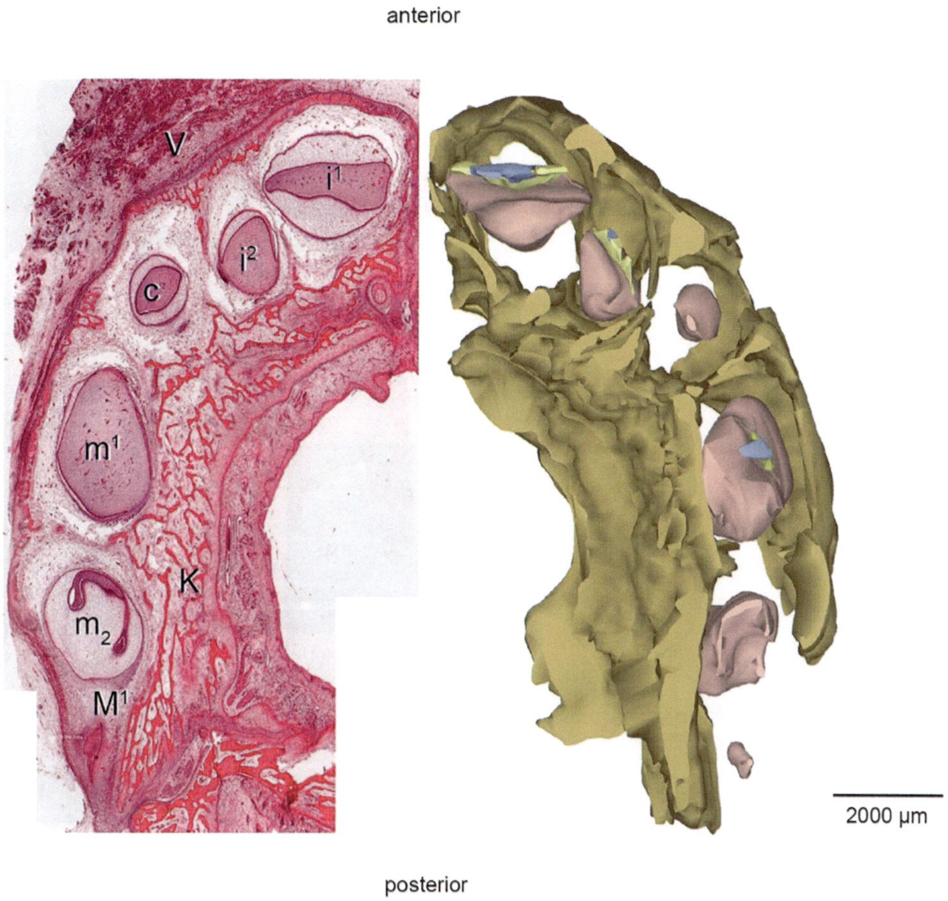

Fig. 32.5 Human fetus, 160 mm CRL, horizontal section (staining: Hematoxylin-eosin), right half of maxilla, and on the right 3D-reconstruction of the left half of the maxilla with dental primordia, from the serial sections, caudal view

Peridental Bone Formation

The teeth are all surrounded by bone. From the diagnosis in the histological section with regard to bone apposition and bone resorption, the apposition zones were transferred to the reconstructions in red and the resorption zones in green. This shows that also at this stage of development, the bony enclosure of the dental primordia represents an active bone formation (Fig. 32.6).

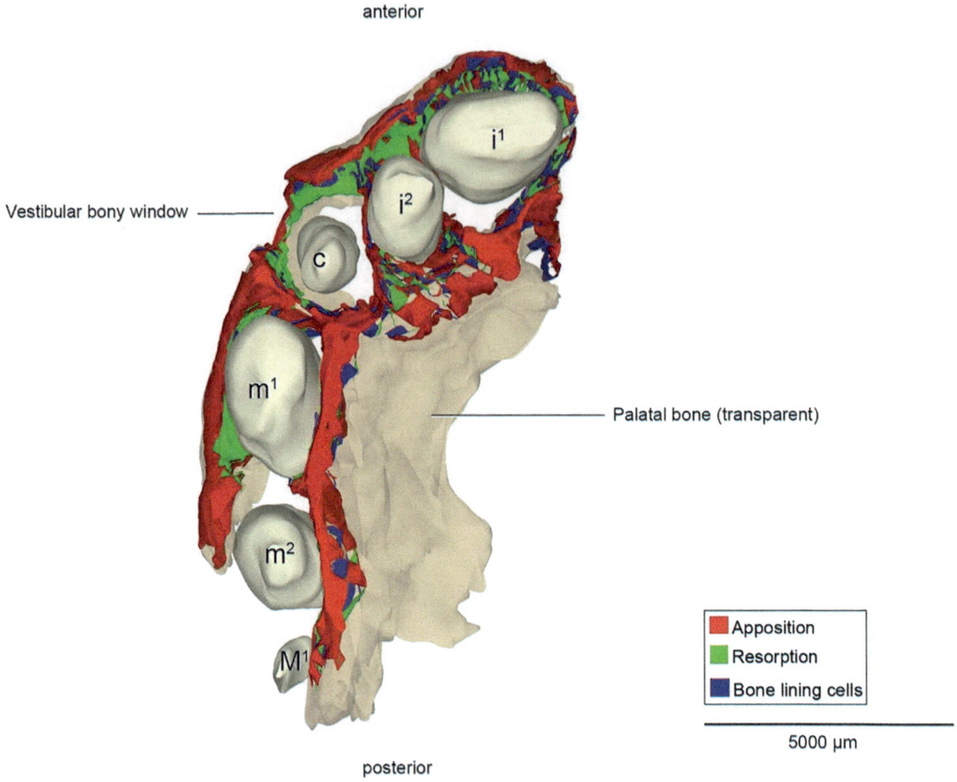

Fig. 32.6 Human fetus, 160 mm CRL, right half of the maxilla with dental primordia and bone remodeling pattern superimposed, caudal view

Further Reading

Kalinke U. Embryologische Untersuchungen zur Entwicklung des Alveolarknochens im Oberkiefer beim Menschen zur Zeit der vorgeburtlichen Entwicklungsstadien von 19–68 mm SSL. Med. Diss. Charité-Universitätsmedizin Berlin; 2014.

Konietzny M. Zur Entwicklung des Kieferknochens im Bereich der Zahnanlagen—Histologische und morphologische Untersuchungen an der fetalen Maxilla des Menschen während des Glockenstadiums der Zahnanlagen (160–270 mm SSL). Med. Diss. Charité-Universitätsmedizin Berlin; 2014.

Radlanski RJ. Contributions to the development of human deciduous tooth primordia. Chicago: Quintessence Publishing Co, Inc.; 1993.

Radlanski RJ, Renz H, Tsengelsaikhan N, Schuster FP, Zimmermann CA. The remodeling pattern of human mandibular alveolar bone during prenatal formation from 19 to 270 mm CRL. Ann Anatomy. 2016;205:65–74.

Schuster FP. Zur Entwicklung des Kieferknochens im Bereich der Zahnanlagen. Histologische und morphologische Untersuchungen an der fetalen Mandibula des Menschen während des Glockenstadiums der Zahnanlagen (68–160 mm SSL). Med. Diss. Charité-Universitätsmedizin Berlin; 2013.

Zimmermann C. Zur Entwicklung des Kieferknochens im Bereich der Zahnanlagen. Histologische und morphologische Untersuchungen an der fetalen Mandibula des Menschen während des späten Glockenstadiums der Zahnanlagen (160–270 mm SSL) Med. Diss. Charité-Universitätsmedizin Berlin; 2013.

Fetus 190 mm CRL (ca. 20th wk): MIN101293

33

Overall Remarks

At this stage of development, the additional cartilage centers, which are unrelated to Meckel's cartilage, are still present in the mandible. They can be recognized in the area of the incisors both vestibularly and lingually. Meckel's cartilage is already degraded in the anterior third of its course although a remnant remains in the very anterior part. Its posterior end extends into the middle ear, where the malleus with the incus continues to form the articulation of the primary temporomandibular joint (Figs. 33.1, 33.2, and 33.3).

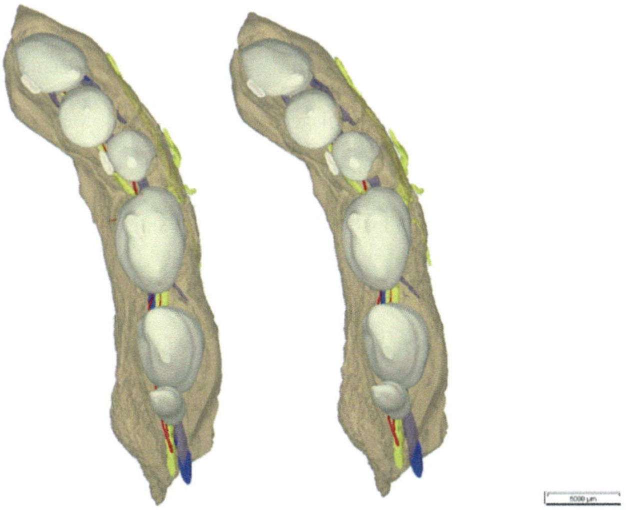

Fig. 33.1 Stereoscopical pair of images. Human fetus 190 mm CRL (ca. 20th week), mandible (right half) with dental primordia and alveolar inferior nerve and vessels. Scale bar: 5000 µm

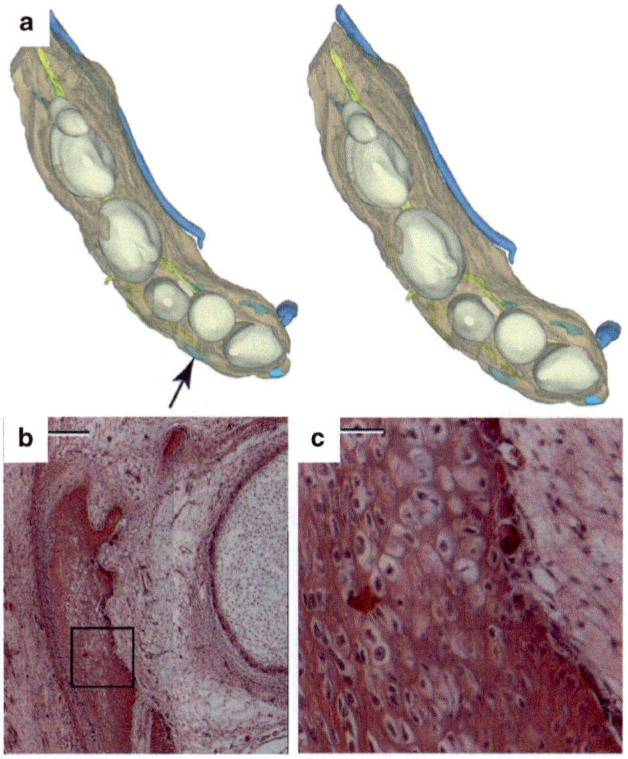

Fig. 33.2 (**a**) Stereoscopical pair of images. Human embryo 190 mm CRL (ca. 20th week), mandible (right half) with dental primordia, Meckel's cartilage. Structures made transparent in order to visualize the single cartilaginous structures, located at the brim of the crestal bone buccally at dental primordia i_1 and i_2 (arrow) and lingually at the central incisor primordium, cranial view. Scale bar: 2000 µm. (**b**) Human fetus, 190 mm CRL, horizontal section (staining: Hematoxylin-eosin) at level of dental primordium i^1 with cartilaginous tissue in the buccal brim of the surrounding bone. Scale bar: 400 µm. Box: Location of (**c**). (**c**) Region of box in (**b**). Cartilage cells in abundant matrix, covered by a thin layer of bone toward the dental primordium. Scale bar: 100 µm

R. J. Radlanski, *Atlas of Human Prenatal Craniofacial Development*, https://doi.org/10.1007/978-3-031-88812-0_33

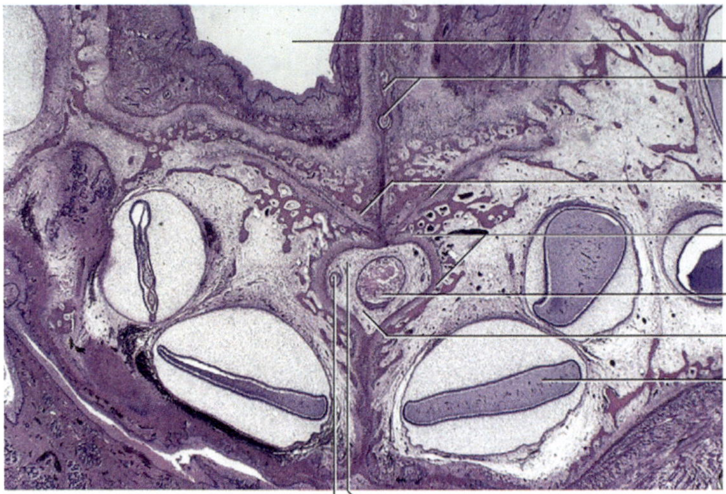

Fig. 33.3 Human fetus, 190 mm CRL, horizontal section at level of the maxillary dental primordia (staining: Hematoxylin-eosin)

Incisive Canal

The incisive canal is enclosed in the caudal section rostrally, laterally, and dorsally by bone of the primary palate. Rostrally and dorsally, a narrow, approximately 150 µm wide interruption is recognizable between the left- and right-sided bone, in which compacted mesenchyme runs, which also lines the canal wall. The nasal openings of the nasopalatine nerves have no bony border, but are separated medially by bone of the primary palate. The nasopalatine nerves lie at the level of the incisive foramen rostrolateral in the loosely filled lumen directly on the lining of dense mesenchyme. The nasopalatine arteries lie dorsally. The nasal openings of the nasopalatine nerves are separated from the surrounding area on each side by dense mesenchyme. The nasopalatine arteries run in the nasopalatine duct in the dense mesenchyme.

Interestingly, several epithelial pearls can be found in the suture between the lateral palatal processes, even a larger one in the incisive canal.

Further Reading

Emmerich S. Die pränatale Morphogenese des humanen Canalis incisivus. FU Berlin: Med. Diss; 2004.

Radlanski RJ, Emmerich S, Renz H. Prenatal morphogenesis of the human incisive canal. Anat Embryol. 2004;208:265–71.

Radlanski RJ, Renz H, Tsengelsaikhan N, Schuster FP, Zimmermann CA. The remodeling pattern of human mandibular alveolar bone during prenatal formation from 19 to 270 mm CRL. Ann Anatomy. 2016;205:65–74.

Schuster FP. Zur Entwicklung des Kieferknochens im Bereich der Zahnanlagen. Histologische und morphologische Untersuchungen an der fetalen Mandibula des Menschen während des Glockenstadiums der Zahnanlagen (68–160 mm SSL). Med. Diss. Charité-Universitätsmedizin Berlin; 2013.

Zimmermann C. Zur Entwicklung des Kieferknochens im Bereich der Zahnanlagen. Histologische und morphologische Untersuchungen an der fetalen Mandibula des Menschen während des späten Glockenstadiums der Zahnanlagen (160–270 mm SSL) Med. Diss. Charité-Universitätsmedizin Berlin; 2013.

34 Fetus 225 mm CRL (ca. 21st wk): AMA-2_160294

Additional Cartilaginous Islands

Even at this stage, the additional cartilaginous islands in the bone of the mandible in the region of the incisors and on the vestibular and lingual canine are still present. Due to their location, it is clear that they have nothing to do with Meckel's cartilage [1]. This is because the anterior part of the cartilage is completely degraded at this stage—its dorsal end, however, continues to extend into the middle ear, where it articulates with the incus as the primary jaw joint (Figs. 34.1, 34.2, and 34.3).

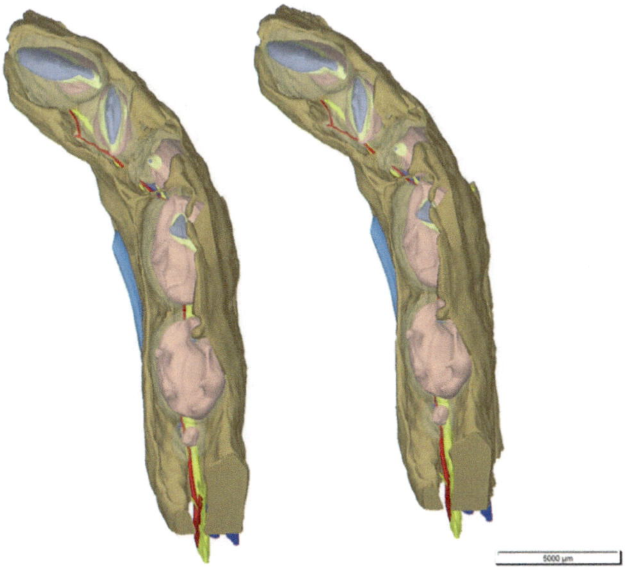

Fig. 34.1 Stereoscopical pair of images. Human fetus 225 mm CRL (ca. 21st week), mandible (right half) with dental primordia, inferior alveolar nerve and vessels, and Meckel's cartilage, cranial view. The condylar process and the coronoid process are not included in the reconstruction. Scale bar: 5000 µm

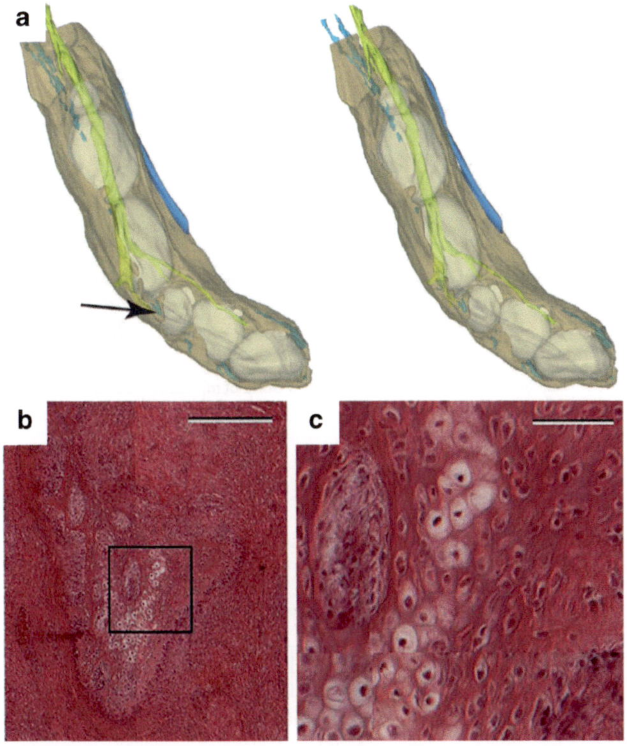

Fig. 34.2 (**a**) Stereoscopical pair of images. Human fetus 225 mm CRL (ca. 21st week), mandible (right half) with dental primordia, Meckel's cartilage. Structures made transparent in order to visualize the single cartilaginous structures, located at the brim of the crestal bone buccally at all deciduous dental primordia. Cartilaginous tissue is also within the mandibular bone posterior to m_2 and occlusal of M_1. Cranial view. Scale bar: 200 µm. The arrow indicates the section plane in (**b**). (**b**) Human fetus, 225 mm CRL, horizontal section (staining: Hematoxylin-eosin) at level between dental primordia i_2 and c with cartilaginous tissue in the buccal brim of the surrounding bone. Scale bar: 500 µm. Box: Location of (**c**). (**c**) Region of box in (**b**). Cartilage cells in matrix, surrounded by bone. Scale bar: 100 µm

R. J. Radlanski, *Atlas of Human Prenatal Craniofacial Development*, https://doi.org/10.1007/978-3-031-88812-0_34

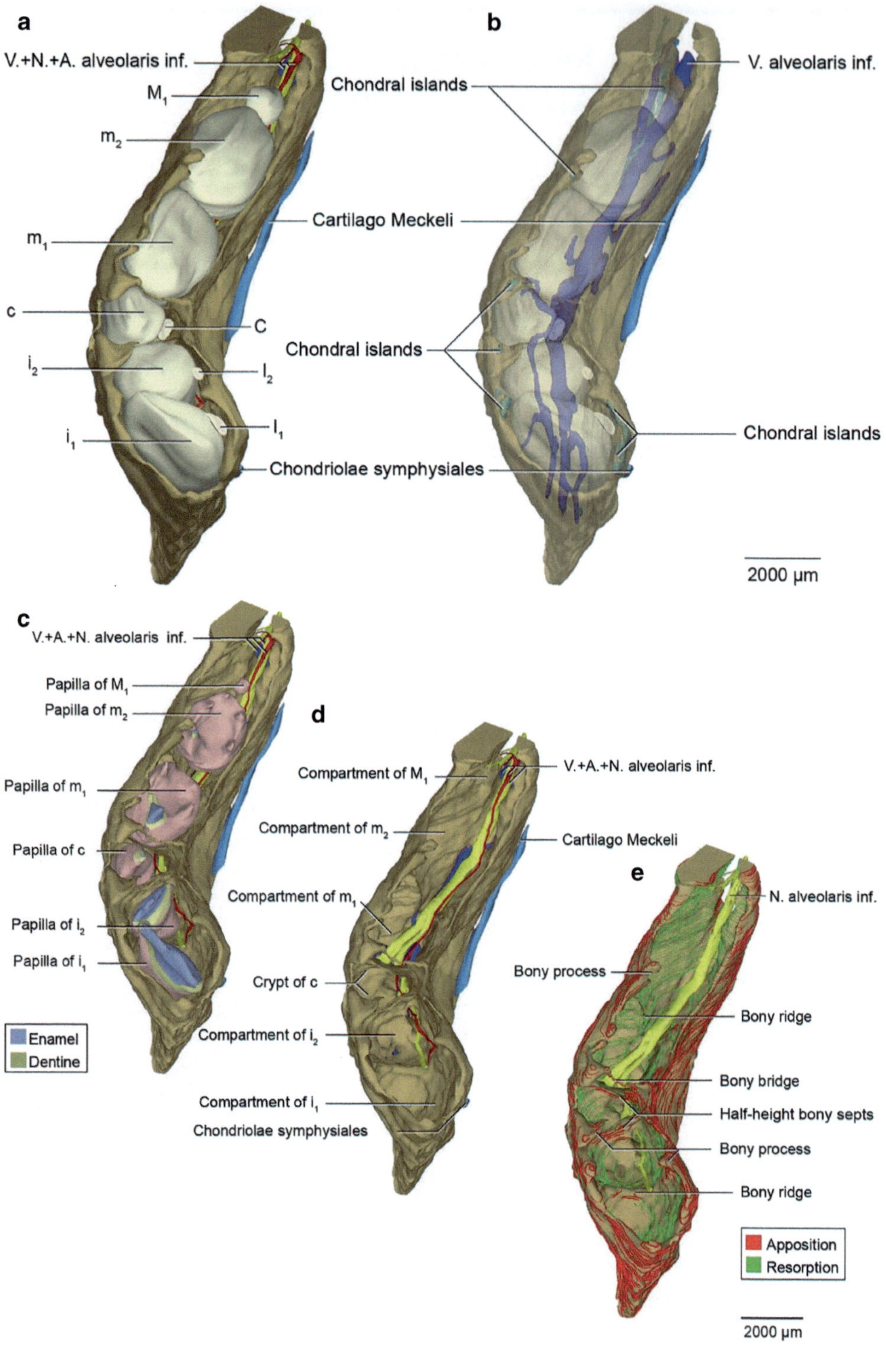

Fig. 34.3 (**a**) Human fetus 225 mm CRL (ca. 21st week), mandible (right half) with dental primordia, inferior alveolar nerve and vessels, and Meckel's cartilage, cranial view. The condylar process and the coronoid process are not included in the reconstruction. (**b**) The same reconstruction, structures made transparent to visualize the inferior alveolar vein. (**c**) The same reconstruction, dental follicles removed to visualize the developmental status of the dental primordia. (**d**) The same reconstruction, dental primordia removed to visualize the bony compartments of the dental primordia. (**e**) The same reconstruction, bone remodeling pattern superimposed

Dental Primordia and Surrounding Bone

All primordia (of the first dentition) of the mandible have significantly increased in volume compared to the previous stage and are now closer together. They are in their late bell stages and now all show an incipient or increasing formation of the hard substances dentin and enamel. The full width of the incisal edges of the deciduous incisors is already covered with dentin and enamel. The tip of the canine also shows hard substance formation. The same applies to the molars although not all of their cusps have reached this stage. All dental primordia are surrounded by a dental follicle proper, as a dense, fibrous coating and can thus be easily separated from the loose mesenchyme of the surrounding area. The 3D-reconstructions show the bony compartments of the dental primordia and at the same time the bone remodeling processes: Bone growth dominates on the buccal and lingual surfaces of the mandible, as well as on the interdental septa of the alveolar process bone. However, the two incisors and the molars are each located in common, enlarged crypts; only the canine has its own alveola, which is surrounded by bone on all sides. The inner surfaces of the alveolae show predominantly bone resorption. This is an indication of the space-occupying growth of the dental primordia. The conditions in the maxilla are comparable. At the dorsal end of the dental arch, the first permanent molars are recognizable in the reconstructions, whereas for the permanent incisors and canines only the replacement dental laminae are recognizable; and there are no indications of the formation of the bicuspids yet (Figs. 34.4, 34.5, and 34.6).

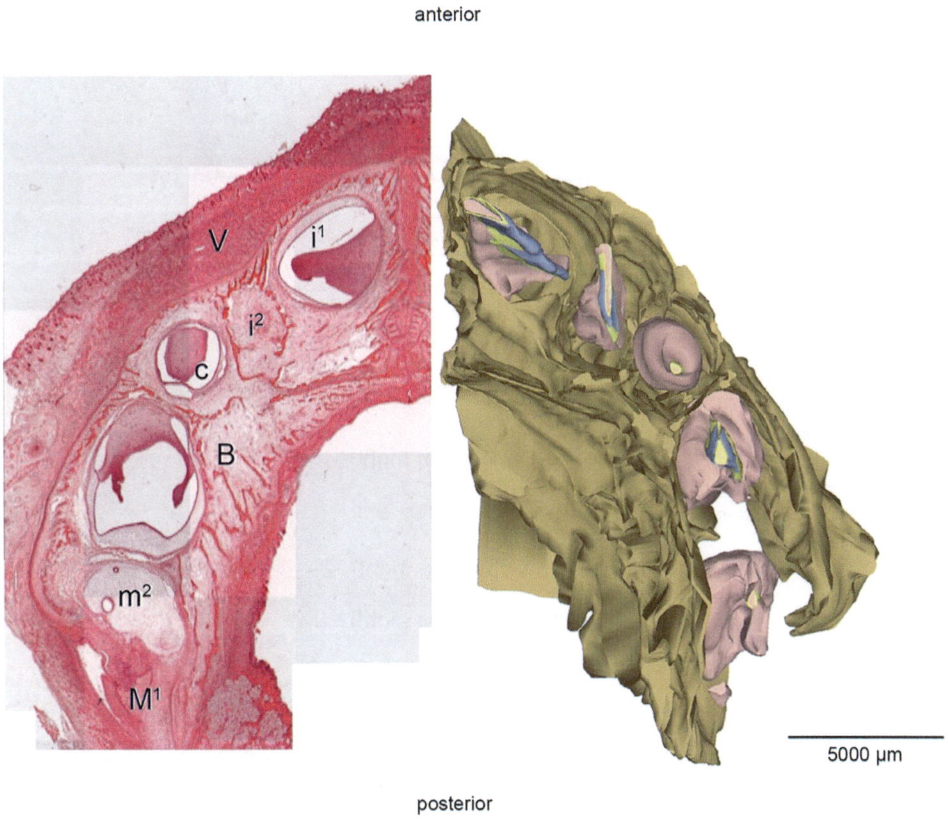

Fig. 34.4 Human fetus, 225 mm CRL (ca. 21st week), horizontal section (staining: Hematoxylin-eosin), right half of maxilla, and on the right 3D-reconstruction of the left half of the maxilla with dental primordia, from the serial sections, caudal view. Dental primordia i^1, i^2, c, m^1, m^2, and M^1; *B* bone of maxilla; *V* vestibulum

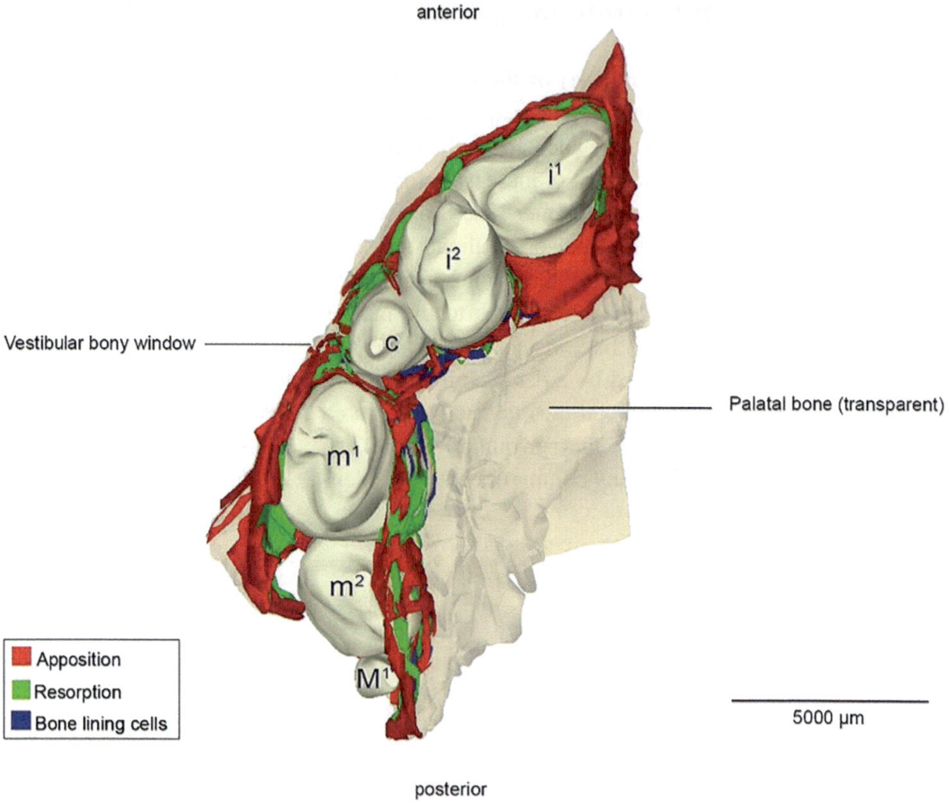

Fig. 34.5 Human fetus, 225 mm CRL, 3D-reconstruction of the right half of the maxilla with dental primordia and bone remodeling pattern superimposed, caudal view

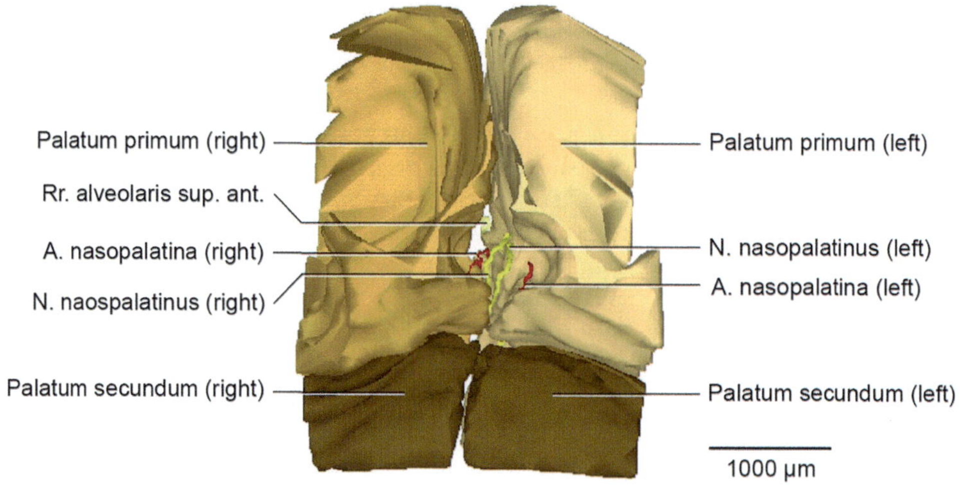

Fig. 34.6 Human fetus, 225 mm CRL, 3D-reconstruction, overview from caudal on the incisive canal; lower border of the figure is pointing in dorsal direction. (*From* Radlanski et al. [2]; *with permission*)

Incisive Canal

The incisive canal forms a funnel-shaped palatal exit, runs uniformly through the caudal part of the primary palate to divide nasally into two separate canals. The nasopalatine nerves on both sides lie directly adjacent to each other palatally. In the caudal section of the incisive canal of about 1.5 mm, there is no bony separation between the left and right nerves. Only in the direction of the nasal cavity does bone intervene, forming the maxillary nasal crest. With the division into right and left incisive canal in the direction of the nasal cavity, the nasopalatine nerve is displaced into the border region of the primary and secondary palate in order to open into the nasal cavity at the junction with the secondary palate. The nasopalatine artery runs lateral to the nasopalatine nerve and is separated from the nasopalatine nerve in the middle section by a bony septum. It runs in a gusset-like protrusion of compacted mesenchyme in the primary palate and divides cranially into two arteries.

References

1. Radlanski RJ, Renz H, Zimmermann CA, Schuster FP, Voigt A, Heikinheimo K. Chondral ossification centers next to dental primordia in the human mandible. A study of the prenatal development ranging between 68 to 270 mm CRL. Ann Anat. 2016;208:49–57.
2. Radlanski RJ, Emmerich S, Renz H. Prenatal morphogenesis of the human incisive canal. Anat Embryol. 2004;208:265–71.

Further Reading

Konietzny M. Zur Entwicklung des Kieferknochens im Bereich der Zahnanlagen—Histologische und morphologische Untersuchungen an der fetalen Maxilla des Menschen während des Glockenstadiums der Zahnanlagen (160–270 mm SSL). Med. Diss. Charité-Universitätsmedizin Berlin; 2014.

Radlanski RJ, Renz H, Tsengelsaikhan N, Schuster FP, Zimmermann CA. The remodeling pattern of human mandibular alveolar bone during prenatal formation from 19 to 270 mm CRL. Ann Anat. 2016;205:65–74.

Radlanski RJ. Oral structure and biology. Batavia: Quintessence Publishing; 2018.

Schuster FP. Zur Entwicklung des Kieferknochens im Bereich der Zahnanlagen. Histologische und morphologische Untersuchungen an der fetalen Mandibula des Menschen während des Glockenstadiums der Zahnanlagen (68–160 mm SSL). Med. Diss. Charité-Universitätsmedizin Berlin; 2013.

Zimmermann C. Zur Entwicklung des Kieferknochens im Bereich der Zahnanlagen. Histologische und morphologische Untersuchungen an der fetalen Mandibula des Menschen während des späten Glockenstadiums der Zahnanlagen (160–270 mm SSL) Med. Diss. Charité-Universitätsmedizin Berlin; 2013.

Fetus 230 mm CRL (ca. 24st wk): ULF270592

35

Temporomandibular Joint

For the first time, in this developmental stage, only the secondary jaw joint, and no longer any primary jaw joint, is present [1].

There is also no longer any connection between Meckel's cartilage and the malleus. As a consequence, the secondary temporomandibular joint, now well established, can move freely without any impairment from the former primary jaw joint. The condylar cartilage is prominent well rounded.

The glenoid fossa now has a concave contour at its inferior margin and the distance between it and the condylar process is clearly decreased. Also, the articular eminence can be seen now for the first time.

The articular disc has developed into a mesenchymal structure with a high proportion of fibrous and collagenous connective tissue. It shows a distinct biconcave shape with a thin central part and thicker peripheral parts that are confluent with the surrounding mesenchyme. Its extension and position is, from now, on very similar to the situation after birth (Figs. 35.1, 35.2, and 35.3).

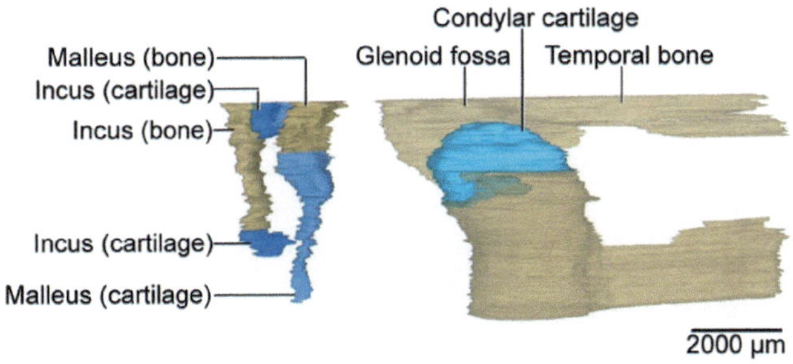

Fig. 35.1 Human fetus 230 mm CRL (ca. 24th week). 3D reconstruction of selected parts of the right temporomandibular joint and middle ear, lateral view. Mandible and temporal bone made transparent. (*From* Fernández Rubio and Radlanski [1]; *with permission*)

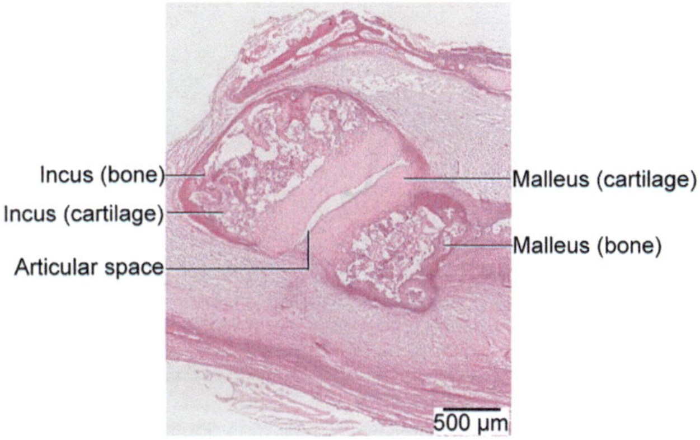

Fig. 35.2 Histological section of the right middle ear. Sagittal plane, staining: Hematoxylin-eosin

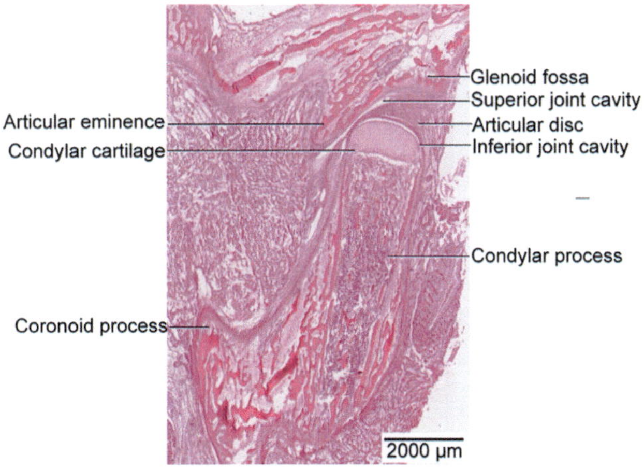

Fig. 35.3 Histological section of the left temporomandibular joint. Sagittal plane, staining: Hematoxylin-eosin

R. J. Radlanski, *Atlas of Human Prenatal Craniofacial Development*, https://doi.org/10.1007/978-3-031-88812-0_35

Reference

1. Fernández Rubio EM, Radlanski RJ. Development of the human primary and secondary jaw joint. Ann Anat. 2024;251:1–8.

Further Reading

Hinrichsen KV. The early development of morphology and patterns of the face in the human embryo. Berlin Heidelberg New York Tokyo: Springer Verlag; 1985.

Radlanski RJ, Lieck S, Bontschev NE. Development of the human temporomandibular joint. Computer-aided 3D-reconstructions. Eur J Oral Sci. 1999;107:25–34.

Radlanski RJ. Oral structure and biology. Batavia: Quintessence Publishing; 2018.

Sperber GH. Craniofacial embryology. Hamilton: B.C. Decker; 2001.

Fetus 250 mm CRL (ca. 26th wk): LUI010293

36

Dental Primordia

In this fetus, the dental primordia of the first dentition have developed into the bell stage and the formation of dental hard substance has begun. It is therefore very worthwhile to take a closer look at the dental primordia. After an overview of the entire tooth bell in a very nice histological section, which shows all the important features, the details are shown in higher magnification. And, of course, the 3D-reconstructions of the dental primordia are also shown, exemplified by an incisor and a molar. Finally, the surrounding bone is also included in the illustrations (Figs. 36.1, 36.2, 36.3, and 36.4) [1].

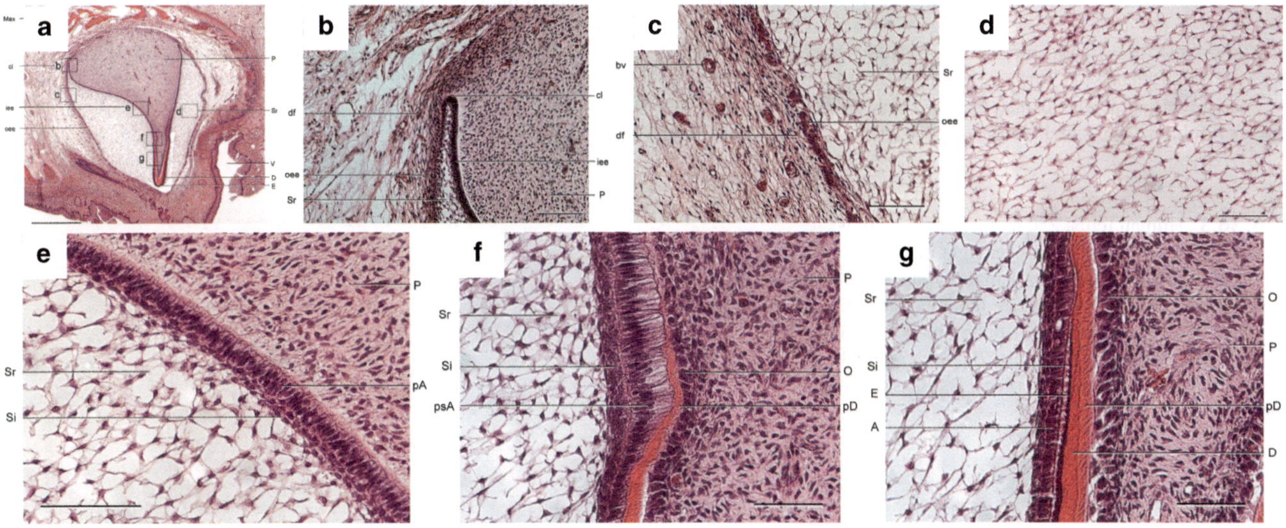

Fig. 36.1 (**a**) Human fetus, 250 mm CRL (ca. 26th wk), vertical section through the upper right deciduous incisor primordium (staining: Hematoxylin-eosin). Scale bar: 1000 µm. *Max* maxilla; *cl* cervical loop; *iee* inner enamel epithelium; *oee* outer enamel epithelium; *P* pulp; *Sr* stratum reticulare; *V* vestibule; *D* dentin; *E* enamel. (**b**) Human fetus, 250 mm CRL (ca. 26th wk), vertical section through the upper right deciduous incisor primordium (staining: Hematoxylin-eosin), close up of region b in (**a**). Scale bar: 100 µm. *df* dental follicle; *oee* outer enamel epithelium; *Sr* stratum reticulare; *cl* cervical loop; *iee* inner enamel epithelium; *P* pulp. (**c**) Human fetus, 250 mm CRL (ca. 26th wk), vertical section through the upper right deciduous incisor primordium (staining: Hematoxylin-eosin), close up of region c in (**b**). Scale bar: 100 µm. *bv* blood vessel; *df* dental follicle; *Sr* stratum reticulare; *oee* outer enamel epithelium. (**d**) Human fetus, 250 mm CRL (ca. 26th wk), vertical section through the upper right deciduous incisor primordium (staining: Hematoxylin-eosin), close up of region d in (**a**). Stratum reticulare. Scale bar: 100 µm. (**e**) Human fetus, 250 mm CRL (ca. 26th wk), vertical section through the upper right deciduous incisor primordium (staining: Hematoxylin-eosin), close up of region e in (**a**). Scale bar: 50 µm. *Sr* stratum reticulare; *Si* stratum intermedium; *P* pulp; *pA* pre-Ameloblasts. (**f**) Human fetus, 250 mm CRL (ca. 26th wk), vertical section through the upper right deciduous incisor primordium (staining: Hematoxylin-eosin), close up of region f in (**a**). Scale bar: 50 µm. *Sr* stratum reticulare; *Si* stratum intermedium; *psA* presecretory Ameloblast; *P* pulp; *O* odontoblasts; *pD* pre-Dentin. (**g**) Human fetus, 250 mm CRL (ca. 26th wk), vertical section through the upper right deciduous incisor primordium (staining: Hematoxylin-eosin), close up of region g in (**a**). Scale bar: 50 µm. *Sr* stratum reticulare; *Si* stratum intermedium; *E* enamel; *A* Ameloblasts; *O* odontoblasts; *P* pulp; *pD* pre-Dentin; *D* Dentin. (**a–g**, *From* Radlanski [1]; *with permission*)

R. J. Radlanski, *Atlas of Human Prenatal Craniofacial Development*, https://doi.org/10.1007/978-3-031-88812-0_36

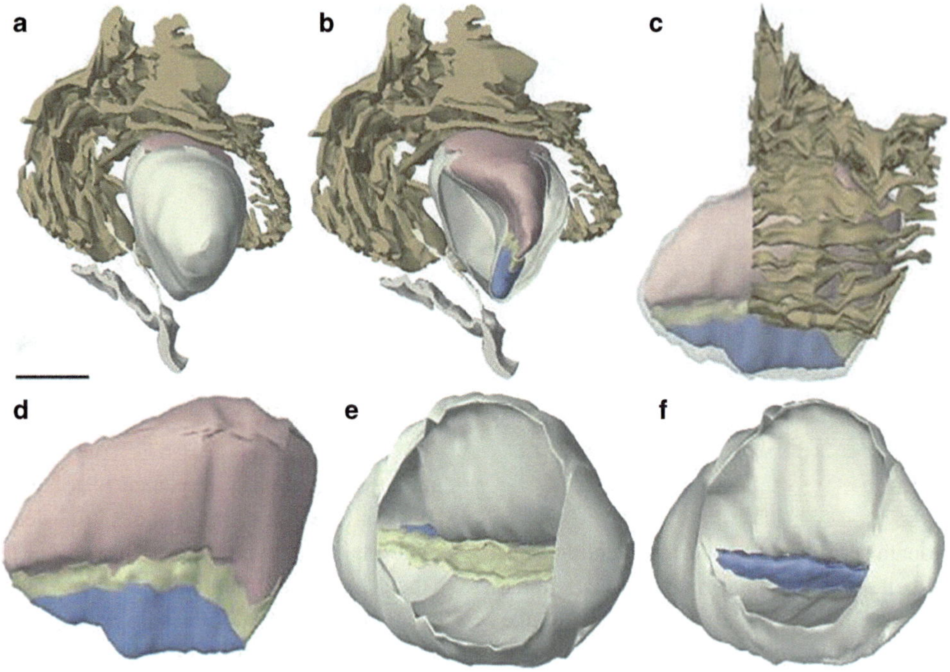

Fig. 36.2 (**a**–**f**) Human fetus, 250 mm CRL (ca. 26th wk), 3D-reconstruction of the upper right deciduous incisor primordium of the based on the serial sections (as in Fig. 36.1). Scale bar: 1000 µm. (**a**) dental follicle (gray) with surrounding maxillary bone (ochre), medial view (*From* Radlanski [1]; *with permission.*). (**b**) Dental follicle partly opened to visualize the dental papilla (pink), dentin (yellow), and enamel (blue). (**c**) Dental primordium with follicle removed, bone partly removed, anterior view. (**d**) Dental primordium, bone removed. (**e**) dental follicle with dental papilla removed to visualize the extent of hard substance formation (dentin: yellow; enamel: blue), cranial view. (**f**) Dental follicle with dental papilla and dentin removed to visualize the extent of enamel formation (blue), cranial view. (*From* Radlanski [1]; *with permission*)

Fig. 36.3 Human fetus, 250 mm CRL (ca. 26th wk), vertical section through the upper right deciduous first molar primordium (staining: Hematoxylin-eosin). Scale bar: 500 µm. *cl* cervical loop; *iee* inner enamel epithelium; *Sr* stratum reticulare; *oee* outer enamel epithelium; *oe* oral epithelium; *Max* Maxilla; *P* pulp; *df* dental follicle; *D* Dentin; *E* enamel. (*From* Radlanski [1]; *with permission*)

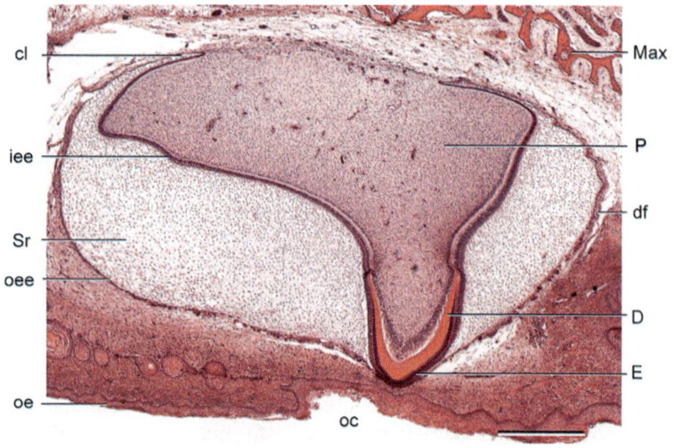

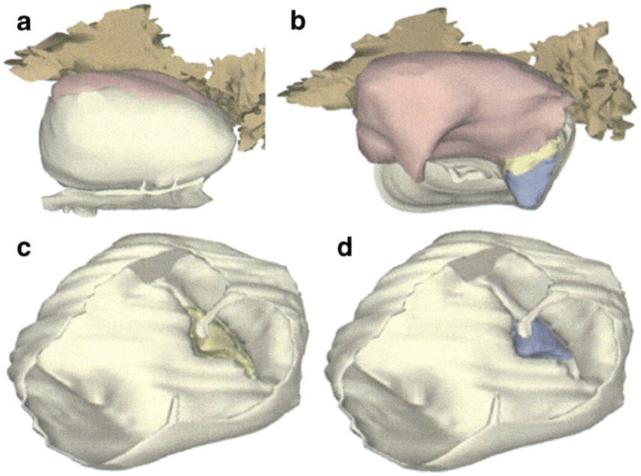

Fig. 36.4 (**a**-**d**) Human fetus, 250 mm CRL (ca. 26th wk), 3D-reconstruction of the upper right deciduous first molar primordium of the based on serial sections (as in Fig. 36.3). Scale bar: 500 µm. (**a**) Dental follicle (gray) with surrounding maxillary bone (ochre), medial view. (**b**) Dental follicle partly opened to visualize the dental papilla (pink), dentin (yellow), and enamel (blue). (**c**) Dental follicle with dental papilla removed to visualize the extent of hard substance formation (dentin: yellow; enamel covered by extent of dentin), cranial view. (**d**) Dental follicle with dental papilla and dentin removed to visualize the extent of enamel formation (blue), cranial view. (*From* Radlanski [1]; *with permission*)

Reference

1. Radlanski RJ. Oral structure and biology. Batavia: Quintessence Publishing; 2018.

Fetus 270 mm CRL (ca. 28th wk): DAG-7_020392

37

Overall Remarks

At this stage, a 3D reconstruction of the mandible was only possible in its anterior part because the fetal material was divided into different segments by macro-preparation. Some parts of Meckel's cartilage are still recognizable although it has been largely degraded in its anterior course. Irrespective of this, individual separate cartilage islands are still detectable in the mandible, as in the previous stages. The compartmentalization of the bone for the dental primordia has remained largely the same: general bone apposition, especially at the inter-dental septa, but also bone resorption inside the crypts, which compensates for the growth of the dental primordia (Figs. 37.1, 37.2, and 37.3).

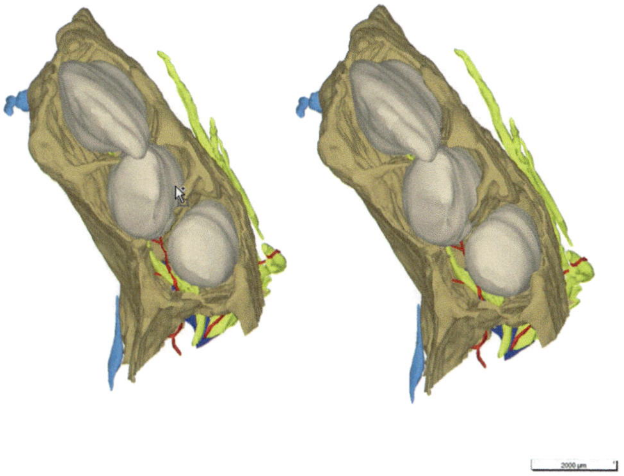

Fig. 37.1 Stereoscopical pair of images. Human fetus 270 mm CRL (ca. 28th week), partial reconstruction of the right half of the mandible with dental primordia i_1, i_2, and c, vessels and alveolar inferior and mentalis nerve, remains of Meckel's cartilage, cranial view

R. J. Radlanski, *Atlas of Human Prenatal Craniofacial Development*, https://doi.org/10.1007/978-3-031-88812-0_37

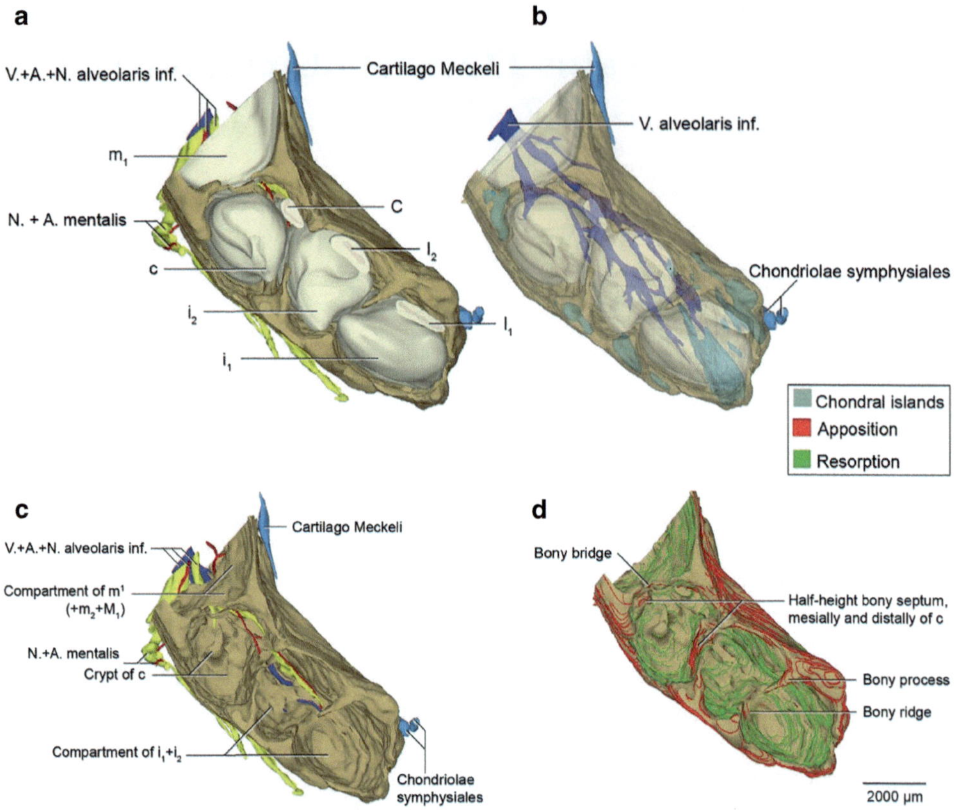

Fig. 37.2 Human fetus, 270 mm CRL. (**a**) Human fetus 270 mm CRL (ca. 28th week), part of the mandible (right half) with dental primordia, inferior alveolar nerve and vessels, and remains of Meckel's cartilage, cranial view. The condylar process and the coronoid process are not included in the reconstruction. (**b**) The same reconstruction, structures made transparent to visualize the inferior alveolar vein. (**c**) The same reconstruction, dental primordia removed to visualize the bony compartments of the dental primordia. (**d**) The same reconstruction, bone remodeling pattern superimposed

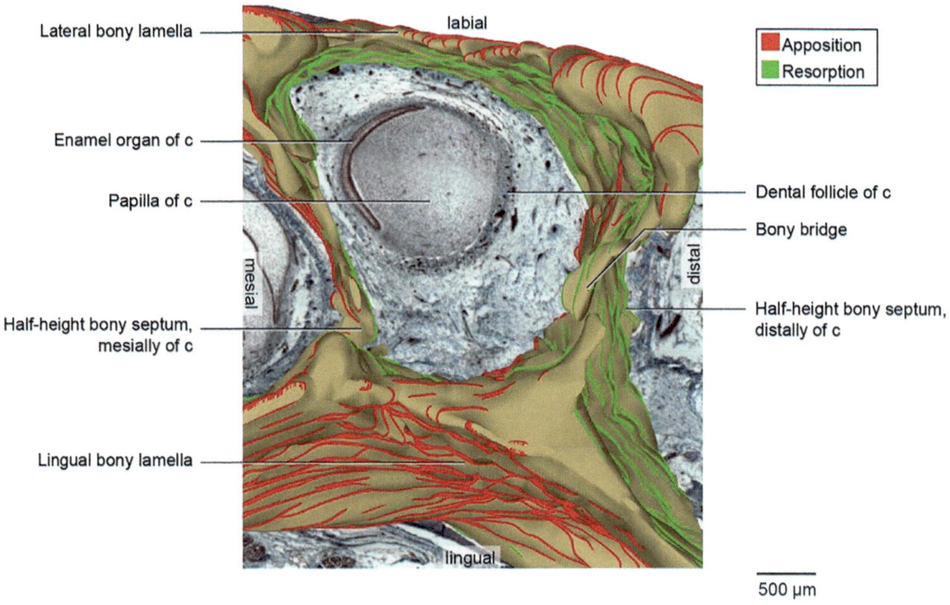

Fig. 37.3 Human fetus, 270 mm CRL. Histological section (staining: Trichrome) with partially reconstructed structures and bone remodeling pattern superimposed next to crypts of lower deciduous canine primordium, and bone remodeling pattern superimposed, cranial view

Mental Symphysis

The mental symphysis is still open in fetuses at this stage of development and does not close until the child is between 2 and 4 years old. There are still some cartilage remnants in this area, which are regarded as remnants of Meckel's cartilage.

Degradation of Meckel's Cartilage

Meckel's cartilage is degraded, but textbooks have always been somewhat reluctant to describe this process as enchondral ossification. This would have contradicted the view that the mandible is a purely desmal bone. However, it has been shown that a considerable number of other cartilage islands are also temporarily present in the mandible, which is why enchondral ossification occurs in many places in the mandible anyway (Figs. 37.4 and 37.5).

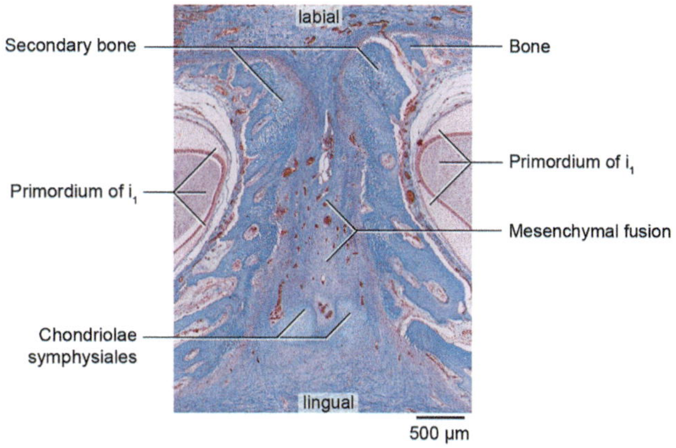

Fig. 37.4 Human fetus, 270 mm CRL, horizontal section (staining: Trichrome, Masson-Goldner) at level of lower deciduous incisor primordia and mental symphysis

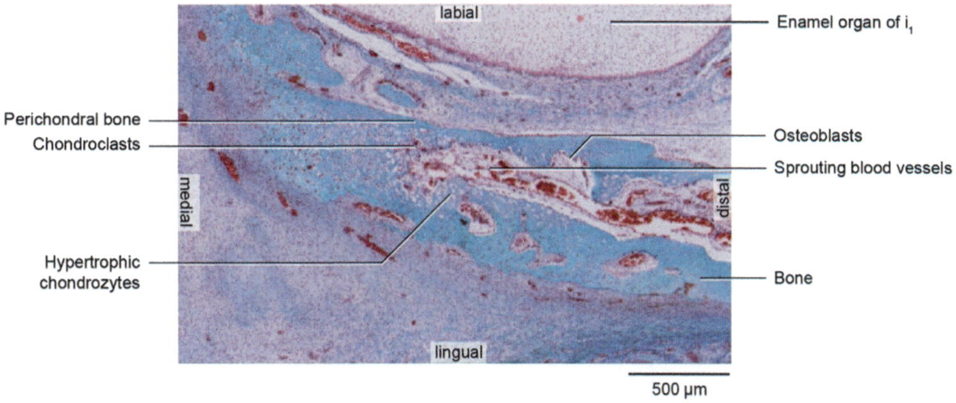

Fig. 37.5 Human fetus, 270 mm CRL, horizontal section (staining: Trichrome, Masson-Goldner), at level of lower deciduous incisor primordium and degrading Meckel's cartilage

Tooth Position, Dental Crowding

This 3D reconstruction shows the position of the teeth in their bony compartments and even at this stage of development it can be seen that the incisors are so closely together that they overlap at their edges. Further growth of the maxilla, during further fetal development and especially in early childhood, usually leads to the resolution of this dental crowding—although an arrangement of the teeth like here appears rather similar in the permanent dentition of quite a few adults (Figs. 37.6 and 37.7).

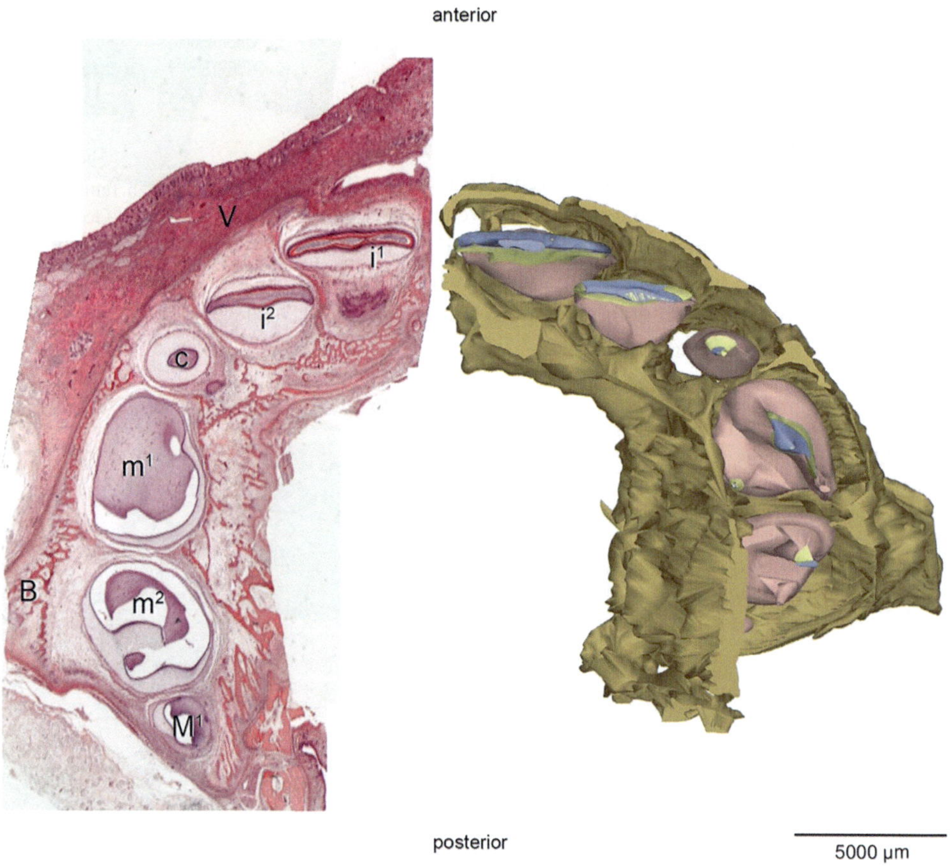

Fig. 37.6 Human fetus, 270 mm CRL, horizontal section (staining: Hematoxylin-eosin), right half of maxilla, and on the right 3D-reconstruction of the left half of the maxilla with dental primordia, from the serial sections, caudal view. Dental primordia i^1, i^2, c, m^1, m^2, and M^1; *B* bone of maxilla; *V* vestibulum. (From Radlanski (2018))

Fig. 37.7 Human fetus, 270 mm CRL, 3D-reconstruction of the right half of the maxilla with dental primordia and bone remodeling pattern superimposed, caudal view

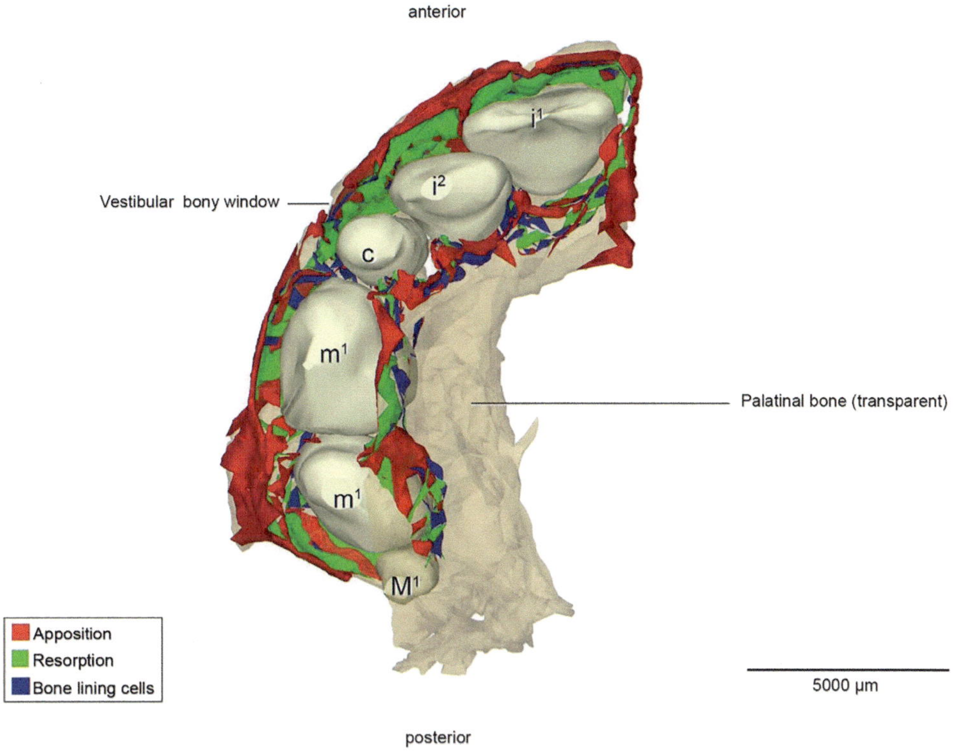

Bone Remodeling

Bone remodeling in the maxilla in the area of the dental primordia behaves similarly to bone remodeling in the mandible although the bone of the maxilla initially develops in a fundamentally different way: there are individual bone islands that form in earlier stages and have become already largely confluent at this stage of development. But now the bony margins of the alveolar process bone are preferably formed by apposition, and the inner walls of the crypts are undergoing bone resorption accordingly depending on the growth of the dental primordia.

Further Reading

Konietzny M. Zur Entwicklung des Kieferknochens im Bereich der Zahnanlagen—Histologische und morphologische Untersuchungen an der fetalen Maxilla des Menschen während des Glockenstadiums der Zahnanlagen (160–270 mm SSL). Med. Diss. Charité-Universitätsmedizin Berlin. 2014.

Radlanski RJ. Oral Structure and Biology. Batavia: Quintessence Publishing; 2018.

Schuster FP. Zur Entwicklung des Kieferknochens im Bereich der Zahnanlagen. Histologische und morphologische Untersuchungen an der fetalen Mandibula des Menschen während des Glockenstadiums der Zahnanlagen (68–160 mm SSL). Med. Diss. Charité-Universitätsmedizin Berlin; 2013.

Zimmermann C. Zur Entwicklung des Kieferknochens im Bereich der Zahnanlagen. Histologische und morphologische Untersuchungen an der fetalen Mandibula des Menschen während des späten Glockenstadiums der Zahnanlagen (160–270 mm SSL) Med. Diss. Charité-Universitätsmedizin Berlin; 2013.

Part III

Conclusions

Prenatal Craniofacial Development: The Whole Picture

38

Overall Facial Growth

Of course, there are individual variations as to whether someone is tall or short, but even this compilation of 3D reconstructions of the head specimens on the same scale shows very impressively how size increases. The interocular distance may serve as a reliably comparable measure, and it can be seen, that it definitely increases overall, albeit the individuals with crown-rump lengths of 22–25 mm in particular also show that there can be specific variations. There are also individual variations with regard to the vertical dimension of the face, if you take the distance between the height of the eyes and the chin region. Interestingly, the musculature is already quite well developed in the first stage of development here, which is at 19 mm CRL. Apart from that, however, all embryos and fetuses show that there is a clear increase in the maturation of the cartilaginous and bony structures of the face.

Growth of the Maxilla

If we look at the maxilla alone, in each case from the anterior view, we can also see the clear increase in size in relation to the increase in crown-rump length. Over the depicted period of the development, the maxilla remains paired, which means, it does not grow together in the middle during the observation period, instead, there is a recognizable distance between the two halves of the maxilla. In our specimens, we were able to observe a sutural convergence of the two halves of the maxilla for the first time at the developmental stage of 190 mm CRL, but these fetuses were already too large for a complete 3D reconstruction of the maxilla. At the beginning, at 25 mm CRL, the maxilla is rather flat and describes a shallow arch. Already at the 37 mm CRL stage, the frontal processes are clearly recognizable, which have increased in height again at 68 mm CRL. They nestle against the nasal capsule, as can be seen in the reconstruction at 87 mm

R. J. Radlanski, *Atlas of Human Prenatal Craniofacial Development*, https://doi.org/10.1007/978-3-031-88812-0_38

CRL. The bone expands into the space between the nasal capsule and the eyes. Only one half of the maxilla was reconstructed from the developmental stage of 117 mm CRL; as to enable us to compare it with the others, the image has been duplicated and mirrored (Figs. 38.1, 38.2, and 38.3).

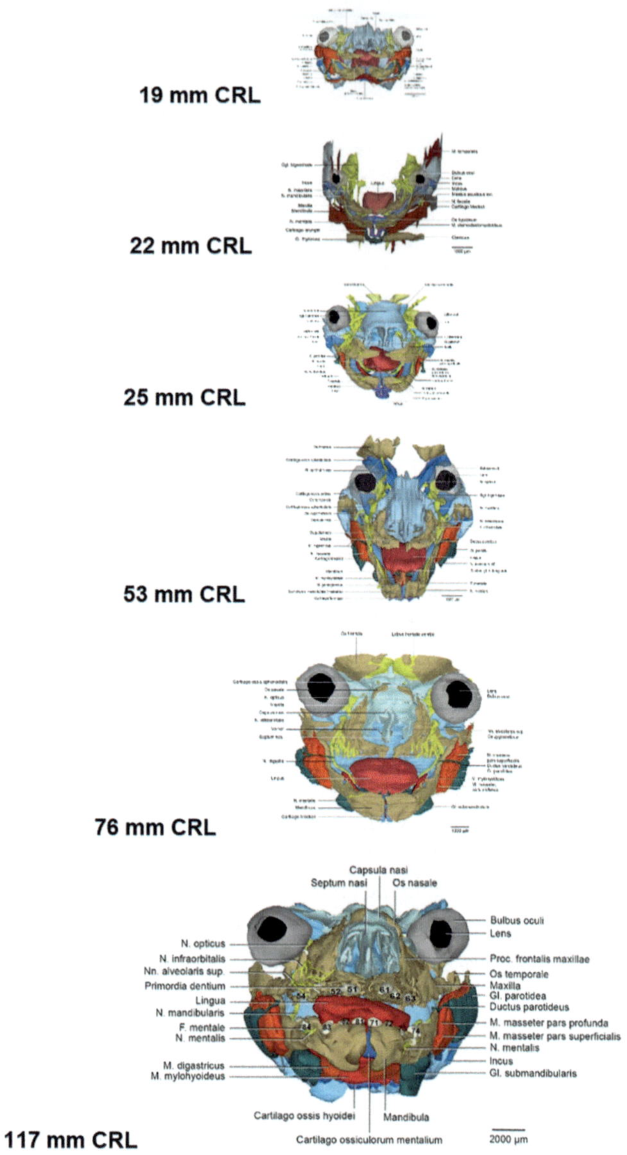

Fig. 38.1 3D-Reconstructions of the facial regions brought to scale. Anterior view

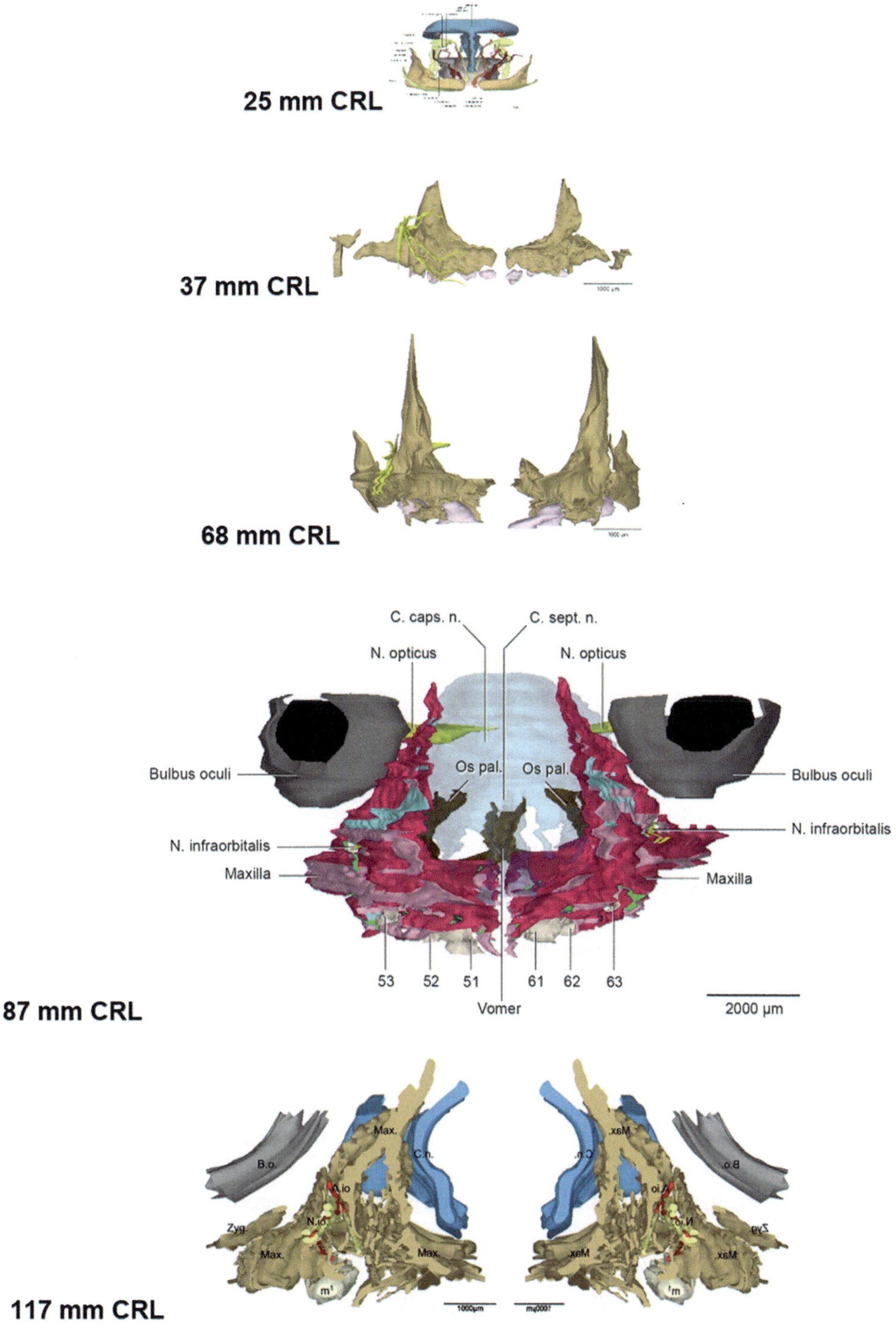

Fig. 38.2
3D-Reconstructions of the maxillae brought to scale. Anterior view. Red color code: remodeling pattern, see Chap. 26

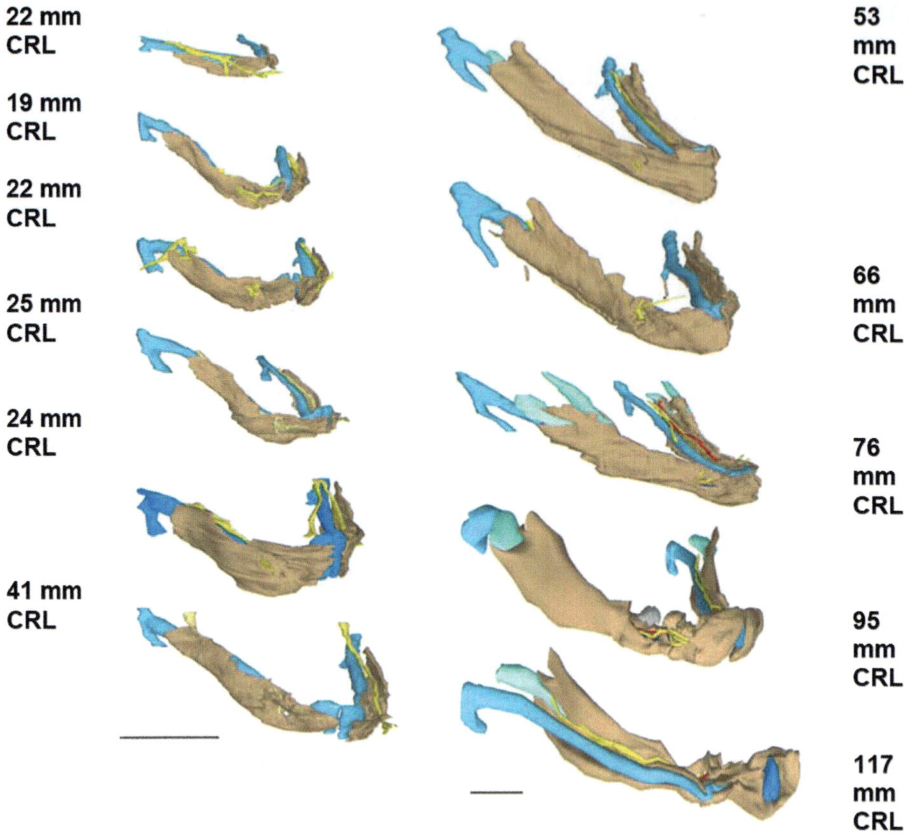

Fig. 38.3 3D-Reconstructions of the mandibles brought to scale. Lateral and 45° anterior view. Scale bar: 2000 µm (left column) and 1000 µm (right column). (*From* Radlanski et al. [1]; *with permission*)

Growth of the Mandible

First Ossification and Mental Foramen

Development of the mandible starts with the formation of Meckel's cartilage as the cartilaginous structure of the first visceral arch, and it runs almost straight in the embryo of 22 mm CRL. The first mandibular ossification was observed in our collection in the embryo of 15 mm CRL at the branching of the mandibular nerve and the future region of the mental foramen. The formation of the mental foramen starts as a bony notch through which the inferior alveolar nerve crosses into lateral direction. It was found completely encircled by bone in our specimen of 19 mm CRL, but we also have specimens in our collection, in which its circumference remains cranially open until a developmental stage of 29 mm CRL. However, as we could show in a more detailed study [2], it was reopened in later stages because of heavy resorption due to the expansion of the adjacent dental primordia in a fetus of 76 mm CRL and in a fetus of 117 mm CRL. After that, it was completely formed again in later stages.

Coronoid and Condylar Processes

The maximum vertical height of the mandibular bone moves from the region of the mental foramen in younger embryos to the posterior third of the mandibular bone, and the formation of the condylar and coronoid processes at the end of the mandible can clearly recognized in the embryo DOR 37 mm CRL although there were several smaller specimens where one could interpret bony lingulae such as primordia. Fawcett [3] and Kitamura [4] described the first formation of the coronoid process at day 47 and the condylar processes at day 60. Lee et al. [5], however, did not recognize the coronoid process before the 14th week. The formation of the condylar process with its cartilaginous core has been described by us elsewhere [6]. As regular formations, we see cartilaginous cores in the condylar as well as in the coronoid processes. This happens from a size of 76 mm CRL on. As it is the case in many of the cartilaginous tissues, rapid expansion leads to a significant elevation of both of the processes. The cartilaginous core within the condylar processes allows for a tremendous growth potential of the mandible in that region, even after birth.

Mandibular Body Formation

Our reconstructions revealed the first signs of mandibular bone penetrating underneath and above Meckel's cartilage in an embryo of 29 mm CRL, in a region directly posterior to the mental foramen.

The attachment of the mylohyoid muscle was at Meckel's cartilage in earlier stages, and it changed to a bony attachment when bone emerged underneath Meckel's cartilage. We could see specimens that had an attachment of this muscle to Meckel's cartilage and to mandibular bone next to each other [7]. This embracing of Meckel's cartilage spreads out longitudinally into both directions, but never covers the lingual aspect of Meckel's cartilage, except in the anterior region where resorption of Meckel's cartilage takes place in the fetus 76 mm CRL and in later developmental stages. Lee et al. [5] reported that after a close contact between bone and Meckel's cartilage in earlier stages, bone detaches from Meckel's cartilage at 12 weeks into a lateral direction.

The remodeling pattern is quite simple in earlier stages where almost ubiquitously apposition can be found [7, 8]. In later stages, however, an elaborate pattern of apposition and resorption is found, which can be matched with the 3D outline of bone in most cases. Resorption fields are found at the posterior margins of the mental foramen and at the lateral and medial posterior bony planes at concave surfaces. The bulging of bone underneath and over Meckel's cartilage is also recognized as a region with active appositional growth. Meckel's cartilage itself lies in a trough with can be characterized by less apposition and even resorption from the 53-mm-CRL stage onwards. It might be true that Meckel's cartilage acts as a frame for mandibular bone [9] to spread out along its lateral borders, but bony formations extend further away from the framing cartilage, and the remodeling pattern leads to the typical—and within certain limits—individual appearance of the mandible.

The part of the mandible known as the alveolar process bone is already clearly recognizable in the embryonic stage or at the latest in the fetal stage. It is characterized by the fact that it houses the teeth, which are located in their own crypts. The edges of the crypts—which are later, i.e., postnatally, comparable to the alveolar crest—are already formed in the fetal stage as typical appositional bone. We only see zones of bone resorptions on the inside of the crypts, this is due to the fact that the dental primordia enlarge and therefore take up more space. The bone surface reacts to this by activating its osteoclasts.

Today, we know much about the chemical processes that lead to bone formation at the cellular level [10–16], but even today it remains unknown, what information do the bone forming cells receive to create those different forms of the mandibles, as seen in our 3D reconstructions with some being asymmetrical with varying shapes, even deviating from the typical schematic pattern.

Changes in Form of the Mandible

The mandible not only becomes larger, it also changes its shape quite clearly. At the beginning of its development, the mandible is a broad U-shape and from 53 mm CRL this broad U-shape changes to a V-shape: The mandible thus becomes more pointed. In the older fetuses in our collection, we were then able to observe how Meckel's cartilage in the anterior part develops a clear S-curvature before it is also broken down and bone is built up in its place. A detailed description, also with morphometric data can be found elsewhere [1].

Temporomandibular Joint

Primary Jaw Joint

The temporomandibular joint of mammals—including humans—shows a peculiar development prenatally: Meckel's cartilage has a hook-shaped, thickened structure at its dorsal end, which represents the malleus. This is the first of the three in the chain of auditory ossicles. And long before a condylar process develops on the mandible, the entire lower jaw can already move. Surprisingly, the articulated connection between the malleus—the posterior end of Meckel's cartilage—and the stapes serves as the primary jaw joint during mouth opening. At this stage of development, the ossicles are not used for sound conduction, but function as a jaw joint. The official term for this is "the primary temporomandibular joint" (Fig. 38.4).

In this context, it becomes clear that nothing at all develops in the embryo for a certain "purpose." There are always physiologically understandable reasons why a tissue or organ differentiates. The "purpose" is only added to it by the astonished observer with human standards. And it is often the case that the search for and the question of purpose obscures our view of what actually only develops for purely physiological reasons. Philosophical and also embryological thoughts on this topic can be found in Aristotle [18], Peter [19], Meyer [20], Blechschmidt [21–23], and also Radlanski [24].

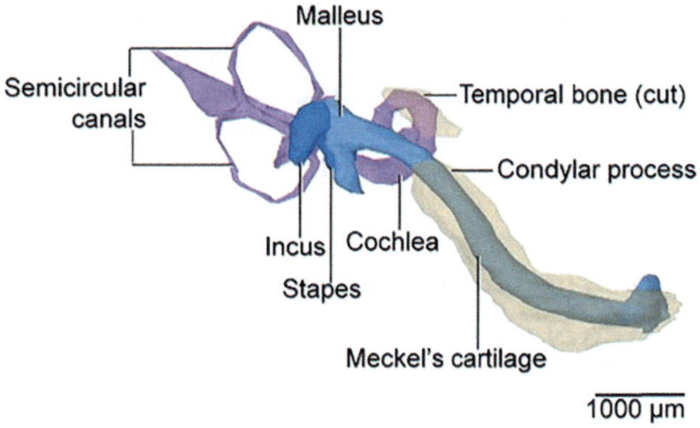

Fig. 38.4 Human Embryo, Region of the right jaw joint, human embryo, 25 mm CRL, 8th week. Lateral view, mandible made transparent. From Fernández Rubio and Radlanski (2024). (*From* Fernández Rubio and Radlanski [17]; *with permission*)

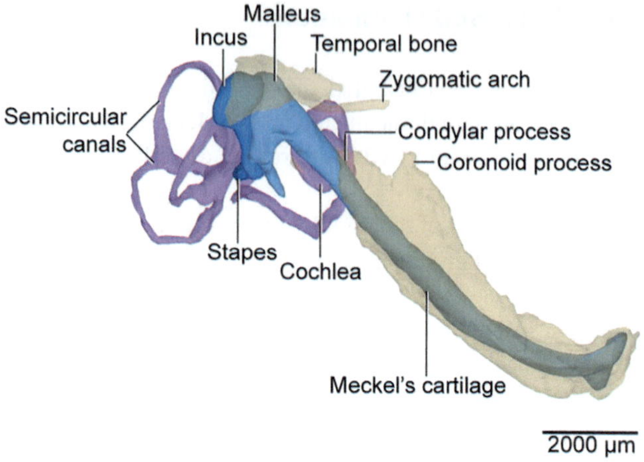

Fig. 38.5 Human Embryo, Region of the right jaw joint, human embryo, 53 mm CRL, 9th week. Lateral view, mandible made transparent. (*From* Fernández Rubio and Radlanski [17]; *with permission*)

Secondary Temporomandibular Joint

Only later, when the mandible, which develops next to Meckel's cartilage, develops a condylar process (and also a coronoid process), can a new, secondary temporomandibular joint develop somewhat further anterior to the primary temporomandibular joint (Figs. 38.4, 38.5, 38.6, 38.7, and 38.8).

The condylar process grows rapidly—probably because it contains this carrot-shaped cartilage core—and in this way it pushes itself further and further dorso-cranially until it has made contact with the temporal bone in the mandibular fossa. Along the way, it pushes the connective tissue in front of it together, and compresses it to form the articular disc. In this way, the secondary temporomandibular joint is formed as an appositional joint, and not by separation, like most other joints in the human body. (Other appositional joints are the sternoclavicular joint, the sacroiliac joint and the intervertebral joints—just for the sake of completeness).

This development in the temporomandibular joint region leads to the peculiar situation that both joints exist simultaneously in the developmental stage around 117 mm CRL. The primary temporomandibular joint is therefore still fully functional, and the secondary temporomandibular joint, a little further anterior to it, is now also physiologically fully developed and could also function as a joint. However, Meckel's cartilage still has a completely stable connection with the body of the mandible. It is therefore completely incomprehensible how humans at this time (and other mammals, for example, the mouse even around birth) can perform any jaw movements: Purely mechanically, both joints would have to block each other (see Fig. 38.7) [25]! Only when Meckel's cartilage has degraded to such an extent that there is no longer a connection between the malleus and the mandible can the secondary temporomandibular joint move freely and the ossicular structures can then finally be used for sound conduction [17]. This is the case with our specimens at the 230 mm CRL stage of development (see Fig. 38.8).

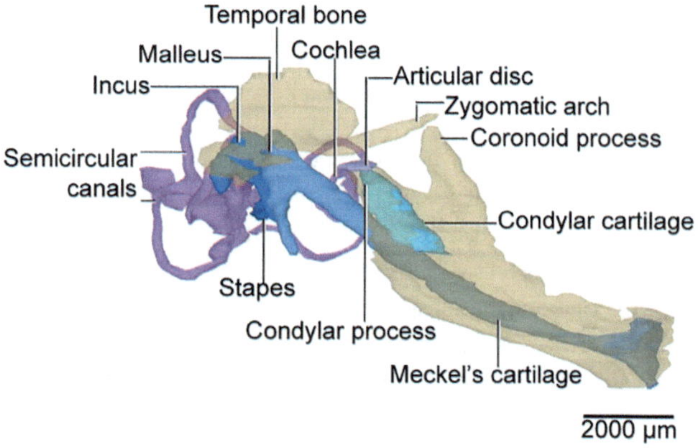

Fig. 38.6 Human Embryo, Region of the right jaw joint, human embryo, 76 mm CRL, 11th week. Lateral view, mandible made transparent. (*From* Fernández Rubio and Radlanski [17]; *with permission*)

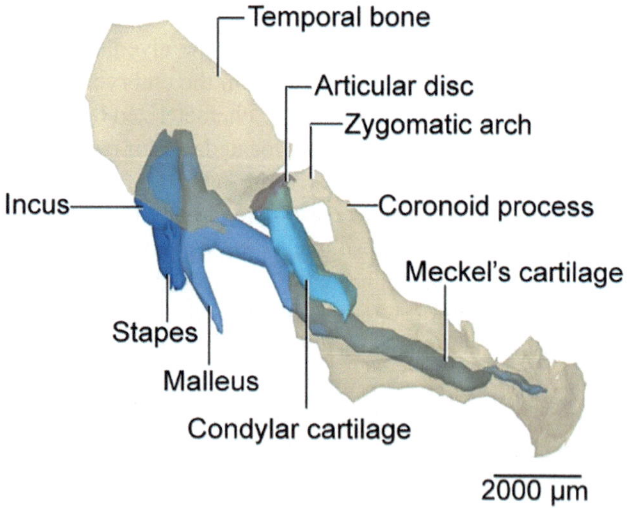

Fig. 38.7 Human Fetus, Region of the right jaw joint, human embryo, 117 mm CRL, 14th week. Lateral view, mandible made transparent. Both jaw joints are fully functional at this stage of development. (*From* Fernández Rubio and Radlanski [17]; *with permission*)

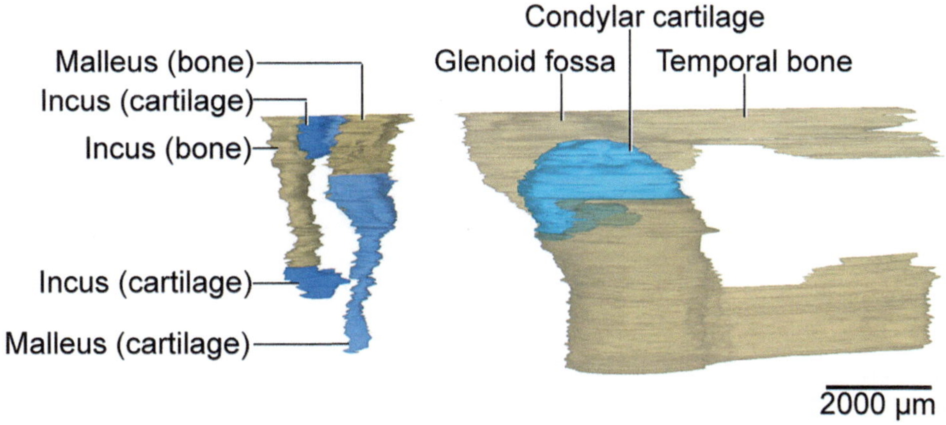

Fig. 38.8 Human Fetus, Region of the right jaw joint, human embryo, 230 mm CRL, 24th week, lateral view. (*From* Fernández Rubio and Radlanski [17]; *with permission*)

References

1. Radlanski RJ, Renz H, Klarkowski MC. Prenatal development of the human mandible. 3D reconstructions, morphometry and bone remodelling pattern, sizes 12–117 mm CRL. Anat Embryol (Berl). 2003;207:221–32.
2. Radlanski RJ, Renz H, Müller U, Schneider RS, Marcucio RS, Helms JA. Prenatal morphogenesis of the human mental foramen. Eur J Oral Sci. 2002;110:452–9.
3. Fawcett E. Description of a reconstruction of the head of a 30 mm human embryo. J Anat. 1910;44:303–11.
4. Kitamura H. Embryology of the mouth and related structures. Tokyo: Manzen; 1989.
5. Lee SK, Kim YS, Oh HS, Yang KH, Kim EC, Chi JG. Prenatal development of the human mandible. Anat Rec. 2001;236:314–25.
6. Radlanski RJ, Lieck S, Bontschev NE. Development of the human temporomandibular joint. Computer-aided 3D-reconstructions. Eur J Oral Sci. 1999;107:25–34.
7. Radlanski RJ, Klarkowski MC. Bone remodeling of the human mandible during prenatal development. J Orofac Orthop. 2001;62:191–201.
8. Mauser C, Enlow DH, Overman DO, McCafferty RE. A study of the general growth of the human face and cranium. In: McNamara JA, editor. Determinants of mandibular form and growth. Proceedings of a sponsored symposium honoring Prof. Robert E Moyers (Monograph No 4; Craniofacial Growth Series). Ann Arbor, MI: The University of Michigan; 1975. p. 243–75.
9. Sperber GH. Craniofacial embryology. Hamilton: B.C. Decker; 2001.
10. Helms JA, Kim CH, Hu D, Minkoff R, Thaller C, Eichele G. Sonic hedgehog participates in craniofacial morphogenesis and is down-regulated by teratogenic doses of retinoic acid. Dev Biol. 1997;187:25–35.
11. Olsen BR, Riginato AM, Wang WB. Bone Development. Ann Rev Cell Dev Biol. 2000;16:191–2020.
12. Depew MJ, Lufkin T, Rubinstein JLR. Specification of jaw subdivision by dlx genes. Science. 2002;298:381–5.
13. Cobourne MT, Sharpe PT. Tooth and jaw: molecular mechanisms of patterning in the first branchial arch. Arch Oral Biol. 2003;48:1–14.
14. Helms JA and Schneider RA. The cellular and molecular origins of beak morphology. Science. 2003;299:565–8.
15. Shui C, Spelsber TC, Riggs BL, Khosla S. Changes in Runx2/Cbfa1 expression and activity during osteoblastic differentiation of human bone marrow stromal cells. J Bone Miner Res. 2003;18:213–21.
16. Thorogood P, Sarkar S, Moore R. Skeletogenesis in the head. In: Proceedings of the conference oral biology at the turn of the century—misconceptions, truths, challenges and prospects. Interlaken, Switzerland: S. Karger; 1989.
17. Fernández Rubio EM, Radlanski RJ. Development of the human primary and secondary jaw joint. Ann Anat. 2024;251:1–8.
18. Aristotle. Περὶ ζῴων γενέσεως, Latin: De generatione animalium. English: Generation of Animals, translated by Devin, H. In: Anagnostopoulos G, editor. A companion to Aristotle. Chichester, UK: Wiley-Blackwell; 2009.
19. Peter K. Die Zweckmässigkeit in der Entwicklungsgeschichte. Eine finale Erklärung embryonaler und verwandter Vorgänge. Berlin: Springer; 1920.
20. Meyer A. Logik der Morphologie. Berlin: Springer; 1926.
21. BlechschmidtE.MechanischeGenwirkungenFunktionsentwicklung I. Göttingen: Musterschmidt; 1948.
22. Blechschmidt E The stages of human development before birth. An introduction to human embryology. - Die vorgeburtlichen Entwicklungsstadien des Menschen. Eine Einführung in die Humanembryologie. Karger, London, New York, Basel, Freiburg i. B: Karger; 1960.
23. Blechschmidt E. The ontogenetic basis of human anatomy: the biodynamic approach to development from conception to adulthood, edited by Brian freeman. Berkeley, CA: North Atlantic Books; 2004.
24. Radlanski RJ. Mein Gesicht: Haare, Haut, Augen, Nase, Ohren, Knochen, Mund und Zähne. Warum sehen wir so aus? Anatomische Erklärungen. Berlin: Quintessenz; 2016.
25. Fernández Rubio EM, Radlanski RJ. Development of the primary and secondary temporomandibular joint in the mouse. Ann Anat. 2023;249:1–15.

Discussion: Morphology as the Foundation

39

Facts and Possibly Errors

I find it quite impressive that it was Aristotle who observed the beating heart as the "salient point" in a fertilized bird's egg [1]! This would actually have been a good start for research into embryogenesis, but as we all know, the history of science continued in a rather mystical direction [2]. Apart from the fact that a very mysterious, fatalistic, and religious attitude prevented scientific research in the European Middle Ages [3], evolutionary biology research ultimately remains enigmatic when it comes to understanding why and how the different species have formed. There is certainly a change in organism and in species [4], but embryologists are naturally interested in how it is possible for a small (or even a large) change to arise prenatally in the first place, which can then be subjected to the Struggle for Life. After all, this problem was discussed very thoroughly by Gavin R. de Beer [5]. And we can be glad that the so-called basic biogenetic law of Ernst Haeckel (1874) [6] has finally been refuted (Blechschmidt 1976 [7]). It really would have been too fantastic if the history of evolution actually had an influence on the physiological processes during embryogenesis! This is how Ernst Haeckel really imagined it. Not only is it physiologically impossible, as explained by scientific theory [8], but it was also the first known case of comprehensive scientific fraud [9]. Today, the theory of evolution is even described as the fallacy of the century [10] and debunked as a tautology in terms of scientific theory [11]. And another thing: many people believe that "purpose" is the cause or the driving force behind the development of body shape or organs. This view also goes back to Aristotle, quoted at the beginning, and was even attempted to be comprehensively explained by the embryologist Karl Peter [12]. But this view is also easy to refute: "purpose" is an anthropocentric view that engineers must certainly take into account in their designs if their construct is to function, i.e., fulfill its purpose. But physiologically and in terms of scientific theory, the question arises again: How should the later purpose, the later, postnatal functional necessity influence the prenatal processes of morphogenesis? Meyer already explained this impossibility very comprehensibly in 1926. These questions are fundamentally important, but we cannot go into them any further here.

Let us devote ourselves to the visible findings, or the findings made visible!

Pioneers in 3D-Embryology

As early as 1869, the anatomist Emil Dursy [13] carried out studies on the prenatal morphogenesis of the human head in Tübingen (Germany) using very elaborate methods. He used the method of spatial reconstruction from series of sections, which was introduced by Gustav Born in 1833 [14]. This was based on uninterrupted histological section series that were optically enlarged and projected onto wax plates. When these wax plates were cut out and stacked according to the tissue contours that one wanted to see spatially, a physical, enlarged, reconstructed model was created that could be touched and examined by hand. Ultimately—albeit computer-graphically—we also used this method to create the images in this atlas. This method was used extensively by the anatomist Wilhelm His (the elder) (1874) [15] and he has only recently been rightly recognized as the pioneer of scientific embryology Richardson and Keuck [16]. I would not like to end this paragraph without mentioning the embryologist Gerd Steding, who has presented a voluminous atlas of scanning electron microscope images of human embryonic development [17]. This is also 3D.

Knowledge in 3D and Mechanical Effects on the Genome

Erich Blechschmidt is regarded as a pioneering human embryologist who created a very comprehensive human embryology collection of section series and 3D models in

R. J. Radlanski, *Atlas of Human Prenatal Craniofacial Development*, https://doi.org/10.1007/978-3-031-88812-0_39

Göttingen (Germany), which can still be viewed and studied today in the Anatomical Institute of the University of Göttingen. Based on these spatial models, Erich Blechschmidt was able to analyze the differentiations of the tissues in connection with their spatial positional relationships without any misinterpretations due to possibly obliquely cut individual histological sections preventing a view of the spatial relationships. There are pitfalls, if only two-dimensional histology is applied for understanding a spatial problem. With his physical models, Blechschmidt then came to the conclusion that although the differentiation of tissues takes place and is controlled by the living cell, the positional relationships of the tissues, their spatial allocation, their relationships to neighboring tissues and then also effective forces, fluid shifts, and migrations must be additional causes and triggers for the differentiation of cells and tissues: Cells have their potential for differentiation in the genetic code, but ultimately differentiation is only triggered by stimuli that act on the cell (on the cell membrane) and then cause a reaction (due to the ways in which they are stored in the genome in the cell nucleus). Blechschmidt described these processes as *"Mechanische Genwirkungen"* [18]. (English: "Mechanical effects on the genome"). In a letter, Erich Blechschmidt told me [19] that he no longer wanted this term to be used because there are people who might misunderstand it.

Some people think that he would have seen only pressure and traction in the developing embryo as the only cause for the development of the body shape of man and the tissue differentiation. But this is not the case. It is important to note that it is each specific cell that responds to influences from outside. This is an achievement of life, it is a reaction of a cell, which is alive. This is how Blechschmidt saw it [20]. And more research is currently being carried out into the mechanical interactions of tissues that play a role in their differentiation [21, 22].

Bridging the Gap Between Morphology and Molecular Signaling

This brings us to the present time, and we have achieved a perspective that combines both the mechanical aspects of morphogenesis and the molecular signals, which of course also play an active role. This was not the case for a long time. Although the mechanical aspect had dominated since His [12] and Blechschmidt [18] (after the mysticism of the evolutionary biology view had been overcome), this view was increasingly displaced with the emergence of questions from molecular biology. It is certainly highly commendable that the multitude of molecular signals and their pathways have been studied and uncovered in increasingly complex ways. Francis-West et al. [23], for example, have illustrated this very well for the head region. However, this does not yet explain how morphogenesis (i.e., the actual *formation of the shape*) takes place. Davies [24] acknowledges both molecular biology and the question of real morphogenesis. And he sums it up so well that I am willing to quote him verbatim: "…the study of morphogenesis is one of the oldest of all sciences, dating back to ancient Greece where Aristotle (384–323 BC) described … morphogenesis in birds. Until very recently, however, the molecular revolution in biology has almost ignored … (the) … most fundamental aspect of animal and plant development and has instead concentrated on differentiation and communication. It is easy to understand why. The immunologist and philosopher Peter Medawar once defined science as "the art of the soluble": … problems were simply too difficult … it is much more complicated to study at a mechanistic level than either differentiation or communication…" [24].

And we also see it as our task to set the morphological stage with our histological studies and our spatial reconstructions across the various stages of development and to describe the places where the signaling cascades can take place at all and the distances, they have to overcome to unfold their effects. This is also a purpose for which this atlas can serve as a reference. We have actually already done this at least once (see Chap. 40).

Advantages and Challenges of 3D-Reconstruction

However, even today it is still not so easy to create beautiful, valid, and in all aspects correct reconstructions from histological section series. It starts with the preparation of the section series. By the time a histological section is finally placed under a microscope, it has been through a veritable ordeal of chemical processes. Fixation causes all proteins to coagulate, which in itself has an impact on the dimensional fidelity of the tissue. The bone-containing tissue then has to be decalcified and the ascending alcohol series has to be run through for embedding in paraffin. Cutting the paraffin wax blocks under the microtome develops such high forces that the sections are naturally significantly compressed and this compression should then be stretched again in the water bath. Finally, the individual sections come to rest on the glass slide, but here they still have to be deparaffined, stained, and covered. It is therefore clear that it cannot be assumed that the sections are really true to size. Despite all the care taken, these histological sections are not always uniformly stained because the tissue is at different stages of development even on the same section, and therefore, has different affinities for accepting dyes. The tissue boundaries, which are of course important for our spatial reconstructions, must then be recognized and recorded by experienced examiners, i.e., segmented on the computer image. This is often met with objections from mathematicians or computer specialists who hope that artificial intelligence can be programmed to recognize tissue borders just as well as an experienced member of my department. The automated image analysis could also perform a Procrustes analysis and bring the compressed and stretched individual sections into the correct dimension. Of course, we tried this with the mathematicians at the Technical University of Berlin and tested it intensively. In the end, it was safer and more correct when a person from my department did it.

And another problem arises when there is a lack of experience: How should the individual sections, or the contour lines from the individual sections, be stacked up again so that they line up correctly? So, if you cut a banana crosswise into slices and put the slices back together, how do you make sure that you get the true curvature of the banana back (if you do not know what a banana actually looks like)? This is exactly the problem that has arisen often enough in the history of reconstructions of serial sections. So indeed, in history of embryology, there were reconstructed models of embryos where the curvature and also the torsion are not correct. We have calibrated our staff and postgraduate students by having them assemble different embryos several times. As we have always presented each student with the same data sets over the years, we have also been able to study ourselves what sources of error there are and how to correct them. There is also the method of embedding fiducial markers [21], but this did not prove to be accurate for our purposes. It was safer and more error-free to reconstruct the eyes, Meckel's cartilage and other reference structures whose true shape we have come to know as correct over the decades. With this orientation, we have to assume that the other unknown structures are also correct.

And yet, the course of nerves and blood vessels in particular could not always be traced across all the individual sections of the series. We can be sure that nerves really do spread continuously, even with aberrant courses, but in the case of blood vessels it is known that they also develop as individual islands, which only later confluent or even obliterate to form pathways [25, 26]. In bone formation, we also find individual islands, especially in the maxilla, which only become confluent in the course of further development, whereby the maxillary bone is more trabecular in structure anyway [27, 28]. And although the mandible essentially begins as a flat bone plate lateral to Meckel's cartilage, there are also individual bone islands on the mandible that ultimately confluent, such as the ossification island at the mandibular angle. This could conceal sources of error that make the continuity of the 3D reconstruction uncertain.

Visualization in μCT

One could argue that why are we doing this complicated and error-prone reconstruction of section series at all? A μCT technique could be much easier to automate and could save a lot of time-consuming manual labor. In fact, we have examined several fetuses that we did not section histologically using μCT, see Chap. 40 [29].

In order to also visualize soft tissue in the μCT, especially the course of the blood vessels, we contrasted our specimens with phosphotungstic acid beforehand. This naturally made tracing the blood vessels more reliable compared to 3D reconstruction using section series, as described above. Nevertheless, we went to the great effort of sectioning and then reconstructing these many specimens because we wanted to do both: To see the shape and also the differentiation of the cells and tissue on a histological level. This is not possible with μCT although we have achieved a spatial resolution of 10 μm in some areas. The μCT has a further problem: as our specimens were also placed in formalin for a long time, decalcification of individual bone regions could have occurred (although the formalin was buffered). Furthermore, we do not know to what extent the bone tissue in the embryos and fetuses is already mineralized during their normal development. These two reasons make it unsafe in μCT to determine the bone threshold for imaging bone. In addition, there is the Hounsfield effect at the very thin bone edges. Nevertheless, we were at least able to use our histological method to calibrate our μCT images [30].

References

1. Aristotle: Περὶ ζῴων γενέσεως, Latin: De generatione animalium. English: Generation of Animals, translated by Devin, H. In: Anagnostopoulos G. (Ed.) A companion to Aristotle. Chichester, UK: Wiley-Blackwell; 2009.
2. Wiener PP, Noland A, editors. Roots of scientific thought. A cultural perspective. New York: Basic Books Publishers; 1957.
3. Asher R. Evolution and belief. Cambridge University Press; 2012.
4. Darwin C. The origin of species by means of natural selection or the preservation of Favoured races in the struggle for life. London: John Murry; 1859.
5. deBeer GR. Embryos and ancestors. Oxford: Clarendon Press; 1951.
6. Haeckel E. Anthropogenie oder Entwicklungsgeschichte des Menschen. Keimes—und Stammesgeschichte. Leipzig: Engelmann; 1874.
7. Blechschmidt E. Der Irrtum des Biogenetischen Grundgesetzes. Fortsch Med. 1976;94:465–6.
8. Meyer A. Logik der Morphologie. Berlin: Springer; 1926.
9. Freeman B. Haeckel's forgeries. Am Biol Teach. 2001;63:230.
10. Illies F. Der Jahrhundertirrtum. Würdigung und Kritik des Darwinismus. Frankfurt (Main): Umschau Verlag; 1983.

11. Goll R. Der Evolutionismus. Analyse eines Grundbegriffs neuzeitlichen Denkens. München: Verlag C.H. Beck; 1972.
12. Peter K. Die Zweckmässigkeit in der Entwicklungsgeschichte. Eine finale Erklärung embryonaler und verwandter Vorgänge. Berlin: Springer; 1920.
13. Dursy E. Zur Entwicklungsgeschichte des Kopfes des Menschen und der höheren Wirbelthiere. Tübingen: Laupp'sche Buchhandlung; 1869.
14. Born G. Die Plattenmodelirmethode. Arch Mikr Anat. 1833;22:584–99.
15. His W. Unsere Körperform und das physiologische Problem ihrer Entstehung. Briefe an einen befreundeten Naturforscher. Leipzig: Verlag von F. C. W. Vogel; 1874.
16. Richardson MK, Keuck G. The revolutionary developmental biology of Wilhelm His, Sr. Biol Rev. 2022;97:1131–60.
17. Steding G. The anatomy of the human embryo. A scanning electron-microscopic Atlas. Basle: Karger; 2009.
18. BlechschmidtE.MechanischeGenwirkungenFunktionsentwicklung I. Göttingen: Musterschmidt; 1948.
19. Blechschmidt E: Letter to the author (RJR) march 25; 1985.
20. Radlanski RJ, Renz H. Genes, forces, and forms: mechanical aspects of prenatal craniofacial development. Dev Dyn. 2006;235(5):1219–29.
21. Helms JA, Kim CH, Hu D, Minkoff R, Thaller C, Eichele G. Sonic hedgehog participates in craniofacial morphogenesis and is down-regulated by teratogenic doses of retinoic acid. Dev Biol. 1997;187:25–35.
22. Shroff NP, Xu P, Kim S, Shelton ER, Gross BJ, Liu Y, Gomez CO, Ye Q, Drennon TY, Hu JK, Green JB, Campas O, Klein OD. Proliferation-driven mechanical compression induces signalling centre formation during mammalian organ development. Nat Cell Biol. 2024;26:519–29.
23. Francis-West PH, Robson L, Evans DJR. Craniofacial development: the tissue and molecular interactions that control development of the head. Advances in anatomy, embryology and cell biology 169. Heidelberg: Springer; 2003.
24. Davies JA. Mechanism of morphogenesis: the creation of biological form. Burlington: Elsevier; 2003.
25. Ongaro I, Sperber GH, Machin GA, Murdoch CA. Fiducial points for three-dimensional computer-assisted reconstruction of serial light microscopic sections of umbilical chord. Anat Rec. 1991;229:285–9.
26. KV. The early development of morphology and patterns of the face in the human embryo. Berlin Heidelberg New York Tokyo: Springer Verlag; 1985.
27. Radlanski RJ. Contributions to the development of human deciduous tooth primordia. Chicago: Quintessence Publishing Co, Inc.; 1993.
28. Tsyhykalo OV, Kuzniak NB, Palis SY, Dmytrenko RR, Makarchuk Capital IU. Peculiarities of the sources of origin and morphogenesis of the human mandible. Wiad Lek. 2022;75:824–30.
29. Tsyhykalo OV, Popova IS, Skrynchuk OY, Dutka-Svarychevska TD, Fedoniuk LY. Pequliarities of Morphogenesys and topography of Infrahyoid triangles in human Prefetuses and fetuses. Wiad Lek. 2021;74:102–6.
30. Radlanski RJ, Weinhold JK, Jerichow S, Mehner M, Schwandt H, Keilig L, Bourauel C. A method to calibrate fetal μCT scans using histological sections from the same specimens. Ann Anat. 2021;237:151725.

Outlook: From Serial Section Histology to µCT-Imaging

40

Morphology, Morphogenesis, and Mechanics

In this final chapter, we want to take another look at where research in the field of morphological embryology might be heading.

For me, it is certainly necessary that we continue to focus on the elucidation of the shape, I mean, really the *μορφή* (*morphe*, form, outline, *gestalt*) and how it changes in the course of development. After all, embryology is a spatial problem. I also think it is important to look at what happens at the cellular level.

In Fig. 40.1, we see a sagittal section through the anlage region of the mandible at the level of the first milk incisor. The vestibular lamina, Meckel's cartilage and the first signs of bone formation of the mandible are also shown. The arrow points to the mesenchymal cells at the cranial-anterior edge of Meckel's cartilage. These cells are arranged at the edge of Meckel's cartilage like the bow waves on a moving ship. And indeed, Meckel's cartilage expands exactly in the direction in which the mesenchymal cells are displaced like the bow waves on a ship due to the growth of the cartilage cells, which are obviously gaining volume and lead to expansion of the cartilage bar. The dynamics can be recognized directly although it is a still image. Whenever cells move against each other, forces are generated. The compression of the mesenchyme can be recognized by the flattened cells at the arrow in Fig. 40.1. The mesenchymal cells on the side of the elongated expanding Meckel's cartilage are located in a field in which shear forces are very likely present if they could only be measured directly on the embryo. In any case, Erich Blechschmidt described and explained that desmal bone formation is always observed in a region in which shear forces should be effective [1, 2].

Using mouse embryos, we were able to show a correlation between bone formation and shear forces [3]: To this end, we reconstructed the tooth-bearing parts of the mandible in three dimensions using the familiar technique on the basis of histological section series, whereby we primarily also included the bone remodeling processes—i.e., bone apposition and bone resorption—in the reconstructions on the basis of the morphologically detectable presence of osteoblasts and osteoclasts. From a purely morphological observation, it can be

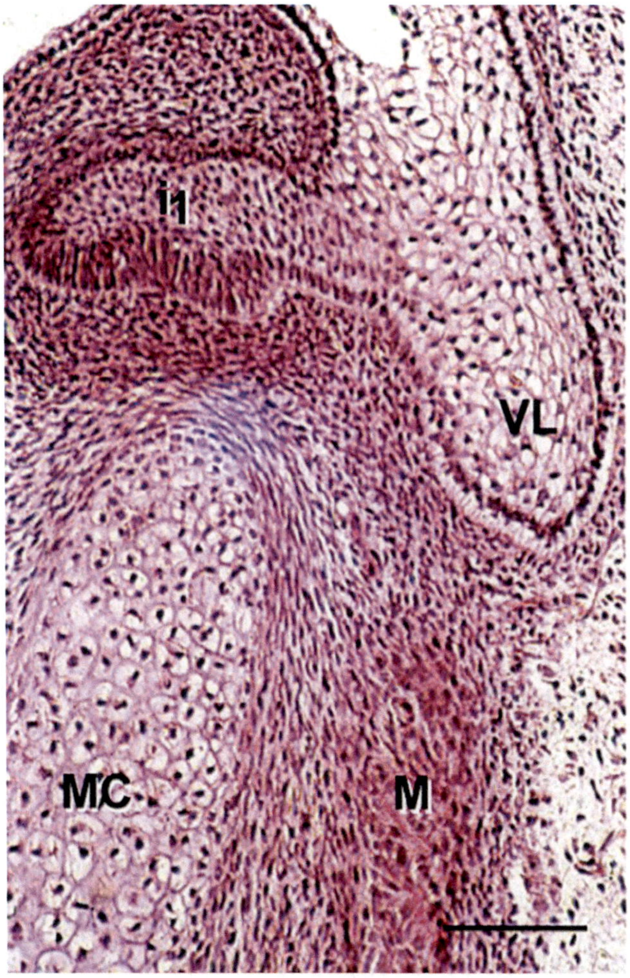

Fig. 40.1 Human fetus, 37 mm CRL, vertical section (staining: Hematoxylin-eosin) at level of the mandibular dental primordium of the first deciduous incisor i_1). Additionally labeled are the vestibular lamina (VL), Meckel's cartilage (MC), and the mandibular bone (M). The arrow points at the arrangement of the mesenchymal cells. Scale bar: 250 µm. (*From* Radlanski [4]; *with permission*)

R. J. Radlanski, *Atlas of Human Prenatal Craniofacial Development*, https://doi.org/10.1007/978-3-031-88812-0_40

Fig. 40.2 (**a**) Mandible of a mouse P10, right half, molar region. Anterior and 45° oblique cranial view. (**b**) The same data set, 3D-shear stress analysis. Bone apposition at alveolar crests, where shear stress is highest. (*From* Chen [3]; *with permission*)

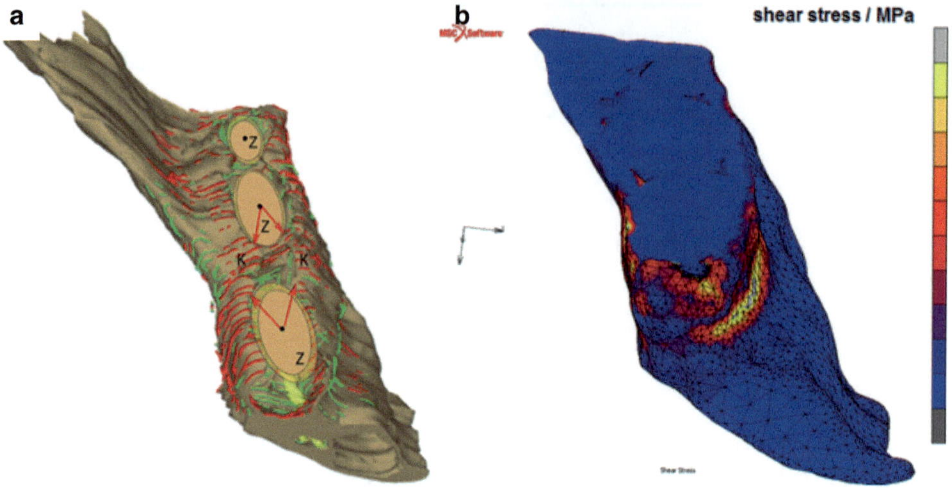

seen that there are regions, particularly at the alveolar crests, where bone formation clearly predominates (Fig. 40.2a). We examined various stages of development in this way and then used a finite element analysis to examine these spatial data sets for the distribution of shear forces. And indeed, the shear forces determined physically in the growth simulation were found exactly where we also found histological bone growth (Fig. 40.2b). We can see this finding as a further indication that Erich Blechschmidt's assumption, according to which desmal ossifying bone forms in regions of shear forces, has received further strong support. Since the data sets of the three-dimensional reconstructions are available as series of sections, they should be subjected to a simulation of the growth processes and the determination of the possible effective forces on a much larger scale.

Bridging the Gap Between Morphology and Signaling

Chapter 39 (Discussion) has already pointed out the need to build a bridge between purely morphologically oriented embryonic research and research in the field of molecular developmental biology. In any case, it is very obvious that cells crosstalk by means of molecular signaling cascades and that these are also triggers for differentiation processes. At the same time, however, all this also takes place in space and it is of course interesting to see the stage on which all this takes place. We have already taken a step in this direction, but we have not published it as a paper. Together with Richard A. Schneider (University of California at San Francisco), we reconstructed not only the plain morphology of a quail mandible (Fig. 40.3a) but also the signals for bone (Runx-2) (Fig. 40.3b), for cartilage (Sox-9) and for muscle (Myf-5) on the basis of a series of sections of a quail mandible in 3D and superimposed them on the morphological findings (Fig. 40.3c). This showed that the expansion of the cartilage and the musculature really is congruent in terms of both shape and signaling molecules, but that the mandible itself was found inside a cloud of Runx-2 expression. In my lectures on orofacial structure and developmental biology, which can be viewed on YouTube, you can see more of these impressive results if you wish at: https://www.youtube.com/@Lectures-Radlanski, Lecture 3: From Gene to Body Form, TC 55′:22″ (Chapter: Signaling Proteins). Runx-2, therefore, has nothing to do with the actual shape of the mandible, but is—as is now also known—more of a master regulator. However, how the actual formation of the mandible really occurs should be investigated further and, above all, the expression of the relevant signaling molecules in space should also be mapped.

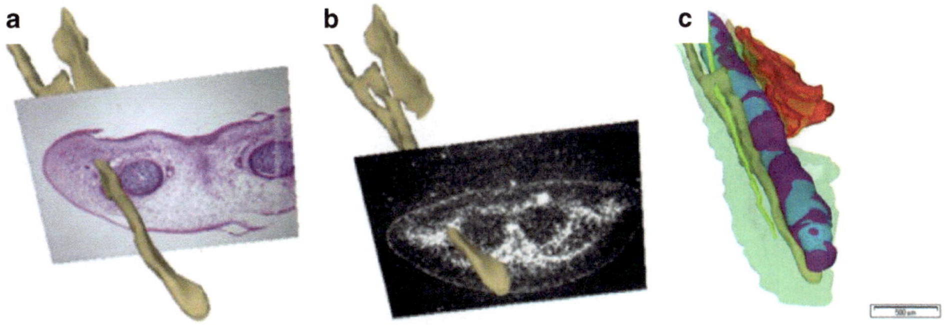

Fig. 40.3 Mandible of a quail, HH stage 35, right half, anterior and 45° oblique cranial view. (**a**) Superimposition of one of the histological sections (staining: Hematoxylin-eosin). (**b**) Superimposition of one of the histological sections with Runx-2 in situ hybridization. (**c**) Mandible of a quail, HH stage 35, right half, anterior and 45° oblique cranial view (the same specimen as in (**a**, **b**). 3D-reconstruction of the serial sections showing the outlines of Runx-2 (light green), bone (ochre), alveolar inferior nerve (yellow), Meckel's cartilage (light blue: cartilage; purple: Sox-9-signaling), mylohyoid muscle (orange: muscle tissue; red: Myf-5-signaling)

µ-CT Imaging

Digital imaging based on µCT has now become so good that we can obtain almost realistic-looking images of our embryonic and fetal preparations (Fig. 40.4b). However, contrast is a prerequisite if we also want to clearly visualize soft tissue with X-rays (Fig. 40.4a). In addition, as already mentioned in Chap. 39 (Discussion: Morphology as the Foundation), it is not so easy to visualize the extent of weakly mineralized bone in a reliable way and, if necessary, automatically. It is certainly also necessary to calibrate the image data, as the bone structures of embryos and fetuses are also mineralized to different extents at different stages of development [4]. It remains to be hoped that a sufficient number of suitable specimens will be available worldwide for examination using µCT. The results are not only instructive, but also very impressive.

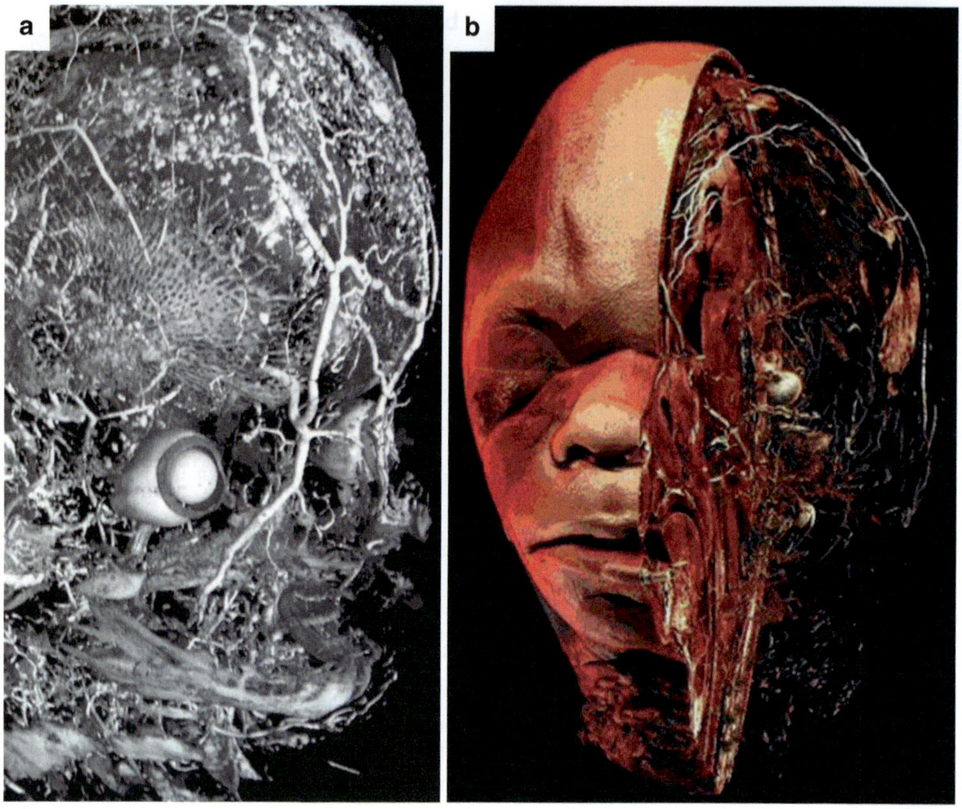

Fig. 40.4 (**a**) Human fetus, 155 mm CRL (17th week) µCT image, contrast staining with phosphotungstic acid. Right view. (**b**) The same specimen, computer-graphic revision, performed by Gery Weber, Volume Graphics, Heidelberg, 2018. (**a** and **b**, *From* Radlanski [5]; *with permission*)

References

1. Blechschmidt E. Mechanische Genwirkungen Funktionsentwicklung I. Göttingen: Musterschmidt; 1948.
2. Blechschmidt E. The stages of human development before birth. An introduction to human embryology.—Die vorgeburtlichen Entwicklungsstadien des Menschen. Eine Einführung in die Humanembryologie. London, New York, Basel, Freiburg i. B: Karger; 1960.
3. Chen J, He Y, Keilig L, Reimann S, Hasan I, Weinhold J, Radlanski RJ, Bourauel C. Numerical investigations of bone Remodelling around the mouse mandibular molar primordia. Ann Anat. 2019;222:146–52.
4. Radlanski RJ. Contributions to the development of human deciduous tooth primordia. Chicago: Quintessence Publishing Co, Inc.; 1993.
5. Radlanski RJ, Weinhold JK, Jerichow S, Mehner M, Schwandt H, Keilig L, Bourauel C. A method to calibrate fetal µCT scans using histological sections from the same specimens. Ann Anat. 2021;237:151725.

Further Reading

Radlanski R.: From Gene To Body Form. Lecture 3, on Youtube: https://www.youtube.com/@Lectures-Radlanski.

GPSR Compliance

The European Union's (EU) General Product Safety Regulation (GPSR) is a set of rules that requires consumer products to be safe and our obligations to ensure this.

If you have any concerns about our products, you can contact us on ProductSafety@springernature.com

In case Publisher is established outside the EU, the EU authorized representative is:

Springer Nature Customer Service Center GmbH
Europaplatz 3
69115 Heidelberg, Germany

Batch number: 10731524

Printed by Printforce, the Netherlands